WILDLIFE,
FORESTS,
AND FORESTRY

Principles
of Managing Forests
for Biological Diversity

WILDLIFE, FORESTS, AND FORESTRY

*Principles
of Managing Forests
for Biological Diversity*

MALCOLM L. HUNTER, JR.

*Wildlife Department
College of Forest Resources
University of Maine
Orono, Maine*

Illustrated by Diane Bowman

PRENTICE HALL, Englewood Cliffs, N.J., 07632

Library of Congress Cataloging-in-Publication Data

Hunter, Malcolm L.
    Wildlife, forests, and forestry : principles of managing forests
for biological diversity / Malcolm L. Hunter , Jr. ; illustrated by
Diane Bowman.
        p.   cm.
    Includes bibliographical references.
    ISBN 0-13-959479-5
    1. Biological diversity conservation. 2. Forest conservation.
3. Forest management.   I. Title.
    QH75.H85   1990
    639.9–dc20   89- 16164
                                                             CIP

Editorial/production supervision and
    interior design: Diane M. Delaney
Cover design: Ben Santora
Chapter-opening design: Debbie Toymil
Manufacturing buyer: Laura Crossland

©1990 by Prentice-Hall, Inc.
A Division of Simon & Schuster
Englewood Cliffs, New Jersey 07632

Printed in the United States of America

10   9   8   7   6   5   4   3   2   1

ISBN   0-13-959479-5

PRENTICE-HALL INTERNATIONAL, INC., *London*
PRENTICE-HALL OF AUSTRALIA PTY. LIMITED, *Sydney*
EDITORA PRENTICE-HALL DO BRASIL, LTDA., *Rio de Janeiro*
PRENTICE-HALL CANADA INC., *Toronto*
PRENTICE-HALL OF INDIA PRIVATE LIMITED, *New Delhi*
PRENTICE-HALL OF JAPAN, INC., *Tokyo*
PRENTICE-HALL OF SOUTHEAST ASIA PTE. LTD., *Singapore*
WHITEHALL BOOKS LIMITED, *Wellington, New Zealand*

For my family, especially Henry Hunter (1894–1983), carpenter, logger, and naturalist.

Contents

Preface

HOW IT BEGAN

It's 5:05 A.M., 17 October 1985. I'm lying in bed wrapped in a quilt, with one hand sticking out to write. I always keep a pad of 3″ × 5″ paper next to my bed to jot down nocturnal notes to myself, but it is very unusual to be writing entire sentences. If I had spent last evening at home and cooked supper on my wood stove, it would be warm enough to get out of bed and write at the kitchen table on a proper 8 1/2″ × 11″ pad. I woke up about an hour ago and lay looking out at the river wondering why I had woken. At first I thought I was hungry, but then I realized that I had been thinking in my sleep about the previous evening. A group of about twenty people, roughly half foresters—half wildlifers, half grad students—half faculty, gathered at Bucky Owen's house to discuss wildlife management on industrial forest lands. Dan Keppie, who was just finishing a six-week visit from the University of New Brunswick, opened things up with some moderately controversial statements on the subject, always delivered in his vaguely humble, "I would submit that . . ." style. Three hours of lively debate ensued on topics as specific as the altitudinal distribution of various songbirds and as general as how does one value wildlife. Over the last several months, I have formed many opinions on forest wildlife management, and as I expressed them I noted a number of heads nodding in agreement. I think that tacit approval planted the germ of an idea with me. The group reached a reasonable degree of consensus on the point that managing forests for wildlife diversity was relatively simple and feasible. The big problem was the lack of communication, the absence of a means for getting information out to the practicing land managers. Lying here now, I have decided to try to help bridge that gap by writing a book for forest managers. I have many ideas, and it should be fun to organize them. One page down, ??? to go.

HOW IT TOOK FORM

Readership and Scope

24 July 1986—It has taken some time, but I have found a publisher who will gamble that I can achieve my goal of writing a book on forestry and wildlife that will appeal to a broad audience without falling between the cracks. Although I am writing for a diverse readership, I still have priorities, and my primary audience is the men and women who are out on the ground deciding how to manage forests. Most of them are foresters; some are wildlife, watershed, or recreation managers first, and forest managers second. I particularly want to reach those who work on lands owned by the forest industry and on private, nonindustrial forest lands. These people do not necessarily have a strong mandate to manage forests for a variety of values, unlike their colleagues who work on government-owned forests. An obvious second audience is the students who aspire to become natural resource managers. In most programs, forestry students take a wildlife course and wildlife students take one or more forestry classes, but classes that examine the interface are uncommon.

People who care about forest wildlife, but have no professional expertise in the subject, are also intended readers, and the first three chapters are primarily for them. Most prominent among these are individuals who own forest land, some seven million people in the U.S. alone, and thus control an enormous resource. Then there are all the people who use forests and are interested in them: naturalists, hunters, anglers, backpackers, environmentalists, campers, etc. Finally, and most critically, there are the policy makers who must guide natural resource management from afar, from the halls of government. I am trying to make the book readable for these people without burdening the professionals with too much information they already have. (Perhaps this readability will also prompt professional readers to take the book home from the office and share it with their spouses. In fact, I have a wonderful painting of a forest created by Mary Beth Memmer, age seven, whose father Paul—a Scott Paper Co. forester—took time out from reviewing my manuscript to explain some forestry principles to her, using the book's illustrations.)

Last among my audience are my colleagues—researchers and academics. Although they will already be familiar with much of the material presented, I think they will find it a useful and interesting compendium.

The geographic scope of the book is also broad. Although most of the research on which it is based was conducted in North America, much of the information is applicable to temperate and boreal forests throughout the world. Its relevance to tropical and subtropical forests is rather uncertain because these ecosystems are very different and very complex, and my familiarity with them is relatively superficial.

A Conceptual Focus

The book's geographic breadth is made possible by an emphasis on concepts rather than specific recommendations. This is not a cookbook with recipes for achieving goals in the style: Given stand *A*, in condition B, and with constraints *C*, if you do *X*, *Y*, and *Z* you can maintain wildlife diversity. Because of the complexity of even the simplest forest ecosystems, such a volume would be very hard to produce. Even regional works have not really approached this level of detail, although some, most notably *Wildlife Habitats in Managed Forests: The Blue Mountains of Oregon and Washington*, have a great wealth of information (Thomas 1979, Hoover and Willis 1984, Brown 1985). The Blue Mountain book has over 250 pages of tables that outline the interactions among 326 wildlife species, their habitats, and management options. Ultimately, every situation is unique, and therefore management decisions must be individually tailored by the person who has immediate responsibility. Supervisors, absentee landowners, and scientists can only provide general guidelines and an information base.

I have read many definitions of what is a conservationist, and written not a few myself, but I suspect the best one is written not with a pen, but with an axe. . . . A conservationist is one who is humbly aware that with each stroke he is writing his signature on the face of his land. Signatures of course differ, whether written with axe or pen, and this is as it should be. (Aldo Leopold 1949)

Because this book investigates only the concepts that form the foundation for specific guidelines, it cannot tell a forester exactly how many snags to leave in a stand or how large a clearcut can be. All it can do is give some general guidelines. Perhaps forest managers who have been told to leave two snags per acre will read this book to learn the reasoning behind these instructions, and foresters who have not been told how many snags to leave will be inspired to adapt the ideas of this book to their particular situation.

The conceptual focus and broad geographic scope have caused me to limit the book in two ways. First, there is very little mention of specific laws and policies, although Appendix 1 provides a brief review of some U.S. laws and policies related to managing forests for biological diversity. Second, only two aspects of forest management—wood products and wildlife—are considered in depth; recreation, livestock grazing, and watershed management are largely ignored.

A Caveat About Documentation

Most of the ideas presented here are simple and intuitively obvious. It takes little insight to realize that a forest landscape with young stands and old stands, deciduous trees and conifers, will have a greater diversity of wild-life than a landscape covered by 40-year-old red pines. However, in general, the validity of these ideas has not been exhaustively tested. Most research projects last one to three years; many forest processes operate on scales of decades and centuries, even millennia. Entire landscapes form the setting for many processes, but most studies compare subsections of a few stands. These limitations force ecologists interested in influencing natural resource management to stick their necks out to a degree that often makes them uncomfortable. This is particularly true of zoologists, who often cannot even estimate current animal populations with an accuracy that satisfies them, let alone predict future population changes that will result from various management practices. It is very common to hear scientists say, "We need to do more research; be patient," while the managers reply, "We can't wait; we have to make decisions now." As Jack Ward Thomas (1979) has written: "To say we don't know enough is to take refuge behind a half-truth and ignore the fact that decisions will be made regardless of the amount of information available."

I think it is useful to compare ecologists and economists; both groups construct models of the real world based on an inadequate understanding of how it works and use these models to make predictions that will influence the decisions of politicians and business people. The economists' models are no better than those of the ecologists, but economists are quite happy to make predictions that have multibillion dollar impacts, while ecologists wring their hands over pocket change and the possibility that they will make a mistake that will sully their credibility. This is understandable; scientists are trained to be conservative, and they should be, especially when human lives or the continued existence of a species is at stake. However, I think we err on the side of conservatism too often.

A Word About Metrics

I have used the metric system here, because the book is intended to be global in scope and because most of the data on which it is based are in metric units. Unfortunately, many readers will be more familiar with English measurements. Sometimes I have allowed for this by providing English equivalents in parentheses. However, for the most part I have not done so—in part to save space, but mainly because the book's emphasis is on principles, and the exact numbers are not particularly important. For these same reasons, when describing papers that used only English measurements, I have usually not provided the metric equivalent. Appendix 2 provides some metric and English measurements of familiar things, e.g., a football field, that may help form a visual image of a hectare or an acre.

ACKNOWLEDGMENTS

12 September, 1988. The book is just about finished, and to date the writing has been a source of continuing pleasure—even when I had to turn to sugar and chocolate, in the form of brownies, to get me through a couple weeks of 18-hour workdays. But now a year of compiling indices, proofreading galleys, etc. lies before me; the creative work is done and I want to thank the scores of people who have helped me before my enthusiasm wanes.

Selecting one person for special thanks is easy, because the most thoroughly enjoyable moments in this project came whenever I saw how Diane Bowman had transformed my unimaginably crude sketches into beautiful and informative illustrations. Everytime she came to my office with portfolio in hand, I felt like a child at Christmas.

From this point on, it becomes very difficult to prioritize my gratitude; I'll start with the largest group, all the people who reviewed various manuscripts and thus guided my efforts: Larry Harris, Raymond O'Connor, Hal Salwasser, Ray Owen, Paul Memmer, Andy Whitman, John Barkham, Tom Barron, Rob Bierregaard, Bob Blake, Mark Brinson, Paul Brouha, Evelyn Bull, Dave Capen, Jerry Clutts, Hew Crawford, Dick DeGraaf, Bill Drury, Cathy Elliott, Peter Emerson, John Fay, Jerry Franklin, Phil Gaddis, Bill Glanz, Barry Goodell, Mike Greenwood, Gordon Gullion, Yrjö Haila, Trina Hikel, Leslie Hudson, Peter Hudson, Lloyd Irland, George Jacobson, Dan Keppie, Jiro Kikkawa, Al Kimball, Sharon Kinsman, Bill Krohn, Dick Lancia, John Lanier, Mitch Lansky, Bruce Marcot, Chris Maser, Bill McComb, Janet McMahon, Chuck Meslow, John Moring, Bill Platt, Rory Putman, R. A. Lautenschlager, Harry Recher, Robert Ruff, Ellen Shields, Mary Small, Don Spalinger, Benee Swindel, Jack Ward Thomas, Peter Vickery, Nat Wheelwright, Alan White, Dave Wilcove, Jack Witham, and Mike Zagata. (The first six people named merit special thanks for undertaking the Herculean task of reviewing the entire manuscript.) During trips to Oregon and Florida, I also received wonderful hospitality from many of these people. They, plus many others—notably Hilary Swain, Ellen Hemmert, Ed and Roy Komarek, Mike Balboni, Rand Erway, Kathy Johnson, and Marilyn Morrison—opened my eyes to new landscapes.

All these reviews coming and going required considerable secretarial support, and this was ably supplied by Shirley Moulton, Maxine Horne, and Nancy Kealiher. A series of editorial assistants, Cathy Elliott, Andi Colnes, Selena Tardiff, Martha Wood, and Tina Stillings also provided much logistic support, particularly in the library, and Eric Stefanovich produced several figures. The Fogler Library staff, especially Bruce Leach, greatly facilitated all the requisite information processing. Janet Rourke of University of Maine's Development Office has helped with a coda to this enterprise—setting up an endowment that will use the book's royalties to assist graduate students from developing countries.

Some organizations must be credited here too. The University of Maine in general, and the College of Forest Resources and Department of Wildlife specifically, have given me a stimulating and nurturing professional home off and on for nearly two decades. During the last several years, my efforts to understand forest ecosystems have been generously supported by the Holt Woodlands Research Foundation and the Maine Agricultural Experiment Station.

Finally, I thank family and friends for being my family and friends, the most important part of my life even during this time-consuming exercise.

A postscript: 10 February 1989. I have just returned from back-to-back trips to Zaire and Nepal—in the first case wearing the hat of a research ecologist, in the latter case as a conservation advisor—and I have begun collecting material for a sequel focusing on tropical forests. And so it goes.

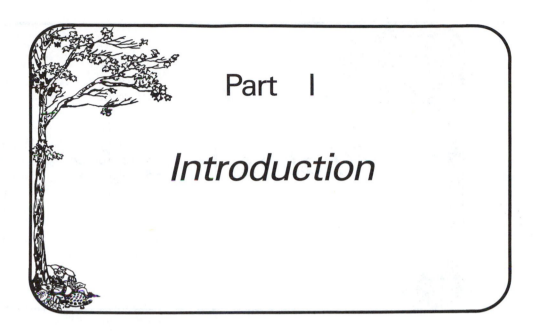

Part I

Introduction

Three chapters with deceptively simple titles, ''What is wildlife?,'' ''What is diversity?,'' and ''What is a forest?'' comprise Part I. The first two chapters define two rather ambiguous words, wildlife and diversity, and provide a brief justification for making maintenance of wildlife diversity an important goal of forestry. In other words, they form the philosophical foundation for the book. Readers who already believe that wildlife and diversity are very important should still read the summary on page 28 to make sure that they know how wildlife and diversity are defined in this book. Chapter 3, ''What is a forest?'' is essentially a brief introduction to forest ecology for readers who have never had an ecology course or whose memory of an ecology course is over a decade old. The latter may be surprised by some current ideas about classic concepts such as ecosystem and succession.

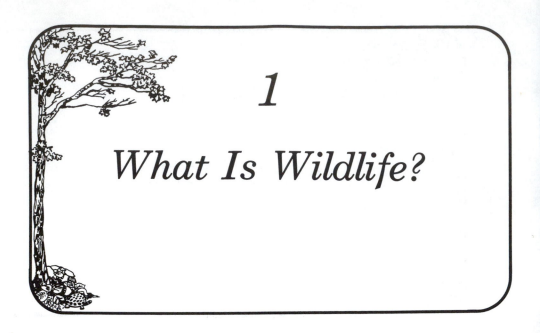

1

What Is Wildlife?

I have initiated a number of wildlife classes by asking the students to write a definition of wildlife. There seems to be a fair degree of consensus, with the majority submitting some variation of "any living thing that is part of a natural ecosystem." I also ask them to list the ten most important kinds of wildlife in the northeastern United States. Their lists almost always begin with deer and go on to include various combinations of moose, coyote, bear, duck, squirrel, hare, grouse, eagle, salmon, etc. Game mammals and game birds dominate the lists; occasionally a game fish or large nongame bird will appear. Nobody ever lists white pine, red spruce, or soil bacteria, even though most of the students are forestry majors. Just once did someone include the spruce budworm, a creature that wreaked havoc on Maine's major industry for over a decade, and she placed it only seventh. Clearly, plants, invertebrates, and microorganisms are living things that are parts of natural ecosystems, and no one can question their critical importance. Why the inconsistency?

An Historical Perspective

Twenty years ago, if students had been asked to define wildlife, a significant number of them would probably have written, "wild animals that are hunted or trapped by man," and thus been far more consistent with their lists. (In my polls, fewer than 1% of the students have given this definition.) This definition is also consistent with some historical precedents. The term wildlife probably originated in 1913 with a book, *Our Vanishing Wild Life*, by William

Hornaday, Director of the New York Zoological Park (Hornaday 1913, Hickey 1974). *Our Vanishing Wild Life* focused on the overexploitation of game birds, mammals, and fishes; it also addressed the harvesting of birds that Hornaday felt were not game, notably the songbirds that recent European immigrants often hunted. By 1937, the term wildlife had been contracted into one word and expanded in scope by an implicit definition on the first page of a new journal, *The Journal of Wildlife Management:*

> Wildlife management is not restricted to game management, though game management is recognized as an important branch of wildlife management. It embraces the practical ecology of all vertebrates and their plant and animal associates. While emphasis may often be placed on species of special economic importance, wildlife management along sound biological lines is also part of the greater movement for conservation of our entire native fauna and flora. (Bennitt et al. 1937)

The publishers of the new journal, a professional organization called the Wildlife Society, designed a logo composed of Egyptian hieroglyphics depicting mammals, birds, fish, *and* plants, thereby symbolizing a fairly broad approach to wildlife management. Unfortunately, this reasonably clear beginning was soon to be obscured by bureaucratic intervention.

In 1940, Franklin D. Roosevelt sat down to sign a presidential proclamation creating a new agency, the U.S. Wildlife Service, by combining a Department of Commerce agency, the Bureau of Fisheries, with Agriculture's Bureau of Biological Survey. Apparently, some fisheries people said something persuasive at the least minute, because before signing the bill the President inserted the words "Fish and," thereby creating the U.S. Fish and Wildlife Service and implicitly declaring that fish were not wildlife. This was probably perfectly acceptable to the American Fisheries Society, an organization that had been around since 1870 and doubtless was reluctant to be upstaged by the fledgling Wildlife Society. Initially, The Wildlife Society tried to bring fish management under its aegis by publishing fisheries papers in the *Journal of Wildlife Management,* but there were never many submitted, and by the late 1970s the trickle had dried up entirely (S. Beasom, pers. comm.). Today, the idea that fish are not wildlife is firmly entrenched, and many people define wildlife as terrestrial vertebrates—succinct enough as long as you do not forget turtles and whales.

Have you wondered how the dictionaries define wildlife? Open a Webster's Dictionary older than 1961 and you will not even find wildlife included; in that year Webster's third edition defined wildlife as:

> . . . living things that are neither human nor domesticated: especially the mammals, birds, and fishes that are hunted by man for sport or food.

The twelve-volume *Oxford English Dictionary* (OED) of 1933 also does not list wildlife; it took until 1986 for the OED to define wildlife as "the native fauna and flora of a particular region."

In practice, the effective definition of wildlife has been constrained by economic realities. Although The Wildlife Society explicitly stated that "Wildlife management is not restricted to game management. . . ." the fact that in many places revenues for wildlife management have come primarily from hunting licenses and fees and taxes on guns and ammunition has weakened that statement. Not only are nongame species largely excluded by this reality, but separate fishing licenses and taxes on angling equipment further support the idea that fish are not wildlife. In the United States this situation is starting to change, primarily because most states now have nongame programs funded by a number of voluntary mechanisms such as income tax check-offs. These programs have embraced all vertebrate species, and some are even venturing into uncharted waters by assembling lists of endangered plants, butterflies, and mollusks. Such activities are still tiny parts of the superstructure of wildlife conservation. For example, Colorado, the first state to have a nongame tax-checkoff, raised $692,000 in 1982, the most of any state; in the same year revenues from hunters amounted to $28,100,000 (Shaw and Mangun, 1984). To take another example, it has been estimated that nearly 2,000 species of plants in the United States are threatened with extinction, yet the U.S. Fish and Wildlife Service has completed the requisite surveys and paperwork to have only 185 species protected by law (Ayensu 1978, USFWS 1988). Similarly, the Service has a list of almost a thousand invertebrate species that are candidates for endangered species status, but fewer than seventy have been listed. All this makes some people impatient, but the important point is that there is a growing trend for wildlifers to diversify the scope of their activities, and with this comes a concomitant expansion in the definition of wildlife. Already, organizations such as the World Wildlife Fund and National Audubon Society define wildlife to include all biota. In this book, wildlife is defined as all forms of life that are wild; this definition at least has the virtue of simplicity.

Incidentally, the American bias of this historical review is not grossly unfair; not only was the world wildlife coined in the United States, but most languages have no close equivalent; words better translated as "nature" are as close as most languages come. Furthermore, wildlife management as a scientific discipline and profession emerged primarily in the United States through combining the emerging science of ecology with the traditional practices of gamekeepers.

The Consequences of Ambiguity

The confusion that surrounds the definition of wildlife is not just of etymological interest. If one accepts a broad definition of wildlife, it changes how one views the interaction of people and wildlife. For example, many natural resource managers talk about various timber management practices that are "good for wildlife," while other professionals and the general public complain about practices (e.g., clearcuts and herbicides) that are "bad for wildlife." Sometimes the claims become very exaggerated. More than one person has asserted that the landscape will be virtually devoid of both trees and wildlife because of mismanagement by foresters. Others will tell you that a forest that is not regularly harvested will become stagnant and bereft of wildlife. Both groups usually make their claims from a perspective that is heavily biased toward game animals.

When someone says that a forestry practice is good for wildlife, they typically mean that it creates the mosaic of young forests that is favored habitat of deer, hare, grouse, and other game species. If they complain about clearcuts or herbicides destroying wildlife habitat, it is usually because large, recent clearcuts are avoided by these same species. When you use a broad definition of wildlife, you cannot talk about creating or destroying wildlife habitat, because when you destroy the habitat of species X you create habitat for species Y. Forestry practices do not destroy ecosystems; they only alter them and thus change the assemblage of species comprising the ecosystem.

Using a broad definition of wildlife also makes it easier to avoid the bunnies and bulldozers syndrome. This syndrome is seen in people who declare that bulldozers do not cause problems for bunnies, because bunnies move faster than bulldozers and will simply run to a new habitat when the bulldozers uproot their homes. As fatuous as this argument is, it is used all the time by people who equate harm to wildlife with killing animals. Shakespeare knew the fallacy of this, for as Shylock, the merchant of Venice, said, "You take my life when you take the means whereby I live." Focusing attention on wildlife diversity and away from the welfare of individual species will weaken the stance of those who suffer from the bunnies and bulldozers syndrome.

"But Some Animals Are Created More Equal than Others"

A clearcut that stretches to the horizon will have very few squirrels and woodpeckers, but lots of butterflies and sparrows. A pine forest will have more pine warblers than a spruce forest, but the spruce forest will have more spruce grouse. Which species should we favor? Which do we value most? Should we have more squirrels or more butterflies, more pine warblers or more spruce grouse? As George Orwell (1946) wrote, "All animals are created equal, but some animals are created more equal than others," and his maxim is readily discernible in the way people view various species. Game animals are created more equal than nongame animals, and vermin are far less than equal. Furry creatures with big eyes are preferred over slimy creatures with small eyes.

Animals are valued more highly than plants, unless the plants happen to be trees or other commercially valuable species.

Even ecologists can be quick to assign values. For example, exotics (alien species that people have introduced to a region such as starlings and cheat grass in North America and rabbits in Australia) are usually scorned by people who promote the protection of native species and natural ecosystems. Only feral animals and escapes (i.e., species of domestic animals and plants now living in the wild such as goats in Hawaii) are held in lower esteem. This attitude toward "foreign intruders" is not just a matter of whimsy. There are numerous examples of exotic and feral creatures causing profound disturbances to native ecosystems, even the extinction of local biota (Elton 1958, Laycock 1966, King 1984).

Although people's attitudes toward wildlife are conspicuous, they are not always consistent, and one person's deer is often another person's rat. (In many places farmers consider deer just oversized vermin; in Nigeria rats are prized food and sell for more than beef and pork [Ajayi 1979].) In the absence of a clear and universal consensus, and because all species have some value, however intangible, one of the most fundamental goals of wildlife management should be to maintain or restore all native wildlife populations, with a particular emphasis on endangered species, and that goal is the focus of this book.

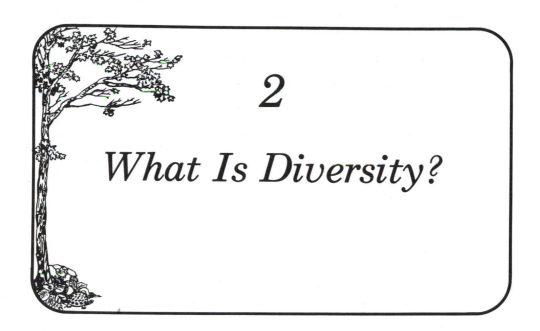

2

What Is Diversity?

Words such as diversity and variety have a positive connotation when people think about work, diets, experiences, and the spice of life. Similarly, the phrase "biological diversity" also elicits positive images—a variety of birds at the feeder, a diversity of flowers in the meadow; these are easily perceived as things to be valued. However, biological diversity is much more complicated than different kinds of birds and flowers. The phrase—often abbreviated biodiversity—refers to the diversity of life in all its forms and all its levels of organization, not just the diversity of plant, animal, and microorganism species. At its most elemental level, biological diversity encompasses the varied assemblages of organic molecules that comprise the genetic basis of life. On the other end of the spectrum there are biomes—the vast stretches of tundra, desert, forest, ocean, etc., that cover a whole region and reflect the planet's diversity of climate and physical form. In between these extremes lies a panoply of other levels of organization—population, race, subspecies, community, ecosystem—which are also components of the larger concept, biological diversity.

Measuring Diversity

When research ecologists use the word diversity, they are not referring to an abstract quality, but rather a quantifiable character used to describe collections of elements, most often communities of species. Table 2.1 illustrates

TABLE 2.1   Tree Species in Three Hypothetical Forests and Their Relative
Abundance (%).

Forest A		Forest B		Forest C	
Douglas-fir	43%	American beech	74%	Sweetgum	24%
Western larch	32%	Yellow birch	13%	Nuttall oak	17%
Ponderosa pine	25%	Sugar maple	13%	Willow oak	17%
				Hackberry	16%
				Green ash	15%
				American elm	11%

the basis of an ecological definition of diversity; based on the compositions listed in Table 2.1, which forest seems most diverse?

Obviously, Forest C, which has the most species, seems the most diverse. Ecologists refer to this characteristic, simply the number of species present, as *richness*. In terms of richness, Forests A and B are identical, but differences in the relative abundance of the trees change the perception of diversity. Ecologists would say that Forest A, where the number of trees is quite evenly distributed among all species, seems more diverse than Forest B, where American beech predominates. The distribution of abundance among different species is referred to as *evenness*; forests with high evenness (A) are generally considered more diverse than ones with low evenness (B). A community with higher richness is not necessarily more even, and vice versa; for example, if Forest B had a fourth species, eastern hemlock, comprising 12% of the trees, it would be more rich, but less even than Forest A. Thus, ecologists often use one of several formulas that combine richness and evenness to determine which forest, the one with the greatest evenness or the one with the greatest richness, is more diverse. Appendix 3 describes some of these in detail. Suffice it to say here that species richness is weighted more heavily in these formulas than evenness and that, according to the most popular formula, adding 12% hemlock to Forest B would make it more diverse than Forest A.

One can also combine richness and evenness conceptually to think of a diverse community as one in which there are many species, all of which are relatively rare (Patil and Taillie 1982). The idea of *average rarity* has led to some new graphical techniques to compare diversity of communities (Swindel et al. 1987); these are also described in Appendix 3.

Problems with Measuring Diversity

There are two significant problems with measuring biological diversity. First, there are no universally accepted classification schemes for elements of diversity, except at the species level of organization. In other words, while biologists can usually (but by no means always) agree that tree species x is distinct from tree species y, there are no general rules for saying sequoia population x is genetically distinct from sequoia population y, or conifer forest x is different from conifer forest y.

The second problem is that even species diversity is effectively impossible to thoroughly measure. As far as I know, no one has ever attempted to list *all* the species found in a forest. Perhaps the most extensive effort, one for Monk's Woods near Cambridge, England (Steele and Welch 1973), came up with a list of 3,682 species. Even this list, the collective effort of a team of scientists working many years, was deemed incomplete for several groups and completely left out microorganisms. To list all the species for a site would probably involve more than just collecting thousands of creatures and then spending years in a lab with a microscope. Often it would entail describing species previously unknown to science. Entomologists are constantly discovering new insect species, about 20 per day. For example, a Smithsonian entomologist collected 55 kinds of caddis fly on a visit to Amazonia and subsequently determined that 53 of them were new species (Myers 1979). And you do not have to go off to the tropics for such discoveries; in a couple years of trapping spiders in the spruce-fir forests of Maine, an arachnid enthusiast found six new species (Jennings pers. comm.). Estimates for the total number of species on earth range from 5 to 30 million, while the total number of species described is about 1.4 million; thus the problem of undescribed species is likely to be with us indefinitely, especially in tropical forests (Wilson 1988).

In some senses, the near impossibility of measuring total species diversity makes the task easier; i.e., because it is so difficult to measure total diversity no one even tries. Managers who want to ensure that they are indeed maintaining diversity usually measure only a few groups of organisms. Birds are commonly used because they are easy to find and identify and are important to the public. Vascular plants are good too, probably better than birds because most creatures rely on plants for food and shelter. Managers of U.S. National Forests are required by law to maintain viable populations of all native vertebrate species, and they monitor certain vertebrate species, mainly endangered species or those sensitive to forest management, to help meet this objective (Barton and Fosburgh 1986).

WHY MANAGE FOR DIVERSITY?

The fundamental reason for managing for diversity is simple: All life forms have some value, economic or ecological, realized or potential, and by managing for diversity we manage for all life forms. Many eloquent arguments have been made to this end. Two classics are *Extinction* by Paul and Anne Ehrlich (1981) and *The Sinking Ark* by Norman Myers (1979); more recently, cases have been built by the U.S. Congress' Office of Technology Assessment (1987) and the U.S. National Academy of Sciences and the Smithsonian Institution (Wilson and Peters 1988). The arguments run from the esoteric to the utterly practical. For example, E. O. Wilson (1984) compares the genetic structure of an organism to the bits that computers use to encode information and then likens an ant to a string of words a thousand miles long, enough to fill a very large bookcase. Robert and Christine Prescott-Allen have written two books: *What's Wildlife Worth?* (1982), which carefully catalogues the thousands of uses people in developing nations make of wild plants and animals, and *The First Resource* (1986), a similar accounting for North America.

These books contain many case histories documenting the importance of maintaining all forms of life. The following is one from another source, *In the Rainforest* by Catherine Caufield (1984); it is the story of two kernels of rice. In the 1960s there was only one major disease afflicting the world's rice, grassy-stunt virus, and scientists at the Philippines' International Rice Research Institute screened 10,000 varieties of rice searching for some that would be resistant to the disease. They found only one resistant variety; it was represented in their collection by only two kernels; and attempts to collect more material from the site of origin in India failed because the population had been destroyed. Fortunately, this was a success story. By the slimmest of threads, two kernels, the world's rice was made resistant to grassy-stunt virus. It is tempting to speculate that a similar story might someday be told about forest species such as chestnut and elm (Zobel and Talbert 1984, Ledig 1988).

Another reason to manage for diversity involves a closely related concept, stability. One of the traditional principles of ecology has been that diverse ecosystems are more stable than ecosystems that are less diverse (Elton 1958). Very briefly, this means that it is less likely that Forest C (Table 2.1) would be profoundly disturbed by a disease that infects willow oaks, than that Forest A would be disturbed by a disease of ponderosa pine. Some ecologists, however, have amassed empirical and theoretical evidence suggesting that there is no predictable relationship between stability and diversity (Watt 1968, May 1974, Kikkawa 1986). Others have turned the argument around and stated that a stable ecosystem is more likely to be diverse than an unstable ecosystem (Goodman 1975). And finally, some ecologists debate the nature of stability—for example, whether it is resistance (i.e., the ability to withstand change) or resilience (i.e., the ability to recover after change) (Pimm 1986). Despite this discord, most ecologists would agree with the basic truth of a metaphor that the Ehrlichs' use in which the Earth is likened to an airplane. If one or two rivets popped off (i.e., species went extinct), you would not worry too much, but everyone knows that if too many parts are lost the whole plane will come apart.

To switch metaphors, there are already some examples of the weave of an ecosystem beginning to unravel because of extinction. Take the case of the dodo, a huge, flightless relative of the pigeon that inhabited a small island in the Indian Ocean, Mauritius, until it was wiped out some three hundred years ago. In the 1970s, Mauritian foresters noticed that a formerly abundant species of trees, the tambalacoque, had become rare and no longer reproduced. The 13 remaining individuals were nearly three hundred years old, and although they still produced fruits, no seedlings ever germinated. Stanley Temple (1977) hypothesized that perhaps the trees' seeds would only germinate after they had passed through the digestive system of a bird, and the only bird large enough to eat tambalacoque seeds, the dodo, was long extinct. Fortunately, this relationship was discovered in time to save the tambalacoque by using force-fed turkeys to play the role once held by dodos.

The critical importance of maintaining biological diversity is poignantly explained in a prediction by Edward Wilson (1984):

> ... the worst thing that will *probably* happen—in fact is already well underway—is not energy depletion, economic collapse, conventional war, or even the expansion of totalitarian governments. As terrible as these catastrophes would be for us, they can be repaired within a few generations. The one process now ongoing that will take millions of years to correct is the loss of genetic and species diversity by the destruction of natural habitats. This is the folly our descendants are least likely to forgive us.

Equally eloquent is a classic quote from Aldo Leopold (1949):

> The last word in ignorance is the man who says of an animal or plant: 'What good is it?' If the land mechanism as a whole is good, then every part is good, whether we understand it or not. If the biota, in the course of aeons, has built something we like but do not understand, then who but a fool would discard seemingly useless parts? To keep every cog and wheel is the first precaution of intelligent tinkering.

What Are the Alternatives?

It would be very naive to assume that maintaining diversity is the only important goal of wildlife management. If you don't believe the Orwellian (1946) maxim about some animals being more equal than others, spend a summer weekend tallying T-shirts emblazoned with whales, puffins, and pandas and then go looking for one with a shrew, a wren, or a snake. As long as people are manipulating ecosystems, it is inevitable that their value systems will influence their goals. Species that are prized by hunters and anglers, or sought because of their aesthetic qualities, will always be accorded more attention by land managers, and there is nothing intrinsically wrong with this. Certainly, species threatened with extinction should receive this kind of special emphasis. In short, maintaining diversity is not the only important goal of wildlife management, but it is one of the most fundamental. Managing for certain preferred species is covered in a single chapter in this book, but this does not reflect the importance of this approach. Rather than give a superficial overview of the voluminous literature covering scores of species, Chapter 13, "Special Species," reviews the basic principles of species management and examines the interaction between species-oriented management and management for diversity. For several species, more specific information is presented in case histories distributed throughout the book.

Enhancing productivity is also often stated as a goal of wildlife management. Usually this refers to maximizing the productivity or sustainable yield of game species so that large crops can be harvested annually. With our emphasis on all biota, enhancing productivity could also be interpreted to mean increasing the rate at which the sun's energy is fixed by photosynthesis and subsequently transformed into the potential energy of plant and animal matter. Enhancing ecosystem productivity is fairly straightforward; farmers have been doing it for millennia primarily by increasing the availability of water and nutrients. However, the costs of irrigation and fertilization are so high that these practices are employed on only a tiny portion of forests and other natural ecosystems.

GOALS FOR DIVERSITY MAINTENANCE

Returning to the definition of biological diversity that opened this chapter—diversity of life in all its forms and levels of organization—it is apparent that the balance of the chapter has largely departed from this broad perspective and narrowed to a focus on species diversity. This is no accident. It has occurred because species are the most easily recognized and defined elements of biological diversity, and this makes maintaining species diversity a feasible, measurable goal. (This should be qualified as relatively measurable, given that probably no more than 20% of the world's species have been described by scientists, and the differences between closely related species are often obscured by hybrid individuals.)

One could argue that genes are the most important elements of biological diversity—the building blocks of life—but with each individual organism containing roughly 1,000 to 100,000 genes, they are costly and complex to measure and understand (Wilson 1988). As a result, genetic diversity is difficult to manage except under controlled circumstances such as labs, zoos, botanical gardens, and farms, or with wild populations that are intensively manipulated, for example by artificially exchanging individuals between genetically different populations (OTA 1987). These are situations that have generally little to do with forest management and lie beyond the purview of this book.

Alternatively, one could go to the opposite end of the spectrum and maintain biome diversity, but biomes are large, coarse things and there are far fewer different kinds of biomes on the whole planet than there are genes in a single insect. Many people would argue that by maintaining biome diversity—some deserts, some tropical forests, some savannas, etc.—almost all the species that comprise the biome would be saved too. This is true, and the same reasoning will be used many times throughout this book to advocate maintaining diverse forests in order to maintain the diversity of forest wildlife species. However, this argument is not based on biome diversity as a goal; it is based on biome diversity as a means by which a somewhat different goal—maintaining species diversity—is pursued.

In subsequent chapters we discuss many ways to make forests diverse; we discuss tree age diversity, forest landscape diversity, the diversity of dead trees versus live ones, and others. These units of diversity have intrinsic value—consider the disease-resistant gene of the wild rice from India or the beauty and productivity of a rich, fertile forest. However, for the practical reasons just described, we generally consider these types of diversity as means to be pursued in the context of maintaining species diversity, not as ends in themselves. By focusing on the maintenance of species diversity, we have a reasonably measurable goal, and in pursuing it, biological diversity at many other levels—genes, populations, communities, and biomes—can usually be maintained as well.

A Diversity of Diversity

It is extremely important that forest managers take a broad-scale perspective toward managing for diversity by striving for a "diversity of diversity" across the landscape, not maximum diversity on each hectare of forest (Samson and Knopf 1982). This point is critical enough to merit both an example and a metaphor.

You cannot expect to grow wild turkeys, ruffed grouse, spruce grouse, ring-necked pheasants, sharp-tailed grouse, greater prairie chickens, bobwhite quail, and gray partridge on the same property, even in Wisconsin where their geographic ranges come together. With careful management, you might have reasonable populations of ruffed grouse and spruce grouse on a hundred-hectare tract of aspen and jack pine stands, but it would be extremely difficult to also find room for a sizable bit of prairie where a prairie chicken could survive. Every species has a unique set of habitat requirements, an ecological niche consisting of its preferred physical environment and its interactive roles with other species in the community. Consequently, there is a limit to how much diversity you can crowd into a finite area. Similarly, if it were determined that one type of ecosystem could support a large diversity of these birds in a limited area—e.g., quail, pheasants, partridge, and prairie chickens in farmland—it would be a mistake to turn much of Wisconsin into farmland in order to maximize the diversity of upland game birds. Significant areas should be managed for spruce grouse and ruffed grouse and turkeys too lest diversity at a regional scale be sacrificed for small-scale diversity.

Here is a metaphor about the same point: Imagine you own two boatyards employing a total of 24 apprentices, and in each boatyard 8 apprentices learn carpentry skills to make wooden hulls, and 4 learn to make sails. (In this metaphor each boatyard represents a piece of land, and the apprentices are the "biota," with 24 individuals representing two "species," carpenters and sailmakers.) Now suppose you decided to increase the diversity of your training program, and you discovered that by building metal-hulled motor boats you could train apprentices with four skills, instead of the two required for wooden sailboats. Would you convert both your boatyards to metal boats? No. You would convert one to metal boats, leave the other for wooden boats, and collectively end up with a program that taught six different skills, not just the four required for metal boat building.

For any given area, that is, for any combination of latitude, longitude, and altitude, there will generally be one type of ecosystem that will have the maximum diversity. It would be a serious mistake for land managers to strive to have this, and only this, type of ecosystem in the area they manage, if they wish to manage for diversity. Assuming they have sufficient land area, they should also maintain a number of other types of ecosystems. Which ecosystems to maintain will depend on the physical environment of their land and the types of ecosystems occurring on neighboring land.

Manage, Maintain, and Maximize

Before leaving this chapter it is important to distinguish explicitly among managing for diversity, maintaining diversity, and maximizing diversity. *Managing* for diversity simply assumes that having a variety of wildlife is a management goal; it is a somewhat vague phrase. To *maintain* species diversity is to ensure that viable populations of all the native species of flora and fauna characteristic of the management area will be present. *Maximizing* diversity should probably not be a goal, because it could lead to the aforementioned prospect of trying to raise eight species of game bird on the same area, or even trying to import exotic species. In cases where the natural diversity has been reduced by human interference, it would be preferable to speak of restoring diversity, rather than maximizing it.

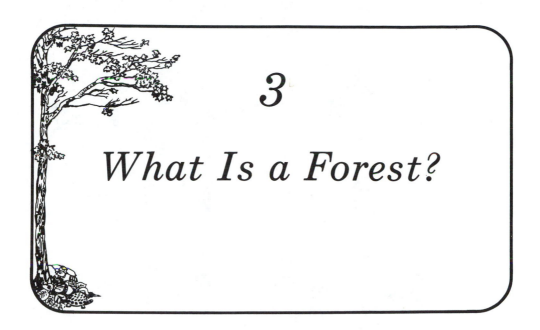

3

What Is a Forest?

In all likelihood, the majority of people reading this book have spent most of their lives within a stone's throw of a forest. However, it is worth remembering that, with only about one-fifth of the earth's land area covered with forest and nearly half the people living in cities, forests are not a central part of everyone's experience. Even for people who live and work daily in forests, perhaps especially for them, it is useful to stop and think, "What is a forest?" The simplest answer, an ecosystem dominated by trees, raises two more questions, "What is an ecosystem?" and "What is a tree?" or, in less simplistic terms, "What special features of trees distinguish them from other plants?"

An *ecosystem* is all the interacting populations of plants, animals, and microorganisms occupying an area, plus their physical environment. The living organisms in an ecosystem are collectively called a *community,* sometimes natural community or biotic community. This definition is very easy to envision with an ecosystem such as a lake: The basin of water and all its inhabitants comprise an ecosystem. Similarly, a forested island in the middle of the lake would also constitute a distinct ecosystem. The problem of defining an ecosystem usually comes in trying to decide where one ecosystem ends and another begins. Would an "island" of deciduous trees in a "lake" of conifers comprise a separate ecosystem? If the deciduous island were reasonably large, many would consider it a separate ecosystem even though there would be extensive interaction between the conifer "lake" and the deciduous "island." There will always be a certain degree of interaction between adjacent ecosystems; even a forest and a lake exchange frogs, swallows, mink, and autumn leaves. Some people define ecosystems in a way that will encompass much larger areas. Biol-

ogists concerned with the grizzly bears of Yellowstone National Park speak of a 20,000 km² area, including the park and most of the land within 50 km of it, as the Yellowstone Grizzly Bear Ecosystem (Servheen 1985). One could argue that the only distinct ecosystem is the planet. However, for the purposes of this book, we can think of forest ecosystems on a scale that makes them comparable to what foresters call a *stand*, a group of trees that are reasonably similar in age structure and species composition, usually occupying at least two hectares (5 acres). However, a forest ecosystem is *not* a stand; stand refers to just the trees; ecosystem refers to all the organisms and their environment.

A tree is a large, woody, perennial plant, usually with a single stem. In most people's vocabulary softwood trees, such as pines, spruces, cedars, and hemlocks, are also known as conifers, evergreens, and needle-leaved trees, whereas hardwoods, e.g., oaks, maples, elms, and birches, can be called broad-leaved, or deciduous trees. Although these terms are nicely familiar, they do not provide a very definitive classification. Most "hardwoods" are relatively slow-growing and thus have high-density wood, but some, aspen for example, are as soft as a "softwood." Indeed, one of the softest woods of all, balsa, comes from what is, in other respects, a hardwood tree. Most "deciduous" trees in temperate latitudes lose their leaves in the fall (this North American name for autumn would be nicely balanced if we called our vernal season "rise"), but some oaks keep their leaves through the winter and some "evergreens," e.g., larches, lose theirs. In tropical areas, many hardwoods drop their leaves during a dry season, but some retain theirs for long periods; the latter are called broad-leaved evergreens. Leaf shape is a fairly consistent criterion, but again some "broad-leaved" trees, e.g., acacias, have leaves that look narrower than some "needle-leaved" trees such as cedar. Plant taxonomy is primarily based on sexual structures of plants, and it is here that the distinction is clearest. Pines, larches, and cedars are gymnosperms, because they have "naked" seeds that are not enclosed in an ovary. These seeds are born in a cone, thus giving rise to the name conifer. Oaks, aspens, and balsa are angiosperms, because their seeds are encapsulated in an ovary. In this book, gymnosperms are referred to as conifers and, for lack of an equivalent term, angiosperm trees are called deciduous trees.

Leaves and sexual structures may be the basis of classifying trees, but the critical adjective in the overall definition of a tree is woody. Wood is extraordinary stuff and, like forests, we tend to take it for granted. It is the essence of what makes a forest, because two of its characteristics, strength and durability, allow trees to be large and long-lived.

Wood and the Forest

When you compare forests to other ecosystems, the first thing you notice is how tall they are. The strength of wood means forests can routinely grow over 25 meters tall; some redwood, Douglas-fir, and eucalyptus trees approach 75 m. Only by going underwater can you find an ecosystem with a greater vertical dimension, and even there below 25 m things generally become quite dark and dismal. Other terrestrial ecosystems such as deserts, grasslands, and tundra lack this three-dimensional quality and thus have much less vertical stratification than forests. Two features shape the vertical stratification of a forest: the structure of the trees and the microclimate (Figure 3.1).

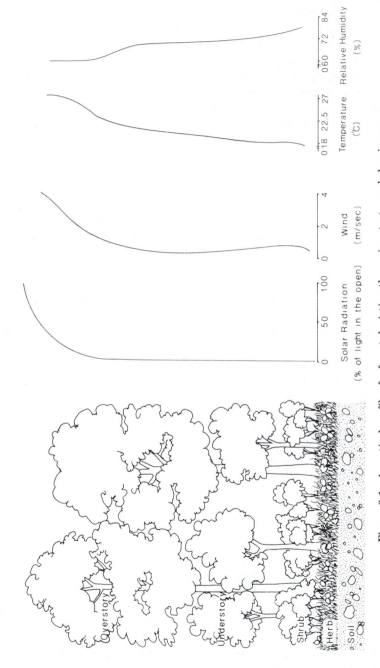

Figure 3.1 A vertical profile of a forest depicting the major strata, and showing some representative microclimate parameters changing from top to bottom in a temperate deciduous forest at mid-day in mid-summer. Briefly, the forest canopy is sunnier, windier, warmer, and drier than the forest floor.

The forest canopy or overstory is composed of the crowns of individual trees, elaborate structures designed to support a mass of leaves in a fashion that will efficiently intercept sunlight. The leaves are borne on twigs, the twigs on branches, the branches on limbs, and the limbs on boles, in a pattern that is found in many natural phenomena such as river drainages, nerves, and mineral structures. This pattern, so commonly associated with trees, is called dendritic from the Greek word dendron, meaning tree. The pattern is repeated in the understory trees and shrubs and to some extent by the herbaceous plants of the forest floors. Below the ground, roots and layers of organic matter and soil continue the vertical stratification. In a humid forest, epiphytes—mainly mosses, lichens, and fungi in temperate climates—grow up and down the trees adding additional variety. The net effect of all this is a great deal of structural diversity from the bottom to the top of a forest.

Everyone has an impression of forests being relatively cool, dark, humid, and windless. Marked differences in these features at different levels in the forest create a stratification of microclimate (Figure 3.1). Solar radiation reaches the top of the canopy relatively unimpeded, but by the time it filters down to the forest floor, many of the wavelengths have been absorbed or reflected and only shifting sunspots of direct radiation remain. Similarly, a large portion of the rain and snow is intercepted and evaporated, or reaches the ground only by flowing down the tree stems. The net loss of water is partly compensated for because the reduced solar radiation means the forest is cool and humidity high. The humidity is high also because forests tend to deflect drying winds up and over the forest. There are many exceptions to this picture depending on forest type, season, and time of day, but the important point is that the interior of a forest is very different from the exterior. Together, the vegetation structure and microclimate gradients make the forest an ecosystem in which vertical stratification provides niches for many species.

The strength of wood, so manifest in the height of forests, is not an ephemeral quality; it lasts; wood is durable. Compare a forest to a salt marsh, an arctic meadow, or a desert creek, and you are immediately struck by how much organic matter, both living and dead, accumulates in a forest (Table 3.1). In most forests, over 95% of the biomass is in woody tissue—boles, limbs, and roots (Packham and Harding 1982). This cornucopia of material is both shelter and food for a large array of organisms: woodpeckers, wood ants, centipedes, millipedes, slugs, sowbugs, etc. Whether the wood is dead or alive, erect and intact, or prostrate and falling apart, myriad creatures make a living tunneling through it, chewing it up, and reducing it to simpler constituents.

Time and the Forest

The height and organic matter accumulation of a forest is not something that develops overnight. It takes time to grow a forest, but not as long as most

TABLE 3.1   Average Standing Crop of Biomass of Various Ecosystems

Ecosystem	Total Grams of Dry Weight per Meter^2	Reference
Temperate forest		
Young oak-pine	10,194	Whittaker and Woodwell, 1969
Mature deciduous	42,500	Whittaker et al., 1974
Tropical rainforest	6,595–48,821	Ovington and Olson, 1970
Grasslands	63– 2,299	French et al., 1979
Salt marsh	3,843–16,012	Gallager, 1974
Desert	400– 4,000	Noy–Meir, 1973
Arctic tundra	19– 1,171	Bliss, 1977, p. 661
English channel	25[a]	Harvey, 1950
Coral reef	846[a]	Odum and Odum, 1955

[a]Includes both producers and consumers.

people think. Images of ancient bristlecone pines and sequoias awaiting the arrival of Columbus are far more pervasive than notions of forests that grow 10 m tall in 10 to 20 years. Relative to the pace of human life, forests grow so slowly that it takes a long-term perspective to avoid thinking of them as static. Even in New England, where stone walls trace memories of former fields through much of the forest, people often express surprise when they see old paintings and photographs showing a landscape dominated by agriculture.

Of course, any good observer of the landscape knows that ecosystems change a great deal through time, primarily through the process called *succession* (Figure 3.2). The idea that one type of ecosystem gradually transforms into another, e.g., that fields become forests, can be found as early as 300 B.C. in the writings of Theophrastus (Drury and Nisbet 1973). Henry David Thoreau, that astute reader of the New England landscape, is credited with naming the process "succession" (Spurr 1952), although a less parochial view would have to give that distinction to a French biologist, Dureau de la Malle, because the French word is identical and he wrote about succession 35 years before Thoreau.

From a classical perspective, succession begins after a disturbance (such as a fire, glacier, or clearcut) removes all or most of the community on a site. Pioneer species invade this vacant area, become established, and reproduce. (Plants that win the appellation "weeds" are usually pioneer species.) After a period, these pioneer species modify their environment—for example, they make a sunny spot shady—and will no longer be able to maintain their populations. New species move in, changing the character of the community and again modifying the environment, until they too can no longer survive and yet another group of species becomes established. In this manner, each group of species facilitates the establishment of the next group (Figure 3.3A). The process

Time →

Figure 3.2 Vegetation changes through time in a pattern called succession. In forests, this usually means a shift from herbaceous plants to shrubs, then shade-intolerant trees, and then shade-tolerant trees.

A

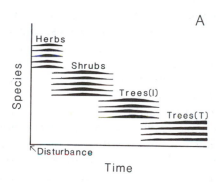

B

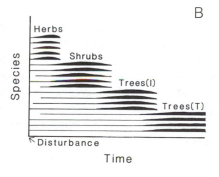

Figure 3.3 Two alternative models of succession in which the lines represent the presence and absence of different plant species and the width of the lines represents their relative abundance. In A, the facilitation model, each group of species alters the environment making it difficult for those species to continue living there and thus facilitating the establishment of a new group of species. In B, the initial floristics model, all the species are present on the site soon after a disturbance, but each group dominates the site at different times, depending on how fast they can grow or how large they can become. (I = shade-intolerant; T = shade-tolerant). (Redrawn from Egler 1954.)

of community replacement and environmental modification is predictable and will continue until it culminates in an ecosystem that is stable and self-perpetuating, a climax ecosystem. The climax ecosystem will persist until another disturbance starts the whole cycle over again.

This model of succession, as expounded by its creator Frederic Clements (1916) and various disciples such as Eugene Odum (1969), has been sharply criticized by many writers (reviewed by McIntosh 1980). They are particularly opposed to Clements' likening a community to a supraorganism, a creature that is hurt by a disturbance and heals itself through succession. They are adherents to Darwin's ideas of natural selection and see competition among individuals, rather than cooperation among the components of a larger entity, as the driving force behind succession. In their alternative model, all the species that will participate in succession become established soon after a disturbance; some are dispersed to the site; some are already there, buried in the soil as seeds or roots (Egler 1954). These species begin competing for the site's resources, primarily light, water, and nutrients, and initially the fastest growing plants, often grasses and other herbaceous plants, become dominant. Later they are outcompeted by slower growing plants that can reach a larger size, typically shrubs, then trees (Figure 3.3B).

The dichotomy between these two models has many ramifications; an especially important one for conservationists is how climax ecosystems are viewed. Followers of Clements tend to view climax ecosystems as representing the natural situation, the state to which ecosystems will eventually return after being disturbed. Clements' critics emphasize that disturbances are natural, frequent occurrences—e.g., every time a tree falls over there is a small-scale disturbance—and that climax ecosystems, if they exist at all, are just another stage of an endless cycle.

Some ecologists would discard the whole idea of succession. However, it seems that the idea has some value as a coarse description of a very prevalent phenomenon, even if it has not yet been developed into a unifying theory. Furthermore, it is reasonable to hope that a coherent theory will eventually be formed (Huston and Smith 1987). One recent attempt synthesizes ideas from both models by recognizing both the importance of competition between species and the positive ways in which one species can facilitate the establishment of another (Finegan 1984). Another model highlights a successional transition from plants that compete well for nutrients to plants that compete well for light, rather than a shift from fast-growing plants to slow-growing plants (Tilman 1985). Despite all the controversy, it is hard to argue about the simplest facts: Most undisturbed fields do become forests in a reasonably predictable manner, and certain kinds of forest often remain relatively stable features of the landscape for periods that are much longer than the time it takes a pasture to become a forest. While ecologists labor to understand this, most people will continue to call such forests climax ecosystems and the process by which they develop, succession.

Clearly, this is all of great relevance to forest managers. The role of different tree species in forest succession is the basis for an important distinction that foresters make between *shade-tolerant* trees and *shade-intolerant* trees (Table 3.2). Shade-intolerant trees are easily dispersed species that soon invade a disturbed site and grow rapidly, forming the first forest community in the successional sequence. Because they cannot reproduce in their own shade, they do not persist on a site and are gradually overtaken and replaced by slow-growing, shade-tolerant trees. The seedlings of shade-tolerant trees are able to survive in the shadow of their parents waiting for an opportunity, an opening in the canopy, to grow up, and thus shade-tolerant trees prevail in climax forests. Forest managers have to understand disturbance, because forests are disturbed frequently, on scales ranging from the toppling of a single tree to the burning of thousands of hectares. The height of trees makes them particularly

TABLE 3.2   Shade Tolerance of Some Common North American Trees

Very Tolerant

Balsam fir	American beech	
Eastern hemlock	Sugar maple	
Western hemlock		

Tolerant

Northern white-cedar	Red maple	
Red spruce	Silver fir	
White spruce	Redwood	
Basswood	Englemann spruce	

Intermediate

Eastern white pine	Hickory spp.	Douglas-fir
Black spruce	Northern red oak	Western white pine
Ash species (spp.)	Southern red oak	Giant sequoia
Yellow birch	White oak	Red alder

Intolerant

Bald cypress	Slash pine	Sycamore
Loblolly pine	Juniper spp.	Paper birch
Pitch pine	Lodgepole pine	Black cherry
Red pine	Ponderosa pine	Yellow-poplar

Very Intolerant

Jack pine	Aspen spp.	
Longleaf pine	Gray birch	
Sand pine	Willow spp.	
Tamarack	Western larch	

Source: Adapted from Hocker, 1979.

susceptible to being blown down or hit by lightning, and all that organic matter becomes fuel during a fire or food during an insect outbreak. On the other hand, most of the organic matter—the wood—is a valuable commodity that makes loggers reluctant to let natural disturbances run their course. Disturbance and succession are recurring topics throughout this book.

Time and the Forest: A Long-term Perspective

It is one thing to look back a few decades and realize that forests are not a static feature of the landscape because of patterns of disturbance and succession; it is another thing to look back a few thousand years and see forests as a dynamic phenomenon on a continental scale. Ecologists have done just this by analyzing the pollen grains contained in layers of sediment deposited on lake bottoms. They have assembled pictures of forests shifting back and forth across Eurasia and North America in response to glaciation and global temperature changes, and of savannas and tropical forests moving across Africa, Australia, and South America with changes in the pattern of rainfall. It is intuitively sensible that 18,000 years ago, when a sheet of ice extended south to New York and Michigan, a boreal forest of spruce trees would dominate the mid-Atlantic states and oak forests would be confined to the southeast (Jacobson et al. 1987). It is much less obvious that these forests may have borne relatively little resemblance to what we now think of as oak forests or boreal forests.

In Figure 3.4 the major distribution of four types of trees are mapped from pollen records; the maps show beech, hemlock, pine, and oak distributions 500 years ago (before European immigration) and 6, 8, and 12 thousand years ago (Jacobson et al. 1987). The 500-year maps bear a good resemblance to current vegetation: The overlap of beech and hemlock corresponds well to a region that is dominated by what is often called the northern hardwoods-hemlock forest (S.A.F. 1967); in the south one can see the region dominated by southern pines and the interface with a more central region where oak forests predominate. However, when you look back in time, you do not see this same pattern simply shifted south in response to the southward advance of the glaciers. You see the distribution of hemlock and beech diverge entirely; you see southern pine forests disappear completely and only a small area of oak-pine remain in southern Florida.

What is the relevance of all this? Although the forest maps derived from pollen analysis on a continental scale are not as precise as descriptions based on direct observations of stands, the underlying message is abundantly clear: Many of the types of forest ecosystem we recognize today may be relatively recent phenomena. Instead of a tightly interwoven ecosystem in which all the constituent species have lived together—and more importantly evolved to-

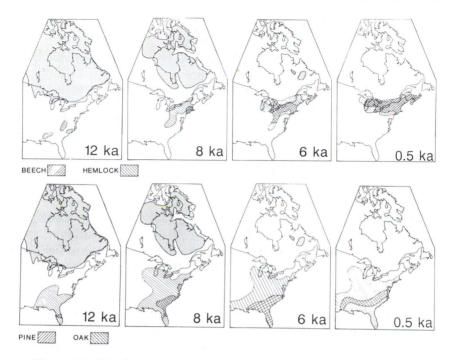

Figure 3.4 Distribution patterns of beech, hemlock, oak, and pine in eastern North America 12,000, 8,000, 6,000, and 500 years ago based on reconstruction from pollen profile analyses. (ka = kiloannum or 1,000 years) (Redrawn from Jacobson et al. 1987.) Note the divergence in geographic ranges as one goes backward through time.

gether—for millions of years, some forests may be more like a loose assemblage that came together just yesterday on an evolutionary time scale (Hunter et al. 1988). This does not mean that these forests are without examples of coevolution; it is probably safe to assume that spruce trees and spruce budworm, pine trees and pine warblers, have been together for more than 10,000 years. However, it does cast further doubt on the idea that all the species of a forest ecosystem combine to form a supraorganism, and thus further weakens the Clementsian model of succession. It also has implications for how one goes about managing for diversity, which are explored in later chapters.

Animals and the Forest

When describing forest ecosystems, it is easy to dwell on the trees because they are the framework, the essence, of a forest. Needless to say, they are not the whole story. Throughout this book, we will explore the relationships of trees, smaller plants, and animals in some detail; here we will briefly outline five of the most important relationships.

To slip toward anthropomorphism for a moment, it is common to think of plants as suffering from the attentions of animals. First and foremost, plants are food for animals; with minor exceptions like carnivorous plants and some microorganisms, they are always the first link in a food chain.

Because essentially all the biomass of a forest is in the plants, they comprise the physical habitat for many animals. Soil invertebrates and some other creatures may not need living plants in which to hide or seek shelter from inclement weather, but most forest animals would be very exposed without a cover of vegetation.

Plants benefit from the presence of animals in some fundamental, if inconspicuous, ways. They use the carbon dioxide that animals exhale and the nutrients, such as nitrogen and phosphorous, that animals help make available through their participation in the decomposition of plant matter. (Admittedly, fungi, bacteria, and other microorganisms have the lead role in decomposition, but larger creatures accelerate the process.)

Plants also benefit from a special feature of animals, the ability to move. Many plant species have evolved ways to exploit this ability in order to move pollen from plant to plant and to disperse seeds away from their parents. Usually the plant rewards the animal with special food, for example nectar or fruit, although some plants, e.g., burdocks, have seeds that cling to animals for a free ride.

There are many examples of plants being protected from one kind of animal by another kind; birds that eat caterpillars are a classic example. The ants that protect certain species of trees by warding off everything from caterpillars to deer and are rewarded with specially produced food, are one of the best examples of plants and animals coevolving for their mutual gain (Janzen 1966).

Why Are Forests Diverse?

Before answering the opening question, we need to consider a more general one: What determines the diversity of an ecosystem? Like many deceptively simple questions, this one has yet to be satisfactorily answered. Most ecologists focus on three factors as likely determinants of ecosystem diversity (Ricklefs 1987):

1. The abundance of an ecosystem's resources (e.g., water, nutrients, light) and competition for those resources from other species can influence diversity (Brown 1981). An ecosystem will have low diversity if a few species can use its resources fully, to the exclusion of

other species, and this is more likely if its resources are limited. If an ecosystem has a heterogeneous environment, wet patches and dry patches for example, it is more likely that species can coexist (Grime 1979).

2. Disturbances that severely alter the community and initiate a new succession may increase diversity by increasing the variability of the environment and thus decreasing the likelihood that a few species will exert long-term dominance (Denslow 1985).

3. Similarly, predators, parasites, and pathogens may limit the ability of any one species to dominate an ecosystem (Paine 1966). This relationship has been well-documented only in aquatic ecosystems, but one can speculate that without caterpillars and other herbivores (plant predators), a few species of plants might come to dominate some terrestrial ecosystems (Janzen 1970).

So why are forests diverse? Firstly, they exist in environments where resources are relatively abundant—not in the dry or cold environments that support grasslands, deserts, or tundra—and thus it is unlikely that a handful of forest species could outcompete all the others. Secondly, they are routinely subject to disturbance, again making it difficult for a few species to predominate. The third factor, limitations effected by animals and disease, may be important to forest diversity, but this has not been demonstrated.

Collectively, these factors may explain how diversity is affected by the environment's effects on interactions of species, but this is not the whole story. If we want to understand why one forest is more diverse than another, it is also important to consider factors of regional and historical scope, specifically the continually changing geographic distribution of species and the evolution of new species (Ricklefs 1987). This point is best explained with an example; consider the mangrove swamps that form along many tropical coastlines. Throughout both Africa and Latin America these forests are comprised of just three or four tree species, each one occupying a different tidal zone. It would be easy to conclude that some fundamental ecological principles are responsible for the similar diversity of these ecosystems—perhaps some basic laws about the competitive ability of tree species that have adapted to having their roots inundated by saltwater (Ricklefs 1987). However, this interpretation falls apart upon further scrutiny; in Malaysia mangrove swamps are formed by 17 principal and 23 less dominant tree species. Apparently, Africa and America have impoverished mangrove swamps because the rich flora of Malaysian mangrove swamps has not dispersed to these continents, nor has a similar flora evolved independently in Africa or America. No doubt, ecologists and conservationists will continue to focus on the local explanations for ecosystem diversity, but the long-term and broad-scale picture should not be ignored.

Part I
SUMMARY

"Wildlife" is a word of recent origin, and although often associated with game birds and mammals, its meaning is gradually expanding. In this book, it is defined as all forms of life that are wild and thus includes all wild animals, plants, and microorganisms. Using a broad definition of wildlife avoids the use of phrases such as "destroying wildlife habitat," and "good for wildlife," because, for example, forest practices that destroy habitat for one species will create good habitat for another. From this perspective, one of the most basic objectives of wildlife management is to maintain or restore all of a region's native wildlife populations, with a particular emphasis on endangered species.

Biological diversity refers to the diversity of life in all its forms and levels of organization: animals, plants, and microorganisms are the three major forms; genes, species, communities, and biomes are among the many levels of organization. Managing for biological diversity is of critical importance because it is essential to the ecological well-being of the planet, and human welfare is ultimately dependent on this. Every element of biological diversity has some economic or ecological value, although in many cases the economic value remains unrealized. Ideally, natural resource managers will maintain all levels of biological diversity; in practice, maintenance of species diversity will often be the fundamental goal because it is a relatively feasible and measurable goal. A broad approach to managing diversity is needed. If you tried to maximize diversity on every woodlot, all the woodlots would end up much the same, the landscape as a whole would be homogeneous, and diversity on a larger, regional scale would be diminished. Therefore, it is essential to think in terms of maintaining a diversity of diversity across the whole landscape, not maximizing diversity on each piece of land.

A forest ecosystem is a community of plants, animals, and microorganisms, and the physical environment they inhabit, in which trees and the dominant life form. Trees are tall, long-lived plants because they are made of wood, a material of great strength and durability. The height of trees, and thus of forests, produces considerable vertical stratification in terms of both the structure of the trees and the microclimate they create. The durability of wood allows large amounts of organic matter to accumulate. Together, vertical stratification and an abundance of organic matter provide niches for a wealth of organisms, making forests especially diverse ecosystems. It takes time for trees to grow tall and for biomass to accumulate and thus succession, the process by which one group of species replaces another, makes forests even more diverse when viewed through a window of time. On a longer time scale, measured in thousands of years, forests are an ever-changing assemblage of species shaped by climate and other physical factors. In terms of spatial scale, we can think of forest ecosystems on a scale that makes them comparable to what foresters call a stand, a group of trees that are reasonably similar in age structure and species composition, usually occupying at least two hectares (5 acres). However, a forest ecosystem is not a stand; stand refers to just the trees; forest ecosystem refers to all the trees and their environment.

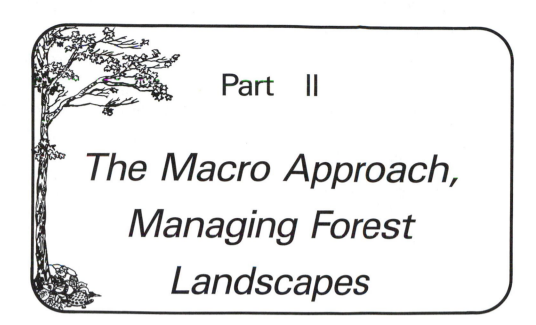

Part II

The Macro Approach, Managing Forest Landscapes

From a forest manager's perspective, a large stretch of forest is naturally subdivided into smaller, relatively homogeneous units called stands. This makes the task of managing a large tract much easier because it can be approached one stand, one group of trees, at a time. Similarly, we can think of those stands of trees as defining the boundaries of ecosystems to better understand the distribution and management of all forest organisms. However, in this section we will view forests from a larger scale, a macro perspective that looks at the whole forest landscape. Stands and the corresponding ecosystems typically range in size from a few hectares to a few hundred hectares; here we will look at the forest in a scale measured in tens or hundreds of square kilometers.

This will remind some readers of the alpha, beta, and gamma diversity referred to by some ecologists (Whittaker 1960, Cody 1975). Alpha diversity is the diversity within a habitat; two lizards that share the same forest by living at different heights would exemplify alpha diversity. If the lizards lived in different, nearby habitats, this would represent beta, or between habitat diversity. Finally, if the lizards were separated geographically, living at different latitudes, for example, this would be a case of gamma diversity. This section is closely akin to beta diversity, although when ecologists think of between habitat differences they are more likely to be referring to a forest versus a meadow than a 20-year-old forest stand versus a 120-year-old stand, or a stand of pine versus a stand of spruce.

The basic assumption of this section is that the most efficient way to maintain biological diversity in a forested landscape is to have a diverse array of stands and thus a diverse array of ecosystems and their constituent species.

The first three chapters explore this idea by considering three key features that define a stand. Two of these features, the stand's tree species composition (Chapter 4) and age (Chapter 5), are standard parameters in stand definitions. However, in Chapter 6 a rather new way of looking at stands based on their area and horizontal structure is described.

The next three chapters examine the way stands are arranged on the landscape. Specifically, Chapter 7 deals with edges, the transition zones between stands, and Chapter 8 details the consequences of a forest stand being isolated from other stands. Together, these chapters outline a topic of growing concern to forest wildlife managers: forest fragmentation. Finally, Chapter 9 concentrates on a special type of edge, the riparian zones that lie between forests and aquatic ecosystems.

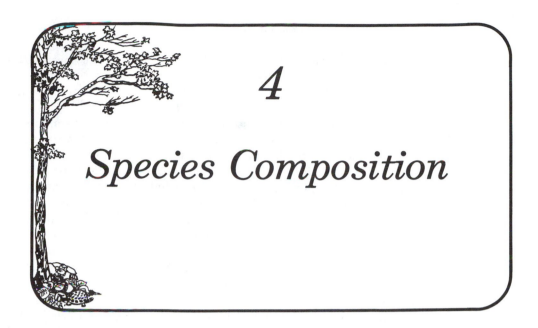

4

Species Composition

Upon entering a forest, the first thing a perceptive visitor will notice is which kinds of trees grow there; trees dominate the ecosystem; they define it. People speak of oak-pine forests and spruce-fir forests, not squirrel-turkey forests or mayflower-moss forests. Understanding the associations of different tree species and their distribution across the landscape is fundamental to forest ecology and management. In some regions a single species will prevail over a large expanse of land. A swath of spruce or birch stretching nearly to the horizon is a common vista in many boreal regions, and fewer than ten major species characterize the forest all the way from Norway to eastern Siberia. Compare this to tropical rain forests where practically every individual in a small area might be a different species. A single hectare of Brazilian forest can have over 200 tree species, and a small forested volcano in the Philippines, Mount Makiliang, is said to harbor more species of woody plants than the entire United States (Myers 1979).

Of course, the assortment of trees that comprises a forest is not simply a matter of how far it is from the equator. A great complex of environmental factors affects forest composition. Some, such as climate and soil, are intrinsic, relatively stable, qualities of a site whereas others, e.g., the presence of competitors or herbivores, are profoundly affected by succession and thus can change quite quickly. To understand what is going on you have to start simply, by considering each species and its niche individually. It is usually easiest to begin by thinking about a species' relationship to the physical environment. How much water does it need? How much potassium? What temperatures can it tolerate? After this picture is reasonably clear, one can start thinking about groups of species and their interactions. Will these two species compete for light? Will these compete for nitrogen? Will species *A* eat species *B,* or will species *B* be infected by species *C?*

To describe all these interactions within an ecosystem rigorously and quantitatively would be nearly impossible. Fortunately, at a more superficial level the relationships are often straightforward. Anyone reasonably familiar with the forests of the Florida coastal plain will know that a forest in a wet depression is likely to be dominated by bald cypress, but that bald cypress would be a very rare sight on a sand ridge. Knowing what type of forest will exist where is much more difficult if you include stands that have been disturbed in the last few decades. People who make predictions on a grand scale, by drawing vegetation maps for a whole region, usually ignore succession and concentrate on mapping climax vegetation. (Actually, to avoid the controversy around climax, terms such as potential vegetation are often used.)

In short, there is a considerable degree of concordance between the distribution of trees and their environment. It is not a rigid equation (exceptions are commonplace), nor a static one (recall the thousand-year perspective of Figure 3.4), but the basic relationship is reasonably sound and predictable. This relationship is of great interest to foresters, because by understanding it they can test its bounds. Through various silvicultural practices they can convert one type of forest to another and, through artificial regeneration, they can even grow trees outside their natural range. The motivation for doing so is simple; trees vary enormously in their profitability and foresters can take advantage of this by replacing low-profit trees with high-profit trees, where feasible. A forester who could figure out how to grow black walnut (a species in which an individual tree can be auctioned off for $10,000) in a swamp where only alders grew before, would garner considerable attention. (Of course, profitability is not just a question of commercial value; the cost of stand establishment, maintenance, and harvest, and growth rates are critical too.)

Around the world, it has proven profitable to convert many millions of hectares of mixed-species forests to single-species plantations (Evans 1982, Sedjo 1983), and this has concerned wildlife advocates who equate these conversions with wildlife habitat degradation. Before exploring this issue, it is important to put it in context. In 1975 there were about 90 million hectares of plantations in the world (Table 4.1; FAO 1978); this is a large area but it represents less than 5% of the world's forests. In other words, the extent of plantations may be a local or regional issue, but is not really a global problem at this time. With careful planning by natural resource managers, hopefully it will never become a global problem.

Tree Species and Wildlife

Every tree in a forest provides food, shelter, or substrate for other organisms, often a diverse and abundant assemblage. Insects consume leaves and

TABLE 4.1 Area of Plantation Forests by Region, Circa 1975 (FAO 1978)

Economic Class and Region	Million Hectares
Developed	
North America	11
Western Europe	13
Oceania	1
Other	10
Total	35
Developing	
Africa	2
Latin America	3
Asia	2
Total	8
Centrally planned	
Europe and the Soviet Union	17
Asia	30
Total	47
Total World	90

lay their eggs in bark crevices; epiphytic lichens festoon the bole and larger branches; mycorrhizal fungi form sheaths around tree rootlets and exchange nutrients for carbohydrates and a substrate; parasitic plants take nutriment and give nothing; birds and mammals come and go in search of insects, seeds, and a place to hide. Some of these relationships are very tight, with a species of insect or parasitic plant occurring only with a single species or genus of tree. Such relationships are often attested by the species' name: Douglas-fir tussock moth, white pine weevil, beechdrops, pinesap, birch bolete, etc. Needless to say, if a forest stand is converted from one type to another, at least a few organisms, those uniquely tied to the eliminated trees, would cease to exist in the stand. In other words, the ecosystem would be altered.

In most cases, the relationships between a tree species and other organisms in the forest are rather looser—spruce grouse are regularly found in jack pine forests, and oak ferns in spruce stands. Nevertheless, the tree species composition of a forest is very important, at least indirectly, to virtually all the organisms living there. The dietary preferences of browsing mammals provide examples familiar to wildlife managers; an especially colorful one comes from Aldo Leopold's (1936) account of the history of Germany's forests and deer. Beginning with the march of Caesar's troops through a vast forest of oak and beech, Leopold traces how politics, economics, and technology combined to shape the interactions of red deer, roe deer, and an array of tree species. Of particular interest is the period (ca. 1810–1914) when the deciduous forests were replaced by spruce, and to a lesser extent pine, on a vast scale. Initially, the production of wood was very high, but in retrospect this period was known as *fichtenomania* or spruce madness, because it led to the deterioration of the soil and greatly diminished yields. Additionally, this same period saw a severe decline in the deer population, because spruce is not very palatable for deer. Later, when a balance between deciduous and coniferous trees was struck to create a *dauerwald* or permanent forest, the deer population rebounded so well that many of their preferred food plants, notably yew, became uncommon except where protected by fencing. (See Wolfe and Berg [1988] for an update on this story.) It is hardly surprising that deer would be dramatically affected by converting deciduous forests to coniferous forests, but it is noteworthy that the diet preferences of herbivores can also be quite subtle; for example, research has shown that to a squirrel an acorn is not just an acorn. One study showed that gray squirrel consumption of seven species of acorn ranged from 20 to 95% in inverse proportion to the acorns' tannic acid content (Harris and Skoog 1980).

Tree species selectivity is also evident in the foraging behavior of insectivorous birds, because most species do not allocate their time among different tree species in proportion to the abundance of the trees (Franzreb 1978, Eckhardt 1979, Holmes and Robinson 1981). Some trees are preferred by certain species and avoided by others—this can be the basis for ameliorating competition between closely related birds (Willson 1970). On the other hand, some trees are favored by practically the whole avian community. For example, in a New Hampshire forest all of the ten commonest foliage-gleaning birds species showed a preference for yellow birches (Holmes and Robinson 1981). The most obvious explanation for such preferences is that some trees have a richer supply of insects for food (Homes et al. 1986), but differences in the trees' structure may also account for bird preferences (Jackson 1979). Perhaps yellow birches are preferred because their leaves have short petioles and are arranged alternately along the twigs in a manner that makes it easy for the birds to spot insects on the leaf surfaces and glean them off without leaving their perch (Holmes and Robinson 1981). Similarly, aspen may be avoided by some birds because they find it hard to forage on leaves that tremble in the slightest breeze (Franzreb 1978), and in winter many Sierra Nevada birds focus their attention on incense cedar, presumably because they can readily find food under its loose, flaky bark (Morrison et al. 1985). In one study, tree architecture was more important than food abundance; black-throated green warblers preferred red spruce over white spruce, even though the insect density was lower in the red spruces, presumably because the red spruces' long, slender boughs probably made it easier to move along them (Morse 1976). Given these relationships between various species of birds and various trees, it is easy to understand why analyses of whole bird communities and tree communities have found some correlations (Rice et al. 1984); in a southern Quebec study there was a highly significant correlation between bird species diversity and the number of tree and shrub species (Menard et al. 1982).

Conifers and Deciduous Trees

Far more important than the differences among individual tree species are the differences between conifers and deciduous trees or, more accurately, gymnosperm trees and angiosperm trees. As a gross generalization, deciduous forests have a richer biota than coniferous forests (Glenn-Lewin 1977). There are a number of possible reasons for this; the most obvious is that conifer foliage is less palatable than deciduous foliage because it is rich in terpenes and other distasteful chemicals (Longhurst et al. 1968). This means that there are fewer animals that consume conifer foliage and this, in turn, means there are fewer predators to consume the consumers (Kennedy and Southwood 1984). Conifer litter is also relatively unpalatable to the invertebrates, fungi, and bacteria that shred and decompose organic matter in the soil. Consequently, decomposition of coniferous matter is often a slow process; in fact, in cool boreal forests the process is so slow that layers of partly-decomposed organic matter accumulate (Pritchett 1979). Furthermore, because conifer litter is acidic, it creates an acidic soil environment in which nutrients are leached out of the top layers of the soil. Thus, available nutrients are often limited in coniferous forests, and this may limit the abundance and variety of both plants and animals.

Two additional, rather speculative, reasons for the relatively rich fauna of deciduous forests have been suggested by Harris and Skoog (1980). First, most deciduous trees are angiosperms, a class of plants that underwent extensive evolution in the Cretaceous period, 135 million years ago. This is also the time when birds and mammals radiated extensively, and it is possible that many new species of plants and animals developed together in relationships involving seed dispersal, pollination, and herbivory (Regal 1977). Second, deciduous trees generally have more structural diversity than conifers because of basic differences in their anatomy. In most conifers the main branches radiate directly from a single trunk, whereas lateral branching is well-developed in most deciduous trees. Moreover, many conifers, especially pines, are self-pruning and thus have few lower branches when growing in mature stands.

The fact that most conifers retain their leaves throughout the year can be counted as an advantage for some forms of wildlife and a disadvantage for others. There are a few herbivores, such as the spruce grouse, that are absolutely dependent on conifer foliage to make it through the winter and, in warmer climes, insects can also continue their work year-round. Where winters are cold, the evergreen quality of conifers is very important to certain members of the deer family. Conifer stands provide crucial winter cover in which wind and radiant heat loss are minimized (Moen 1973), and because less snow accumulates under a conifer stand the deer's mobility is less restricted. As winter becomes spring, the presence of conifer leaves is a distinct disadvantage to many organisms living on the forest floor. In a deciduous forest, this is the time when sunlight reaches the ground unimpeded by leaves, warming the soil to its highest temperatures of the year and providing energy for a flush of photosynthesis and growth by forest herbs. Spring is relatively cool and dark in a conifer forest; in some areas snow persists under a conifer stand after flowers are blooming in nearby deciduous forests.

None of this means that conifer forests are biological wastelands; many support a unique and varied community, and they all support some forms of wildlife (Wiens 1975). Certainly, these differences should not be used to justify the conversion of natural conifer stands to deciduous stands. In northern Michigan, selective cutting and subsequent fires in the late nineteenth century converted magnificent forests of white and red pine to scrubby stands of shrubs and oak sprouts and aspen suckers (Whitney 1987); this was unfortunate from both a timber and wildlife perspective. Conversely, the differences in the communities associated with conifer and deciduous stands should always cause one to give some second thought to any management plan that calls for extensive conversions of natural deciduous or mixed forest into coniferous forest.

Exotic Trees

For over 5,000 years people have gone to great lengths to move plants with commercial value all over the globe. The mountain people of Nepal grow potatoes from Peru, and bananas, a native of Asia, are sold in Andean markets. This phenomenon has been largely restricted to food plants (it is much easier to find a tomato in a local garden than a teak tree in a nearby woodlot), but many trees are of significant economic value far from their native lands. Monterey pine, Australian eucalypts, Scots pine, Norway spruce, and Japanese larch have been planted far and wide for many years. In the early 1980s there was a massive effort to establish plantations of the Leucaena tree, a native of Latin America, in many tropical countries. Considered a miracle tree by its supporters, it can grow 15 meters in five years, produces fuel and building wood, fodder for livestock, seeds for human consumption, improves the soil by fixing atmospheric nitrogen, and is resistant to fire, drought, and pests (Brewbaker and Hutton 1979).

Such wonderful benefits for society do not come without some costs. Whenever a tree is artificially established very far from its natural habitat, it will come in contact with a whole array of new species. (This is the basis for an ecological definition of exotic; whether or not the move crossed some political boundaries is largely irrelevant. A Fraser fir from the mountains of Virginia would be nearly as exotic on the Virginia coast as it would be in Japan.) Almost inevitably, the new tree will displace one or more local species on the planted site, either because it is a successful competitor or, more likely, because its proponents give it a helping hand.

The introduction of an exotic tree may also have an effect on the forest fauna, particularly herbivores. Plants have evolved an impressive array of chemical and physical armaments against herbivores—alkaloids, tannins, spines, hairs, etc.—and the herbivores have had to coevolve means of overcoming these defenses (Coley et al. 1985). Thus, there is often a fairly tight relationship between plants and plant eaters. When a plant is established in a new place, most of the herbivores with which it evolved are left behind. This is obvious in the case of large herbivores (it is easier to move a eucalypt than a koala), but it also applies to the smallest and most numerous herbivores, insects (Strong et al. 1984.) As a result, a plantation of an exotic tree species will support a relatively impoverished fauna until the local fauna is able to adapt to the new plant; this is likely to be a while. Britain's sycamore trees (actually a species of maple) still have a very depauperate set of herbivorous insects even though they have been in Britain for at least 650 years; many claim they arrived during the Roman occupation. More generally, an extensive study of 28 kinds of British trees found that there was a very significant correlation between the length of time a tree had been in Britain (ranging from 350 to 13,000 years) and the number of herbivorous insects associated with it (Kennedy and Southwood 1984). Admittedly, most people could not care less about how many kinds of insect occur on a tree; indeed, if pest insects are left behind in a move this is counted an advantage. However, insects are important in their own right, and of course represent food to more popular forms of wildlife. In a study of tree species selection by English great tits, the only species to be significantly underused (relative to it abundance in a mixed forest) was the most recent addition to the tree flora, sycamore (Hunter 1980).

The effect of exotic trees becomes more dramatic when whole plantations are considered. In Australia, bird populations are larger and more diverse in the native eucalypt forests than in stands of introduced Monterey pine (Disney and Stokes 1976, Driscoll 1977, Friend 1982), but in Africa, Chile, and California it is the exotic eucalypt plantations that have impoverished bird communities (Smith 1974, Cody 1975,1985). As with all generalizations there are exceptions; plantations of exotic pines are poor bird habitat in South Africa (Cody 1975) and New Zealand (McLay 1974), but in Chile bird population densities and species richness are comparable to natural pine stands in the United States (Cody 1985). Interestingly, a study in the mountains of Kenya found plantations of exotic pines to have a generally depauperate bird community compared to native forests, but the pine stands were much more heavily used by winter migrants from Europe and Asia (Carlson 1986). Although best documented for birds, the low diversity of exotic plantations is also known for amphibians, reptiles, and mammals (McIlroy 1978), butterflies and moths (Magurran 1985), and even lichens (Barkman 1958).

Obviously, plantations of exotic trees have an established role in forestry which will not be eliminated in the future. This said, unless an exotic tree offers significant advantages over a native species, the native species should be preferred.

Monocultures

Natural forest stands in which a single species is dominant are moderately common, but natural stands almost entirely composed of a single tree species are rather rare. In contrast, most plantations are nearly pure monocultures (commonly there are some "weed trees" present), and they have a widespread reputation for supporting an impoverished flora and fauna (Figure 4.1) (Moss 1978*a*, Atkeson and Johnson 1979, Hill 1979). This is true whether they are composed of native trees or exotics, deciduous trees or conifers. It is not

Figure 4.1 Plantations of single species have a reputation for supporting a relatively low diversity of wildlife.

always bad—one of North America's rarest birds, the Kirtland's warbler, lives in nearly pure stands of jack pines—but on the whole the replacement of natural forests with monocultures has a decidedly negative impact on wildlife.

The reduction in diversity in monoculture stands could have serious consequences for the trees themselves, because the price of low diversity may be lower resistance to stress. Attempts to grow one of the world's most important nuts, the Brazil nut, in plantations have failed because monocultures do not support adequate populations of carpenter bees, the Brazil nut's major pollinators (Prescott-Allen and Prescott-Allen 1982). The Brazil nut can only be grown in natural forests where its pollinators have a full array of alternative hosts on which to forage throughout the year.

Analogies are often made between forests with a simple structure and the agricultural monocultures that are blamed for the Irish potato famine, boll weevil scourges, and other nightmares for farmers. The growth of insect and pathogen populations can be explosive when a large number of their favorite hosts are crowded into a small area, making it easy to move from one to another (Knight and Heikkenen 1980). For example, outbreaks of spruce budworm, perhaps the most destructive forest insects, are most prevalent in forests with relatively little diversity, especially large tracts of mature balsam fir (Mott 1963). Plantations, usually consisting of a single species of just one age, are among the most vulnerable types of forest (Schmidt 1978). These problems would probably be worse except for an important difference between agricultural and silvicultural monocultures. Agricultural crops tend to be extremely simple genetically, whereas tree geneticists have generally been more careful to maintain genetic heterogeneity (Zobel and Talbert 1984).

To summarize, it is understandable why foresters would want to create forest monocultures; their simplicity makes them much easier to manage and therefore cheaper. However, the economic justification is diminished when the costs of dealing with pests and pathogens are considered and an ecological justification (one predicated on the value of diversity and stability) is very difficult to find. One is reduced to saying: (a) there are some wildlife species that inhabit monocultures; and (b) maximizing timber production on some areas may alleviate the need for intensive utilization of other forests.

Plantations and Moorlands

While on the subject of planting forests, we should briefly return to one of the topics of Chapter 2, creating a diversity of diversities across the whole landscape rather than focusing on one ecosystem at a time. Compared to other ecosystems, forests are relatively diverse, but this should not necessarily be a justification for converting nonforests into forests. Wetlands, meadows, and prairies have a unique biota too, even if it is often not as rich as a forest biota.

The classic example of this comes from Scotland, where the Forestry Commission has drained, fertilized, and fenced extensive areas of moorland to facilitate converting them into forests (Peterken 1981). Increasing the extent of forests in Britain is certainly a laudable goal, and most of the Forestry Commission's efforts are directed toward sites that were forested before sheep and their keepers came to the island. However, wildlife advocates frequently complain about the Commission's work because it is not restricted to former forest sites, because the forests established are usually composed of exotic conifers, and because the wildlife threatened by this activity includes many uncommon species (Marquiss et al. 1985). Certainly there are parts of the world where establishing any type of forest should be counted a positive step—consider the Sahel and the Mediterranean basin. Nevertheless, wildlife interests would be best served if a representative sample of all types of ecosystems were maintained, and if reforestation efforts were directed toward recreating a semblance of the original forest on sites that were once forested (Klopatek et al. 1979).

The Domino Spectre

The aforementioned Brazil nut scenario and the dodo story from Chapter 2 raise a spectre that is of serious concern to tropical forest conservationists: the possibility that, because of close linkages between species, the extinction of one species will lead to the loss of many (Myers 1987). Imagine a tropical forest in which a group of ten tree species share a common set of pollinators. Perhaps five species of bee perform 95% of the pollination for these species, and the flowering times of the trees are spread out enough that throughout the year there is always sufficient food for the bees. Now, consider the potential consequences if a selective logging operation removed all the individuals of just one of the tree species, an extremely valuable tropical hardwood like rosewood, for example. Perhaps the effect would be slight and temporary; after a few decades seedlings left behind may have replaced the cut trees. On the other hand, if the pollinators were very dependent on that tree species at one time of year, their populations might be decimated by its removal, and this could ultimately have a very negative effect on the reproductive success of all ten tree species. In this manner, the elimination of one tree species could have a domino effect on several trees and animals. The likelihood of this latter scenario would depend on many factors: the tightness of the relationship between trees and bees, the proximity of undisturbed forests from which the bees could immigrate, etc. No such major disasters have been documented yet, but scientists who understand the complex dynamics of tropical forests are growing quite alarmed that as more and more tropical forests are harvested, even minor changes in the species composition of a forest could have major reverberations. In other words, the issues surrounding species composition can be far more subtle than monocultures and wholesale conversions.

An old longleaf pine forest in southern Georgia. (Photograph by William Platt)

THE LONGLEAF PINE

When one thinks of biological diversity being compromised by stand-type conversions, the common image is of a mixed deciduous forest being razed to make way for a pine monoculture. Less dramatic changes in tree species composition can be important too, as illustrated by events in the southeastern United States. Historically, a single species of tree, the longleaf pine, was predominant throughout large parts of the region. Along the coastal plain from southern Virginia to eastern Texas, extensive stands of longleaf covered some 25 million hectares, broken only where dry, sandy soils gave way to river bottoms—its uniformity compromised only by a scattering of turkey oaks and other species (Wahlenberg 1946). It was an impressive forest; it was open and park-like with a diffuse canopy of ancient trees, scattered clumps of smaller trees, and a thick lawn of grasses (Croker 1979).

The key to its structure was fire. At short, irregular intervals, perhaps every two or three years, fires started by lightning swept across the ground rejuvenating the grasses and eliminating most of the longleaf's competitors. Such fires are essential to the longleaf's reproductive system, because its large seeds usually get caught up in a mat of grasses unless a recent fire has left the soil exposed (Osborne and Harper 1937). After the seeds germinate and become established, they need six months or so without fire to make it through a sensitive period. Next they enter what is called the grass stage, during which they bear a passing resemblance to grass with only their long needles projecting above the ground. While in this stage the seedlings develop a long, stout taproot and thick bark that allow them to survive occasional fires that scorch their leaves. Only the weaker individuals are killed by these fires, and many benefits are gained: Competition is minimized; a disease called brown-spot is avoided; and a buildup of litter that could lead to a catastrophic fire is prevented. It has even been hypothesized that longleaf pines may actively facilitate frequent fires to gain these benefits (Platt et al. 1988); dead standing trees are often struck by lightning and serve as ignition points for ground fires that are fueled by highly combustible needles that litter the ground. After two to nine years in the grass stage (six is average), the seedlings experience a burst of height growth—made possible by the elaborate root system they have developed—and quickly pass through a stage in which they are vulnerable to fire (Wahlenberg 1946). By the time they are a small tree, 5 m or more tall, their thick bark and propensity for self-pruning their lower limbs make them almost invulnerable to normal fires. With lightning fires setting a dynamic rhythm for the longleaf's biology, the longleaf forest ecosystem was self-perpetuating (Chapman 1932).

The entry of people into this scene has had profound effects. The first inhabitants, native Americans, probably had only a modest influence on the longleaf forest, because they restricted most of their agricultural activity to the richer soils near streams (Croker 1979). They did set fires to drive deer to waiting hunters, but it is difficult to say if this represented a substantial increase in fire frequency. When European settlers arrived, they brought with them two things that would have a significant effect on the forests—livestock and a market for forest products. To provide better forage for their animals, the pioneers set fires too. Sometimes these fires were too hot or too frequent and damaged the pines, but this was probably a minor problem compared to the animals themselves, particularly free-ranging hogs that uprooted seedlings and consumed their taproots. The settlers earned their first revenues by extracting naval stores, initially pitch and tar and later turpentine and rosin, from longleaf pines. (The practice also earned them a nickname, "Tarheels," that is still applied to North Carolinians.) The problems escalated toward the end of the nineteenth century, when commercial harvesting of timber became important as sources of high-quality pine diminished in the northeastern and Great Lakes states and the superior quality of longleaf lumber became widely known. Unfortunately, the style of cutting, extensive clearcuts with little thought given to regeneration, often prevented the reestablishment of longleaf pine. Furthermore, overzealous forest firefighters allowed brush, hardwoods, and other pines to crowd out longleaf over large areas. By the time the relationship between fire and longleaf was understood, and the basis of a sustainable harvest system established, only about a third of the original forest remained.

After World War II, southern forestry was revolutionized by the application of intensive silvicultural practices. Huge areas of commercially unviable forest were clearcut and planted with fast-growing pines that, through careful nurturing, could be grown to merchantable size in 30 years or less. Herbicides, controlled burning, genetic improvement programs, fertilizers, and many other techniques were brought to bear to maximize profitability. Unfortunately for the longleaf, it has been largely left behind in this technological bonanza, because certain characteristics have caused most foresters to reject it for their intensive planting programs. Its most severe limitation is that it produces a good seed crop infrequently and irregularly, roughly only one year in four to seven (Croker and Boyer 1975). A second problem is that because it lies dormant in the grass stage for several years, it does not reach commercial size as soon as some other pines; this deficiency is somewhat compensated for by the superior quality of its wood. It also has a reputation for being difficult to deal with: Stored seeds lose much of their viability; nursery seedlings are easily damaged during transport; fires are needed to control disease and competition; animals consume much of the seed crop. Many of these problems are soluble, but with plenty of suitable alternatives, notably slash pine, loblolly pine, and shortleaf pine, it is not surprising that the area of longleaf forest is less than one-sixth of the original area and continues to shrink (Croker 1979).

Is this a significant problem for wildlife? Perhaps. Longleaf forests have a reputation among hunters as good game habitat; certain southern estates are renowned for producing an abundance of longleaf timber and bobwhite quail. Less well known is the great floral diversity of longleaf forests. Although few species of tree can readily survive an intensive fire regime, smaller plants often flourish and 200–300 species can be found in an area of 100 ha (Gano 1917). Many of these are species more commonly associated with the prairie ecosystems of the coast or the midwest. Whether or not this floral richness is reflected in faunal richness is not known, but one study found a greater density and species richness of birds and small mammals in a mature longleaf stand compared to a mature slash pine forest (Harris et al. 1974). Another study found the diversity, abundance, and biomass of breeding birds to be higher in a natural longleaf pine stand than in slash pine plantations of various ages, but there was little difference in the community of wintering birds (Repenning and Labisky 1985). Various features of longleaf trees themselves may account for some benefits to animals (Harris et al. 1979). The most obvious example involves seeds; longleaf seeds are over three times as large as other pine seeds, making them better food for a number of vertebrates, notably turkeys and bobwhite quail. The seeds also have large wings that probably make them easier to find too. Another potentially important feature is the substantial variability in height growth and mortality among longleaf trees of different sizes (Platt et al. 1988); this leads to considerable vertical diversity within a stand

and could mean there are more niches for various animals. The very shape of the trees, the way leaves and branches are arranged, is quite different from other pines and may facilitate foraging by insectivorous birds. The relatively large amount of heartwood in longleafs may explain why they are the preferred nest sites for the endangered red-cockaded woodpecker.

Given the special nature of longleaf pine forests and the severe reduction they have experienced, it would be unwise to allow the area of longleaf pine forests to diminish further. Fortunately, economically feasible techniques for managing longleaf stands do exist and have been described thoroughly (Walker and Waint 1966, Croker and Boyer 1975, Boyer and Peterson 1983). In brief, they emphasize harvest systems that leave a seed tree to facilitate natural reproduction and, of course, the judicious use of controlled burning. In other words, they are designed to imitate the natural patterns of disturbance and reproduction. If longleaf pine is compared to slash, loblolly, and shortleaf pine as a fast-growing source of inexpensive fiber, it cannot compete commercially. However, when one considers its superior qualities as timber, the beauty of longleaf forests, and its importance to quail, red-cockaded woodpeckers, and other wildlife, it assuredly deserves a significant place in the southern landscape. It is hard to disagree with this statement; the question is, how much is significant? Unfortunately, this is where the wildlife advocate and the timber advocate are unlikely to agree. Ideally, broad social and ecological values would be fairly represented in a reasonable policy, but currently, with longleaf pine forests spread across private and public lands in several states, there is not even a comprehensive forum in which to address this issue.

CONCLUSIONS AND SUMMARY

The trees and other plants and animals that comprise a forest ecosystem are not a random set of organisms; they live together on a site because they are adapted both to the site's physical environment and to living with one another. When people manipulate this situation, by changing the tree species composition of a stand, there are likely to be undesirable consequences for many species of wildlife. To minimize these impacts, foresters have a number of options. From a wildlife perspective, the best option is to avoid the whole problem by maintaining the natural species composition of the stand. This is relatively straightforward (although not necessarily easy) when natural regeneration is relied on following a cut. However, if artificial regeneration is used to reestablish the stand, it will take a specific effort to maintain the original species composition.

When economic goals dictate that biological diversity must be compromised by changing the natural species composition, the following guidelines for selecting tree species should be considered:

1. Choose a species native to the site. Avoid exotics whether they are from the other side of the state or the other side of the world. Exotics that are closely related to native trees are generally preferable, e.g., a Scots pine would be a better substitute for red pine than a Norway spruce would.

2. Try to plant two or more species; for example, a slow-growing tolerant species and a fast-growing intolerant can complement each other both ecologically and economically (Smith 1986). Biological diversity would often be best served if a conifer and a deciduous tree were used.

3. In planning species distributions across a large tract, avoid increasing the extent of conifer stands substantially. If a bias toward conifers is unavoidable, planting a conifer that sheds its leaves annually, e.g., one of the larches, may mitigate the imbalance slightly (Peterken 1981).

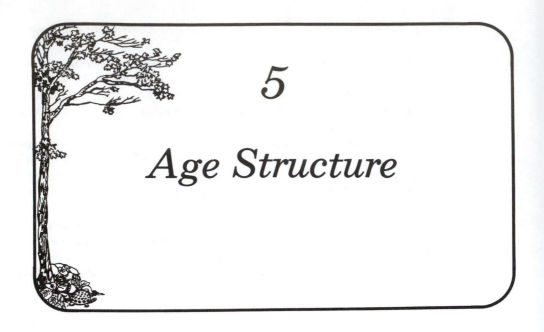

5

Age Structure

Many forests seem ancient from the time-scale of human lifespans, but they are not ageless, immutable features of the landscape. Their age is limited by the amount of time that has elapsed since a significant disturbance—hurricane, fire, logging, glacier, landslide, etc.—last set back the clock of ecological succession. Forest succession is paced by changes in the relative abundance and stature of a handful of conspicuous, dominant plants, but along with these species, thousands of plants and animals come and go too—their populations waxing and waning—as succession proceeds. Plants are influenced primarily by changes in the microclimate and by competition, while animals, especially herbivorous species, change largely in response to transformations in the plant community (MacMahon 1981). Even animals that are relatively unaffected by microclimate and only indirectly influenced by plants (insectivorous birds, for example) are ultimately affected by the modifications wrought through succession (Figure 5.1). Because of all these changes, managing forests for diversity requires managing the patterns of succession that determine the age structure of the landscape.

DIVERSITY AND SUCCESSION

Managing forest landscapes for diversity involves managing patterns of succession for two reasons: (a) Some successional stages have more species than others; and (b) each stage has a different, although not usually unique, set of species. We will explore these two points by considering two questions. *In the*

Time →

Figure 5.1 Succession in spruce-fir forests is dominated by changes in the vegetation, but there are parallel shifts in animal populations as well (Titterington et al. 1979).

hypothetical forest succession portrayed in Figure 5.2, which period will have the greatest number of species?

Given all the controversy that surrounds ideas about succession and climax (see Chapter 3), it is predictable that the answer will not be simple. The traditional view, as expounded by Eugene Odum (1969), is that diversity in-

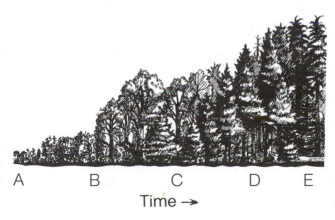

A B C D E

Time →

Figure 5.2 In this hypothetical forest succession, at which point do you think the greatest diversity of species will occur?

creases through succession and reaches its peak in the climax (Letter E in Figure 5.2) because at this stage niche specialization is well-developed and numerous species can share the bounty of biomass that has accumulated. In contrast, some people think that diversity will be greatest soon after a disturbance because all the forest species will persist, perhaps reduced in size and abundance, and new, open-land species will become established (Egler 1954). Finally, one could hypothesize that diversity is greatest in the intermediate stages (Letter C), because enough time has elapsed for a variety of species to invade, but not enough time for any species to dominate and exclude less competitive species. Two patriarchs of forest ecology, Herb Bormann and Gene Likens (1979), predict that diversity will be equally high during both the climax stage (which they prefer to call the shifting-mosaic, steady state) and shortly after a disturbance.

Most of the ecologists who have written about succession are botanists and may be especially influenced by scenarios of invasion and competition among plants. If you consider the enormous number of invertebrates, bacteria, and fungi that depend on dead wood and other detritus (Elton 1966), you can predict that diversity will crest when biomass accumulation does. Here again, there is no unanimity; members of the Odum (1969) school think biomass is maximum at the climax stage, but Bormann and Likens (1979) have developed a model that indicates that biomass will be greater before the climax stage. In their model, biomass peaks when the dominant trees have reached their maximum age, (Letter D), just before they begin to die and fall, creating gaps in the forest canopy. Having these gaps means that there is a bit less biomass in the climax forest, but this is probably more than compensated for by the fact the the the gaps represent a special habitat for certain species. To paraphrase John Lawton (1978), there are more ways to make a living on a maple than on a mayflower, and animal diversity, especially insect diversity, will be greater in a structurally diverse forest. Thus it is easy to conclude that Odum was probably right, albeit for somewhat different reasons, that diversity is maximum at the climax stage.

If climax forests (Letter E) do have the greatest diversity, does that mean that those forest managers whose primary concern is diversity should strive to have as many climax stands as possible? Probably not. In Figure 5.3, a forest is portrayed undergoing succession (Letters A to E) and then remaining in a climax state for an equal period (Letters E to F). *During which span of time, while the forest is undergoing succession or after it has reached a climax, will overall species richness be greatest?*

The answer here is more obvious. All the species coming and going in the first period will add up to more diversity overall. From eucalyptus forests and thornbills in Australia (Loyn 1980) to birches and beetles in Finland (Heliovaara and Vaisanen 1984), the story of species changing through succession has been told and retold and need not be elaborated on here (Figure 5.4). The upshot is very simple: A forest landscape with stands of many ages will, all

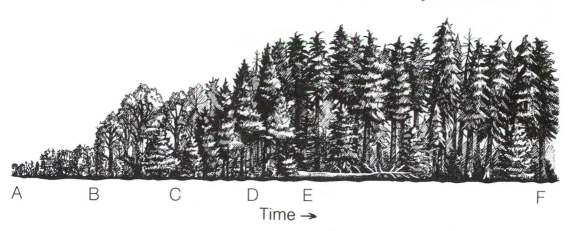

A   B   C   D E   F

Time →

Figure 5.3 In this hypothetical forest, a time span of succession is followed by an equal span of climax forest. Which half of the figure, A through E, or E through F, will have the greatest diversity?

other things being equal, have more kinds of wildlife than a single-age land-scape. In other words, if all the forest stand ages depicted (A, B, C, D, and E) are represented on the landscape, this will be a more diverse situation than a landscape of any single age, even E. Clearly this is good news for foresters, because it coincides well with a basic goal of forest management, providing a continuous supply of wood products. Sustained yield is the operant phrase.

Although it seems obvious that a forest landscape with many succes-sional stages will be more diverse than one at a single stage of succession, research on this subject is often misinterpreted. Imagine a study in which a group of animals is censused in a mature forest and a recently-cut forest and it is discovered that there are fewer species, or lower populations, on the logged site (e.g., Loyn 1980, Bury 1983). Many people would read such a report and conclude that forest harvesting has harmed wildlife. And they are right; cut-ting does often reduce wildlife diversity and abundance—on the cut site—for a finite time. Viewed from a larger time and space perspective, forest cutting will often have little effect on diversity and may actually enhance it. On the other hand, this same narrow approach can also cloud how people think of old forests. Imagine a study in which some plant or animal populations were censused in young forests at various successional stages and in an old forest (e.g., Conner and Adkisson 1975, Middleton and Merriam 1985, Childers et al. 1986). Some people might focus on all the species coming and going through succession and conclude that old forests are not an important type of wildlife habitat, but, as we will see later in the chapter, this is an erroneous perception.

In this chapter we will often refer to managing the age structure of the forest landscape, rather than managing succession, because the age of a forest stand is a reasonable index of its successional state. (It is only an index, how-

Figure 5.4 Succession is often portrayed as progress terminated by disturbance (Figures 5.1, 5.2, and 5.3) but it can also be viewed as a cycle in which a series of plants and animals come and go. Here the cycle is illustrated by four groups of animals. Starting at the upper left are five insects associated with different successional stages of boreal forests in eastern North America (from early to late succession): willow shoot sawfly, white pine weevil, forest tent caterpillar, spruce budworm, and spruce beetle (Dimond, pers. comm.). In the upper right are some birds that characterize taiga forest succession in Finland: pied wagtail, redpoll, willow warbler, chaffinch, and brambling (Helle 1985*a*). In the lower left are some mammals that represent succession in Douglas-fir forests of Oregon: golden-mantled ground squirrel, mountain beaver, dusky-footed woodrat, northern flying squirrel, and marten (Harris and Maser 1984). At the bottom right, a sequence of birds from mixed eucalypt forests of southeastern Australia is shown: buff-rumped thornbill, white-eared honeyeater, brown thornbill, white-browed scrub wren, and striated thornbill (Loyn 1980).

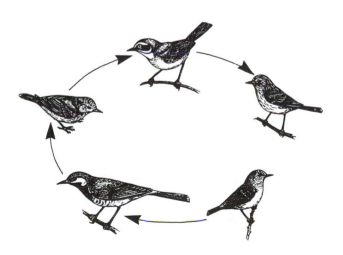

ever, because the rate of succession varies among sites depending on the intensity of disturbance and the overall quality of the site.) Note that the age structure of a forest landscape is not the same thing as the age structure of a stand. The former is based on the ages of all the different stands that comprise the landscape; the latter is based on the ages of the trees that comprise the stand. Stand age structure is discussed in Chapter 6, "Spatial Heterogeneity," and Chapter 11, "Vertical Structure."

BALANCING AGE CLASSES

Despite the incentive to maintain a balance of age classes, failure to do so is quite common and causes major problems. Outbreaks of spruce budworm have defoliated significant portions of the spruce-fir forest of eastern North America, leaving behind a paucity of middle-aged stands and a plentitude of overworked foresters struggling to avoid a timber shortfall in the year 2010. Periods of farm abandonment, such as the 1930s Depression, often create a glut of one age class (Seymour et al. 1986). An uneven age class distribution can also cause problems for any species of wildlife that require forest of a particular age for habitat; this idea—some have called it a temporal bottleneck (Seagle et al. 1987)—is illustrated in Figure 5.5. In the remainder of this section we will take it as given that foresters are trying to stabilize age distributions—an assumption to which there are many exceptions—and concentrate on some general ideas about how different age stands can be arranged on the landscape.

One of the best known examples of balancing age classes is that devel-

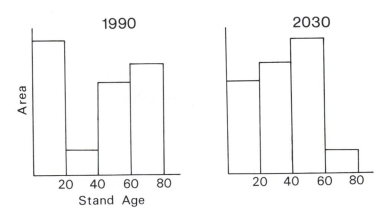

Figure 5.5 Two age class distributions for the same forest at two points in time show how a shortage of habitat for certain species can create temporal bottlenecks: in 1990, there will be a habitat shortage for species requiring stands 20–40 years old, and species requiring stands over 60 years old will experience a shortage in 2030.

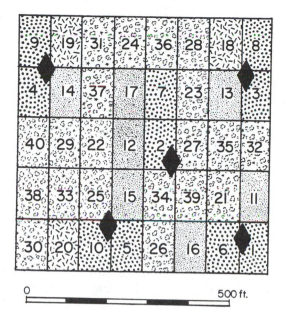

0   500 ft.

◫ Recently Cut

▦ Sapling Stand

▦ Small Pole Stand

▦ Original Forest

◆ Expected Location of a Ruffed Grouse Drumming Site

Figure 5.6 A suggested cutting program for a 40-acre parcel of aspen divided into 1-acre blocks. By cutting one block each year for 40 years in the pattern shown, optimal habitat for ruffed grouse can be created. This pattern provides five potential home ranges for ruffed grouse, each containing a mix of stands that will give a grouse all its life requirements. The pattern's orientation also provides enough southern exposure to facilitate aspen regeneration (From Gullion n.d.).

oped by Gordon Gullion (n.d.) for manipulating aspen stands into a mosaic that ruffed grouse will find attractive (Figure 5.6). His plan calls for cutting an acre a year for 40 years in a pattern that creates a great deal of heterogeneity. In his plan, the age differences between adjacent units range from 2 to 36 years and the average difference is 11 years. As heterogeneous as this looks, it could be much more so. Blocks A and B in Figure 5.7 show two alternative arrangements, and in both cases the average age difference between adjacent stands is 20 years. However, in A the age differences range from 18 to 23 years and the standard deviation is 2 years, while in B the differences range from 1 to 39 years with a standard deviation of 11 years.

The arrangements in Figure 5.7 might be quite reasonable for a landowner who needs to cut wood annually, but many people would prefer to harvest at less frequent intervals and might be inclined toward other arrangements (Figure 5.8). From a biological diversity perspective, how important is the difference between Figure 5.7 and Figure 5.8? That depends on the resolution with which organisms view the age of their habitat. It is doubtful that any species would distinguish between a 28-year-old stand and a 27-year-old stand. But are there any that would discriminate between 27 years and 22 years? Possibly. Between 27 and 13? Probably. And even if the difference between 27 years and 28 years is not important, what about when those two stands were 1 and 2 years old? Existing research is inadequate for addressing these questions. There are many studies in which researchers have surveyed wildlife in stands of several different ages, say 1, 7, 13, 15, 27, and 40 years (Loyn 1980, Helle 1985a, Childers et al. 1986), but none in which there has been adequate replication within different age classes. One study monitored changes in the bird community of a young pine plantation as it aged from 2 to 6 years and found a steady, year after year, increase in bird diversity and abundance (Dickson et al. 1984). However, there were no striking cases of a bird species being absent one year and very common the next, and like other studies, there was no replication. One would have to compare several 27- and 22-year-old stands to determine which, if any, organisms perceive them as significantly different.

Simple systems for arranging stands become a bit more complex when larger forests are considered, because harvesting small stands scattered across the landscape is rather inefficient. Imagine a forest tract that has four separate parcels of 40 acres each. The landowner could repeat the pattern of Figure 5.8 four times, but it would be very tempting to complete all the cutting on one parcel before moving on to the next one (Figure 5.9). As a compromise between these two extremes, a pattern that moved to a new parcel each year would be quite efficient and not reduce diversity very much (Figure 5.10). (There is another obvious possibility: dividing the 160 acres into 40 four-acre stands rather than 160 one-acre stands; the significance of this approach is discussed in the next chapter, "Spatial Heterogeneity.")

This game of arranging 40 cells in a block into interesting patterns is

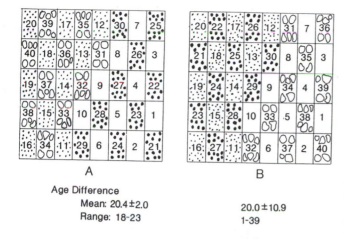

A

Age Difference
Mean: 20.4±2.0
Range: 18-23

B

20.0±10.9
1-39

Figure 5.7 Figure 5.6 has been redrawn to show two patterns which maximize the average age difference between adjacent stands. In A the age differences deviate very little from the overall average (standard deviation = 1.96); in B the age differences range from 1 to 39 years (SD = 10.92).

generally more of an intellectual exercise than a practical one. Still, the patterns do illustrate important principles that can guide management on a real landscape with all its constraints, e.g., stand size, roads, rivers, steep slopes, ownership patterns, and, of course, the age structure of the existing forest. Practical constraints also mean that the process of implementing a balanced age class structure has to be gradual; e.g., if our hypothetical 40-acre aspen parcel were completely occupied by a 40-year-old stand, it would probably take 40 years to change it into the 40 distinct stands portrayed in Figure 5.7. The process has to be adapted to local conditions for equally manifest reasons. For example, if the four parcels in Figure 5.9 and Figure 5.10 were each a different type of forest, say aspen, jack pine, spruce-fir, and maple swamp, it would fundamentally change the picture. For a start, it may well be necessary to extend the rotation from 40 years to 60 or more in a type of forest that grows more slowly (Figure 5.11). Secondly, it would make it all the more important

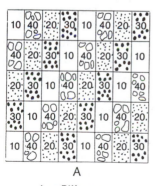

A

Age Difference
Mean: 20.0±7.1
Range: 10-30

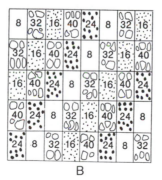

B

19.2±4.0
16-24

Figure 5.8 Figure 5.6 redrawn to show possible patterns for cutting only four (A) or five (B) times per 40 years.

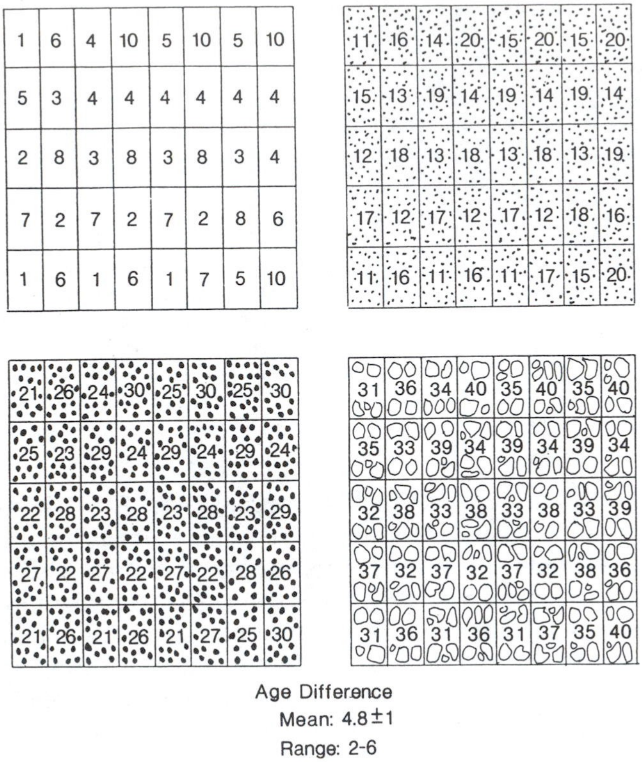

1	6	4	10	5	10	5	10
5	3	4	4	4	4	4	4
2	8	3	8	3	8	3	4
7	2	7	2	7	2	8	6
1	6	1	6	1	7	5	10

11	16	14	20	15	20	15	20
15	13	19	14	19	14	19	14
12	18	13	18	13	18	13	19
17	12	17	12	17	12	18	16
11	16	11	16	11	17	15	20

21	26	24	30	25	30	25	30
25	23	29	24	29	24	29	24
22	28	23	28	23	28	23	29
27	22	27	22	27	22	28	26
21	26	21	26	21	27	25	30

31	36	34	40	35	40	35	40
35	33	39	34	39	34	39	34
32	38	33	38	33	38	33	39
37	32	37	32	37	32	38	36
31	36	31	36	31	37	35	40

Age Difference
Mean: 4.8±1
Range: 2-6

Figure 5.9 If four 40-acre parcels were cut, one parcel for 10 years, then the second parcel, and so on, the average age difference between adjacent stands would be very low.

to not approach each parcel independently and sequentially (Figure 5.9), but to integrate their management (Figure 5.10).

It is important to reiterate here that the age of an ecosystem is only an index of its successional state, and it is probably not especially important, per se, to the organisms that comprise the ecosystem. For a bird, the critical factor might be how much leaf area has grown on which it can search for insects. For a fungus, the accumulation of dead organic matter might be the key issue. Even the trees themselves are often affected more by their size than by their age (Hackett 1985). Age is just a convenient way of keeping track of succession; a forest manager can do it in the office without ever visiting a stand. For many purposes, the height of a stand is probably a better index of its condition than age, and it is much easier to measure in the field. A couple of simple rules of thumb have been suggested for balancing age classes by setting minimum height differences between adjacent stands. In the U.S. National Forests, a stand has to be at least 15 to 25 feet tall (about 4.5–7.5 m; the rule varies regionally) before an adjacent stand can be cut. Some researchers have advocated waiting until a stand is 60 to 70% of the height of adjacent stands (Harris and Skoog 1980).

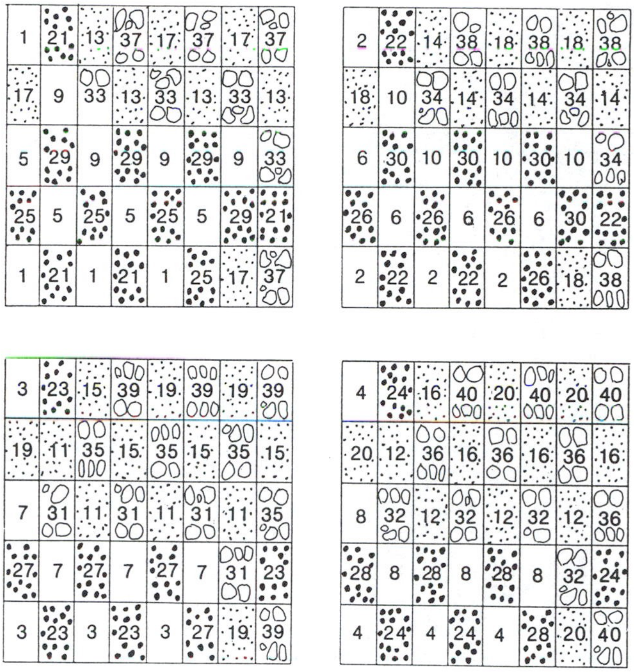

Age Difference
Mean: 19.3±4.3
Range: 8-24

Figure 5.10 In contrast to Figure 5.9, four 40-acre parcels could be cut in a sequence that moved to a new parcel each year and maintained a high average age difference.

OLD FORESTS

Maintaining the continuity of wood production and maintaining diversity go well together but for one significant exception; it does not provide for old forests. This is because a forest becomes economically mature, ready to be cut and rotated for the next crop, long before it is biologically mature. This state, sometimes called the economic climax, is determined by factors such as the annual growth rates of trees, susceptibility to insects and disease, and financial considerations (i.e., the balance between potential revenues and the costs of opportunity loss) (Pearse 1967). To take one of these factors, trees are ready to be cut when their growth rate for a year declines to a point lower than their average annual growth rate; e.g., when the trees grow only 0.0022 cubic meters of wood in their eightieth year having averaged 0.0023 cubic meters per year in the 79 previous years. The economic climax can be reached in as few as 25 years in southeastern pine forests, but in most forests 50–100 years is much more common. Certain European hardwoods that are not cut until they are 300 years old define the far end of the continuum. The age at which a tree or a forest is biologically mature is arguable, but it is certainly much older than

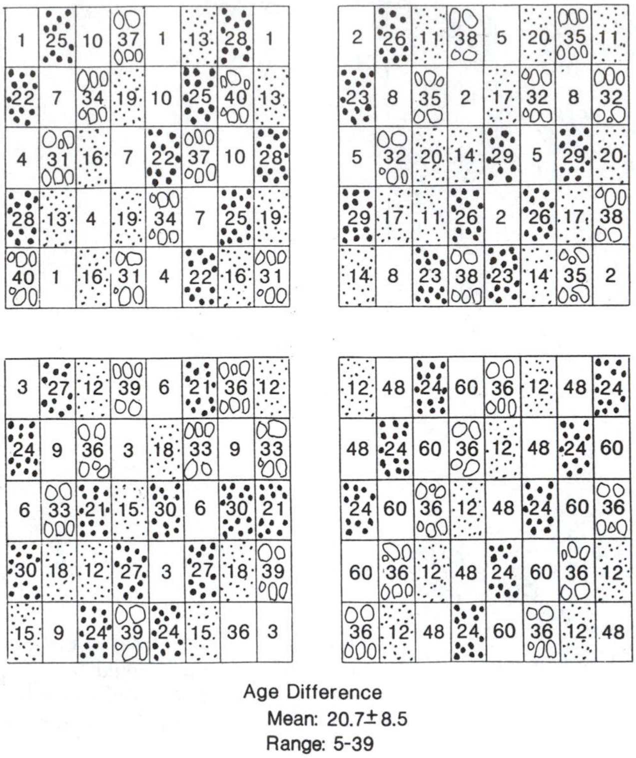

Age Difference
Mean: 20.7±8.5
Range: 5-39

Figure 5.11 If one parcel were managed on a 60-year rotation, then some habitat for wildlife needing older forests could be provided.

the economic climax. Consequently, there are very few old stands remaining in most parts of the world, and the wildlife species associated with them are likely to suffer from a lack of habitat.

This dilemma will increase in the future, because rotation ages are becoming shorter as technological advances allow small trees to be used for purposes that formerly required big trees. The same principle that makes the silk in a spider's web one of the strongest materials known is responsible; a single wood fiber is two to five times as strong as a wooden beam if you correct for relative size (Mark 1943, Wenger 1984). As technologists get better and better at tearing wood apart and gluing it back together again, the demand for big, strong, old trees will diminish. They can already make a panel out of woodchips that is far tougher than a regular board; paper houses may be on the distant horizon. The task has both engineering and biological dimensions. Scientists have found that juvenile wood is weaker than old wood, and although they have learned how to grow trees fast and use them when they are still small, to date they have not figured out how to make them old before their time (Senft et al. 1985). Even without solutions to these issues, old stands will become rarer and rarer as forests are brought under more intensive utilization regimes (Duerr 1986).

A Definition of Old Forests

The adjectives used to describe old forests say as much about the people who use them as they do about the forests. Environmentalists favor words like primeval, ancient, wilderness, virgin, pristine, while in the foresters' lexicon one is more likely to find overmature, decadent, and senescent. Many scientists prefer "old growth"; it is reasonably free of implicit bias and can stand alone as a noun.

There is no generally accepted, or universally applicable, definition of old-growth forest. A simple, more or less idealistic, definition would be "a climax forest that has never been disturbed by humans." However, this definition is too nebulous for practical use and too restrictive if rigorously applied. It is nebulous for the same reasons that the whole idea of climax forests is a controversial issue (Chapter 3). This is especially problematic in places where the frequency of natural disturbance is such that a steady-state, self-sustaining forest never develops. For example, there is evidence that natural fires in pre-settlement northern Minnesota burned the boreal forest every 100 years or less (Heinselman 1973, Clark 1988). The definition is overly restrictive because a comprehensive list of human disturbances would have to include: the activities of primitive people, many of whom regularly set the woods on fire; the pervasive air pollution generated by the rest of us; and the transportation of disease organisms such as the one that has nearly eradicated the American chestnut. If these and similar phenomena are included, it is clear that there is essentially no undisturbed forest left to consider.

Definitions currently in use often avoid the issue of exactly what constitutes human disturbance and tend to be fairly arbitrary about the question of age. For example, a minimum age of 200 years is used by the USDA Forest Service's Old-growth Definition Task Force (1986b) for old-growth Douglas-fir forests of the Pacific Northwest, whereas the Maine Critical Areas Program (1983) uses a cutoff of 100 years.

In fact, it is probably not possible to craft a universal definition of old growth—at least one that would be universally accepted—and it may not be desirable to do so. It would be preferable to have forest ecologists develop a unique definition for each forest type, taking into account both forest structure and development, and the historical and current patterns of human disturbance. Such definitions could be quite specific and include factors such as: (a) species composition; (b) vegetative structure (e.g., size and density of live, dead standing, and dead fallen trees); and (c) minimum area (Thomas et al. 1988). The USDA Forest Service's (1986) definition for Douglas-fir forests approaches this level of detail; for example, it describes old-growth stands as having at least 10 tons of fallen logs per acre on most types of sites, but 15 tons per acre on western hemlock sites. Unfortunately, the current knowledge about old-growth forests is far too limited to construct such detailed definitions for most forest types. Even more unfortunately, such information will be very hard to obtain for the many kinds of forests for which relict old-growth stands are rare or nonexistent.

Even though a broadly applicable definition of old-growth may be impossible to develop, it is still useful to consider the kinds of criteria that are likely to have ecological significance and that could be adapted to form locally appropriate definitions.

Age criteria

1. *Has the forest reached an age at which the species composition is relatively stable; in other words, has it reached a climax?* This criterion will rarely be met in many areas. Even 1,000-year-old Douglas-fir stands are not climax; eventually they are replaced by western hemlock. Moreover, climate changes over millennial time-scales are continuously changing forest communities (Jacobson et al. 1987).

2. *Has the forest reached an age at which average net annual growth is close to zero?* This does not necessarily mean that the trees have stopped growing, but rather that overall tree growth is balanced by death, and thus stand biomass no longer increases and may even decline.

3. *Is the forest significantly older than the average interval between natural disturbances severe enough to lead to succession?* It is simple statistics that roughly half the forests will reach ages over the average (assuming the distribution is not too skewed). Only stands in locations where they are sheltered from natural disturbances (e.g., isolated from fire by being surrounded by swamps; sheltered from wind storms by the surrounding topography) are likely to become much older (Heinselman 1981a).

4. *Have the dominant trees reached the average life expectancy for that species on that type of site?* Determining life expectancy from the scientific literature can be difficult (Table 5.1); it is probably best to visit the stand and see if some of the trees are dying of maladies related to old age. Recognizing that shade-intolerant species generally die at relatively young ages, it may often be desirable to restrict this criterion to shade-intolerant species.

5. *Has the forest's current annual growth rate declined below the lifetime average annual growth rate?* This is a criterion used to deter-

TABLE 5.1   Pathological Longevity (Boyce 1961) and Maximum Longevity (Fowells 1965) for Some Tree Species.*

Species	Pathological Longevity (Years)	Maximum Longevity (Years)
Balsam fir	60–90	200
Pacific silver fir	130–325	540
Subalpine fir	130	250
White fir	150	360
Yellow birch	170	200
Incense cedar	165–210	540
Red spruce	170	400
Sitka spruce	250–300	800
White spruce	160	—
Eastern white pine	160–170	450
Quaking aspen	40–120	200
Douglas-fir	150	1000
Western hemlock	100–275	—

*Pathological longevity is the average age at which the trees begin to suffer from serious decay. Maximum longevity is the greatest life span known. Average life expectancy presumably falls between these limits. It will vary among regions and among sites.

mine rotation ages and is met by virtually any forest that has not been cut at the typical rotation age.

Disturbance criteria

6. *Has the forest been extensively or intensively cut?* This criterion would include not only clearcuts but also selective removal of one or two commercially valuable species that resulted in an altered species composition. Perhaps the most widely used adjective for forests that meet this criterion is "virgin."

7. *Has the forest ever been converted by people to another type of ecosystem?* This criterion would not eliminate forests that have been periodically cut, as long as they have never been converted to another type of ecosystem such as agriculture. Sometimes forests that meet this criterion are called primary, and those that do not are called secondary (Rackham 1980). This criterion could be made less restrictive by allowing: (a) short-term conversions, e.g., traditional slash-and-burn agriculture in which a site is abandoned after a few years; or (b) conversions prior to some cut-off date. In Great Britain, any site that has been continuously forested since 1700 A.D. is referred to as an ancient woodland (Rackham 1980).

With a basic understanding of a given type of forest, it should be possible to select an age criterion and a disturbance criterion that would form the basis for a reasonable definition of old growth. In remote parts of the Amazon basin and in the coastal mountains of southeast Alaska, it might be reasonable to use very stringent criteria (criteria 1 and 6 of the foregoing list), because many forests would meet them. On the other hand, applying criterion 1 to the Douglas-fir stands of Oregon and Washington would eliminate what many consider to be archetypal old-growth stands. Similarly, using criterion 6 in eastern North America would decree that virtually no old-growth forests exist now, or can ever exist in the future. Obviously, both local ecology and the history of human disturbance need to be considered in determining a reasonable working definition.

In many areas, it might be desirable to apply two levels of criteria. For example, the term "old growth" could be reserved for those few stands that meet criteria 2 and 6, while stands that meet criteria 3 and 7 could be called old. The great advantage of recognizing stands that do not meet a narrow disturbance criterion is that they can be created wherever desired, simply by deciding to cease cutting in an area and waiting a century or two, and indeed many are now being created in public parks where timber harvest is prohibited. Unfortunately, it is not known how fully a 200-year-old stand that was once cut would resemble a virgin old-growth forest of equal age, but it is certainly safe to assume that it would be a better facsimile than a stand that was only 80 years old. Recognizing the importance of previously disturbed stands as a surrogate for virgin forests does not diminish the value of undisturbed old forests. They are with us now; we do not have to wait a century for them to develop. They are the best examples of their type of ecosystem and present an unbroken thread of ecological change for scientists to explore. Finally, with a stand that was once cut, one can never know exactly how much adulteration has taken place. Forests cut around Angkor Watt in Cambodia 600 years ago are still distinguishable from uncut forests (Myers 1979).

Exactly what we call older forests that may or may not have been disturbed is not critical; the main thing is that people be as explicit as possible about the kind of forest they are talking about. If they mean a forest that has never been cut, then they should be sure to note that it is virgin; if they mean a forest that has no net growth, they should mention that fact. In the balance of this book the adjective "old" is generally used, because it is less restrictive and can refer to stands that meet any of the five age criteria. "Old-growth" will occasionally be used to refer to forests that meet criteria 1 or 2 plus 6.

The Importance of Old Forests

The exercise depicted in Figures 5.2 and 5.3 demonstrates that it is important to have an adequate number and area of old forests simply because they represent one portion of the successional sequence, and especially because they represent what is likely to be the most biologically diverse portion of the sequence. Ideally, they would occupy a significant part of the landscape, not the minuscule portion they currently represent in most regions. In areas where natural disturbances are fairly uncommon, e.g., where forests are too moist to burn readily and cyclonic winds are a rare occurrence, old forests probably once dominated the landscape (Bormann and Likens 1979). We will not return to that state, but some redressing of the imbalance in in order.

One question that is often asked is, "Are there any species that are dependent on old forest?" Probably the best studied old forests in North America are the Douglas-fir-western hemlock stands of the Pacific Northwest, where the USDA Forest Service has extensive holdings and a mandate for maintaining biological diversity on their lands. Here a number of species closely associated with old forests have been identified (Franklin et al. 1981). These include animals such as northern spotted owls, Hammond's flycatchers, red tree voles, and coast moles; lichens known only by their scientific names, *Lobaria oregena*, *Lobaria pulmonaria*, and *Alectoria sarmentosa;* and saprophytic plants, known by even odder names, phantom orchid, woodland pinedrops, and candystick. (Saprophytic plants, such as fungi, acquire most or all of their energy from organic matter decomposition rather than by photosynthesis.) Across the Pacific and below the equator, biologists studying Australia's eucalyptus forests have produced another list of birds and mammals that are associated with old forest; the list includes gliding possums like the sugar glider and feather-tailed glider, and birds such as the crimson rosella, gang-gang cockatoo, and spotted pardalote (Recher et al. 1980, 1981). Even some Pacific seabirds, the marbled murrelet and ancient murrelet, may prefer old forests for nest sites (Binford et al. 1975, Blood and Anweiler 1984); (the ancient murrelet derives its name from its elderly appearance, not from its ancient habitat).

Spotted owls, spotted pardalotes, and many other species associated with old forests are sometimes found in younger forests. Does this imply that old forest is just optimal habitat for these species and that they could readily survive in younger forests? The only way to find out for sure would be to eliminate all the old forest and see if any species became extinct, but such drastic techniques are not really required. Detailed ecological studies on the requirements of individual species can usually determine the importance of old forest. For example, studies of the northern spotted owl have left wildlife biologists convinced that its extinction would closely follow the disappearance of old forests (Forsman et al. 1984); (the spotted owl story is elaborated below). Furthermore, even if there were no species known to occur only in old forests, this does not mean that none exist. Our knowledge of the total biota, particularly of insects, is far too limited to assume this (Bistrom 1978).

The rationale for having old forests does not rest entirely with saving rare species from extinction. The black-tailed deer and blue grouse of British Columbia and Alaska (Schoen et al. 1981, Doerr et al. 1984) and the capercaillie and marten of Sweden (Bjarvall et al. 1977) are not rare species and are not really tied to old forests. They simply seem to find their optimal habitat there, at least during certain times of the year. The issue of what constitutes optimal habitat is of particular interest to wildlife managers whenever a game species is involved, because the production of a harvestable surplus is more likely under optimal conditions. The relationship between game animals and old forests is further examined later in a section on the black-tailed deer of British Columbia and Alaska.

When discussing the importance of old forests, it is rather misleading to focus on individual species. Generally, the welfare of this species or that should be secondary to the concept of maintaining the entire ecosystem and all the species that comprise it. By most definitions, old forests are a rare commodity, and the ones that have been free of significant human disturbance will always be rare, even in the twenty-second century when the parks set aside during this era are approaching their bicentennials. Current ideas about climax and succession have weakened the image of old-growth forests as paragons of coevolution (Chapter 3), but it is still clear that they are far more than a collection of creatures and the place they live. Old-growth forests are systems with structure and function; they are more than the sum of their parts; they are exemplars of synergism. The people responsible for preserving the Taj Mahal do not think of it as an assemblage of marble blocks and panels inlaid with precious stones. Pieces of the Parthenon, the Gates of Babylon, and the Palace of Persepolis that have been carted off to distant museums are very poor vestiges of the creations they once formed. It should be a fundamental objective of conservation to maintain examples of each of the Earth's ecosystems, and old, virgin forests are the logical choice for selecting examples of forests.

How Much Is Needed?

The preceding paragraph begins to approach some of the purple prose that comes to the fore so readily when old forests are being discussed. In fact, there is nothing particularly controversial about the importance of old forests. Almost everyone is for them. Such pillars of forestry as the Society of American Foresters and the USDA Forest Service have formally recognized the need to protect representative old forest ecosystems for their own sakes, and as areas where baseline research can be conducted (Buckman and Quintus 1972, F.C.E.R. 1977, Heinrichs 1983). The controversy is largely over how much, and even this point is not argued everywhere. In many places such as Europe and the eastern United States, old-growth forests are such rare relics that few people would debate the wisdom of carefully protecting all the virgin forests that remain. On the other extreme, there are remote regions in the Amazon basin and parts of Canada, Alaska, and probably the Soviet Union, where there has been very little forest cutting and old-growth forests measured in thousands of square kilometers could be set aside. People who would argue that such large areas should remain inviolate forever are more likely to be advocates for wilderness than for old-growth forests.

We need to digress briefly to discuss the relationship between old-growth forests and wilderness (large areas free of roads, motor vehicles, and virtually all other forms of human artifact) because they are separate, albeit related, issues. The former is based on ecological arguments; the latter is largely an issue of aesthetics and recreational needs, considerations which are equally legitimate, but quite distinct from, ecological factors. It is true that (a) wilderness areas are important for some species of wildlife, especially large carnivores; (b) old-growth forests are often aesthetic; and (c) there is substantial areal overlap between wilderness and old-growth. However, it is important to realize that many old-growth stands are too small to be considered a wilderness, and many wilderness forests do not meet the more rigorous age requirements for old growth (e.g., they are not climax forests). Furthermore, many old forests are not what the average person would consider beautiful; there may be no huge, magnificent trees; there will certainly be numerous dead and dying ones.

Controversy over how much old growth to maintain is most volatile in regions such as the Pacific Northwest where virgin, old-growth forests are far from rare and quite accessible to loggers. In 1984, there were still about two million hectares of old-growth Douglas-fir forests in western Oregon and Washington, about 19% of the total forest area (S.A.F. 1984), but this figure is steadily decreasing. Here, people who are primarily concerned about biological diversity argue that all the remaining old growth must be saved, while timber companies feel that some designated wilderness areas and a scattering of isolated groves will be adequate.

It is not easy to arrive at an adequate answer for, "How much?" One can construct various ecological and economic arguments, but ultimately it comes down to a subjective, political decision. In *The Fragmented Forest,* an analysis of the Pacific Northwest old-growth controversy, Larry Harris used the maintenance of 5% of the forest in old-growth as a hypothetical objective to illustrate what the consequences of a particular policy would be. This became the most controversial part of the book, with people challenging him to defend his decision (Harris pers. comm.). As Harris knew, there is no right decision, although one can set some reasonable bounds on it. Before settlement, about 50% of the Douglas-fir forests were probably over 300 years old at any one time (Harris 1984); it would be hard to maintain that current forests should exceed this figure. At the other end, it is imperative that enough old forests remain to support viable populations of all species dependent on this habitat. Determining this minimum figure requires careful study and a conservative approach; if not enough old stands are retained, by the time new ones develop it could be too late for some sensitive species.

Focusing on the spatial requirements of old-forest animals with large home ranges is an obvious way to determine the minimum area needed. This is what was proposed in the Pacific Northwest, where a panel of scientists analyzing the viability of northern spotted owl populations determined that nearly two million ha of old forest (ca. 10–20%) would probably be needed (Dawson et al. 1987). To go below that amount would threaten the owl's existence and contravene public policy. In tropical rain forests, the trees themselves are often so rare and widely scattered that maintaining viable populations of them will require large areas of old forest (Hubbell and Foster 1986). For example, in a 23 ha tract of Malaysian lowland forest, there were 377 tree species of which 307 (81%) were represented by only one to ten mature individuals each; for 143 species (38%) a single mature individual was the sole representative (Poore 1968). Again, where to fall between the minimum and maximum is a decision for the public and its policy makers to decide. Given that short-term economic arguments usually persuade more people than long-term, ecological arguments, the decision is likely to be closer to the lower bound than the upper one, and people who care about wildlife should usually add their voices to those calling for the retention of as much old forest as feasible.

This discussion is not irrelevant beyond the regions where virgin forests still remain. Because forests that have been cut in the past can be allowed to grow old and become a surrogate for old growth, the question of how much old forest there should be needs to be considered for every forest landscape. Furthermore, the question has added dimensions, because in areas without virgin forest the scope for choice and positive action is much greater. One can decide where to have old forests and how large each unit should be, and hopefully, the flexibility of creating old forests will move the decision closer to the upper bound; i.e., there will be more habitat for organisms that need old forests.

Despite the possibility of creating old stands in regions where there are none, this idea does not justify the cutting of existing old stands in other regions. As the Society of American Foresters (1984) said:

> With present knowledge, it is not possible to create old-growth stands or markedly hasten the process by which nature creates them. Certain attributes, such as species composition and structural elements, could perhaps be developed or enhanced through silviculture, but we are not aware of any successful attempts. Old growth is a complex ecosystem, and lack of information makes the risk of failure high. In view of the time required, errors could be very costly. At least until substantial research can be completed, the best way to manage for old growth is to conserve an adequate supply of present stands and leave them alone.

Managing for Old Forest Species

Managing for old forest species seems rather straightforward. You provide them with as much habitat as feasible, then "lock it up and throw away the key." Unfortunately, it is not quite that easy. Beyond pernicious problems such as air pollution, deciding not to allow any human disturbance only makes the decision about how to handle natural disturbances all the more sensitive. Ideally, old-growth forest reserves should be subject to natural disturbance regimes; lightning fires should not be extinguished; insect outbreaks should not be suppressed, etc. These are perturbations with which the ecosystem's species have evolved, and ideally, they should continue to be exposed to these events, although this may not be practical unless the reserves are large enough that an entire reserve will not be disturbed by one event.

The effects of short rotations on old forest species can also be mitigated to some extent by a simple process, lengthening the rotations. In a landscape where most of the forests are being cut when they reach 60 years of age, allowing some stands to reach 100 years could make the difference in whether or not some old forest species persist in an area. Even better would be to allow some commercially managed forests to meet climax or zero net growth status (criteria 1 or 2) before being cut. For example, Harris (1984) has proposed that some Douglas-fir stands, which are normally cut after 80 years, be allowed to reach 320 years before being harvested (Figure 5.12). It would take them almost three normal rotations (240 years) to reach old status; then they would have 80 more years to function as old forest ecosystems before being cut. The problem with a long rotation program for providing old forest habitat is that it would occupy more land than a reserve system, because only a portion of the stands would be in an old stage at any one time. In this example, if it were decided that 5% of the forest landscape should be covered by old stands, 20% of the land would have to be allocated to long rotations (i.e., at any given time stands 0–80, 80–160, 160–240, and 240–320 would occupy 5% each).

By combining long rotation systems and uncut reserves, the long rotation stands could be arranged to encircle a core of old-growth forest permanently withdrawn from harvest, thus buffering it and significantly increasing its effective size (Figure 5.12). Other arrangements could be devised; in southeast Alaska it has been proposed that old growth be saved in large blocks occupying entire watersheds (Shoen et al. 1984); in southeastern Australia management guidelines call for old forests to be retained as strips along waterways connecting moderate sized blocks of uncut forest to very large parks and reserves (Recher et al. 1987). Research in the tropical forests of Indonesia (Wilson and Johns 1982) and Malaysia (Wong 1985) indicate the importance of retaining uncut portions of forest as widely distributed refugia from which organisms can recolonize cut areas after harvesting is completed.

The management of forests to produce a facsimile of old growth could

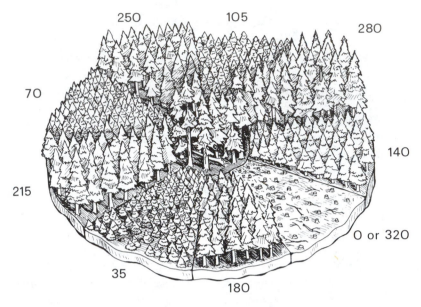

250   105

280

70

140

215

0 or 320

35

180

Figure 5.12   Harris (1984) has proposed a management system for Douglas-fir forests in which a core of old-growth forest is surrounded by a series of nine stands harvested on a long rotation of 320 years.

potentially go far beyond simply lengthening rotation ages (Newton and Cole 1987, Nybert et al. 1987, Lennartz and Lancia in prep.). Foresters often say, "Tell me what kind of stand you want and I can produce it for you." If deer need old forests because they have a dense canopy which will intercept snow, then the forester can grow trees closer together (Nyberg et al. 1987). If woodpeckers need the dead trees that typify old forests in order to excavate nesting cavities, then selected trees can be killed (Mannan and Meslow 1984). The problem with this approach is that it works one species at a time; for each species of concern it is necessary to carefully document what are the critical components of its old forest habitat, then provide those components in younger stands. Needless to say, beyond one or two species this soon becomes a logistical morass. Moreover, if *all* species were being considered, then it would be necessary to provide *all* the components of an old forest, and that would only be possible in an old forest. To put it another way, the special requirements of a few, very important species (e.g., black-tailed deer and red-cockaded woodpeckers) might occasionally necessitate the creation of some special stands to serve as surrogates for old forest. However, this does not diminish the need for maintaining a reasonable amount of old forest ecosystems because, to return to an earlier metaphor, this would be saving a few stones and watching the Taj Mahal crumble.

THE SPOTTED OWL

When the editors of the *Journal of Forestry* put a picture of a spotted owl on the cover with the caption, "The next snail darter?"—thereby invoking the image of a small fish that stood in the way of an enormous dam—they were not exaggerating the controversy that has developed around the most re-nowned bird of the Pacific Northwest (Heinrichs 1984). Four years later the owl made it to the front page of the *New York Times* and the controversy is still growing. The heart of the discord is simply this: A pair of spotted owls requires very large (up to 1,700 ha) tracts of old-growth conifer forest in which to breed, and the timber in these stands is worth about $10,000 per ha (Simberloff 1987). Predictably, many people would ask, is a pair of owls worth over ten million dollars?, and probably answer, no. But, of course, this is reducing the issue to an absurd level of simplicity.

One subspecies, the northern spotted owl, is the focus of interest; its primary range is from Washington to northwestern California, and it has an estimated population of 4,000 to 6,000 individuals (U.S. Fish and Wildlife Service 1982). (There are also a few in southern British Columbia, and some people would argue that the California spotted owls of the Sierra Nevada are not an isolated subspecies.) Home range size of spotted owls varies considerably, but overall there is a decline from an average of about 1,700 ha in Washington to 800 ha in California (Forsman et al. 1984, Sisco and Gutierrez 1984). Large or small, virtually all home ranges are in very old conifer forests, primarily virgin stands over 200 years old (Carey 1985). It is not known exactly which features of a stand are essential to spotted owls, although some observations have been made (Carey 1985, Gutierrez 1985). Large trees that die and have large cavities or broken tops for nesting are likely to be important (Forsman et al. 1984). So are tall trees that produce a multilayered canopy in which the owls can find a benign microclimate (Barrows 1980). Additionally, a thick canopy may shelter juvenile owls from their predators, great horned owls and goshawks. Perhaps most important are the habitat requirements of the owls' primary prey, northern flying squirrels and wood rats, but little is known about their needs (Raphael and Barrett 1984).

The spotted owls' problems are not limited to being tied to an uncommon (and commercially valuable) habitat; data on their demography suggest that they may be ill-equipped to resist, or recover from, catastrophes (Dawson et al. 1987). They do not begin to breed until they are three years old, and the average pair produces only 0.49 fledglings per year (Marcot and Holthausen 1987). Worse yet, juvenile survivorship may be only 11%, because a substantial majority of young birds die soon after they disperse from their parents' territory. Adult survival is probably far better (80 to 96%), but information for adults is very scarce. Using these data to construct a simple life-table model of population change leads to the prediction that the owls will be extinct within a few generations (Barrowclough and Coats, 1985, Dawson et al. 1987, Marcot and Holthausen 1987), although other models are not so pessimistic (Boyce 1987, Lande 1988).

Fortunately, it is apparent that the demographic data available from four years of study are not truly representative of the long-term situation; the values are so low that if true, the owls would have become extinct long before (Dawson et al. 1987). One possible explanation is that there are occasional years in which juvenile survival improves markedly and that these years compensate for the intervening periods in which few young are recruited to the breeding population. Unfortunately, this boom or bust strategy is very fragile when employed by a rare species. It would take only one long bust period (the average life span of an adult or less) to put the species in very dire straits. It is also possible that there is a geographic mosaic of high and low productivity areas with the high-production areas sustaining the overall population.

There is often an interaction between concerns about habitat needs of a rare species and its demography. In the case of the spotted owl, because dispersal mortality of juveniles is a weak link in the demographic chain, the potential effects of cutting old-growth forests are likely to be exacerbated. When continuous blocks of old-growth are fragmented into a few, isolated parcels, presumably it is more difficult for a young owl to locate an unoccupied island of habitat in which it can breed.

Finally, there are issues that confront all small populations. For example, as a population declines there is the likelihood that loss of genetic variation, through inbreeding and genetic drift, could accelerate the slide toward extinction by making individuals less fit, and the whole population less able to adapt to environmental change (Soule' 1980). There is also the possibility that some catastrophe (e.g., disease) could decimate the population, reducing it to a level from which it can never recover. One form of environmental change may be at work now; a close relative of the spotted owl, the barred owl, has extended its range west in recent years and may compete with spotted owls (Hamer et al. 1987). The risks will become more acute if the spotted owl is isolated into small populations; for example, owls north and south of the Columbia River may already be separated from one another, and the Olympic Peninsula owls may be isolated.

All this information and much more was reviewed by a panel of six prominent scientists selected by the presidents of two ornithological societies, the American Ornithologists Union and the Cooper Ornithological Society (Dawson et al. 1987). They recommended that to be reasonably sure of the owls' continued existence, it would be advisable to provide habitat for at least 1,500 pairs, and that these habitat units should be quite large: at least 570 ha in the Sierra Nevada, 1,000 ha in Oregon and northwest California, and 1,800 ha in Washington. The panel did not specify how many owls there should be in Washington, how many in Oregon, etc., so it is not possible to calculate the total area required by their recommendation. However, it is clear that, unless there was a massive shift of owls southward into regions where home range sizes are smaller (obviously an impossibility), between 1.5 and 2 million hectares of old forests—a large majority of all that remains—would be needed.

Understandably, environmental groups were pleased that an assessment by independent scientists called for a major commitment to the spotted owl. However, on the opposite side of the coin is a timber industry that would be quite content to see spotted owls relegated to national parks, designated wilderness areas, and a few other refugia, even though this would substantially increase the risk that the species will become extinct. The timber companies claim that unless they are allowed to continue harvesting old-growth stands,

12,000 jobs will be lost along with about $300 million annually. The impact seems less striking when you realize that it will eventually occur anyway, because in some sense the old-growth forests are being mined; when they are gone they will have to be replaced with a far less profitable system of managing second-growth forests. It has been argued that: (a) the USDA Forest Service, which manages the vast majority of the remaining old growth, is subsidizing the timber industry by building an expensive network of access roads and selling old-growth timber at artificially low prices; and (b) the social costs of an old-growth harvest moratorium could be mitigated by redirecting these government subsidies to employee retraining programs and temporary income supplements (Dixon and Juelson 1987). A greater emphasis on producing value-added wood products would also lessen the economic impact of ceasing old-growth harvest; currently large quantities of raw wood (trimmed to circumvent a law prohibiting export of large logs from federal land) are shipped to mills in Asia.

With the fate of the spotted owl largely in its hands, the Forest Service is caught between the timber industry and the environmental groups (Salwasser 1987). In 1986, it issued a planning document that outlines 12 alternative courses of action for managing spotted owl habitat and describes the likely consequences of each (USDA Forest Service 1986*a*). The alternatives range from doing nothing (this would leave owl habitat only in wilderness areas, parks, etc.) to instituting a moratorium on cutting spotted owl habitat; their preferred alternative calls for 550 habitat areas of 400 to 900 ha each. It is likely that such a compromise position would sustain the spotted owl many more years, if not indefinitely. However, having received 40,000 responses to its planning document, it is clear that the Forest Service's task is far from finished. Ultimately, the resolution will almost certainly be found in the public arena where lobbyists and lawyers for the timber industry and environmental groups will do battle to influence elected officials or persuade judges. Indeed, political considerations are already at work. For example, despite all the concern about its survival, the United States government has not designated the spotted owl an endangered species. It has been speculated that government officials have not wanted to bring the whole Endangered Species Act under attack (thus reenacting events precipitated by the snail darter) for the spotted owl, because it should presumably be protected by the Forest Service's mandate to assure that viable populations of all native vertebrates persist in National Forests (Simberloff 1987).

Is there no solution, no way to save the spotted owl without a significant economic impact? Some very intensive management techniques could be used to deal with the owls' dilemma: moving birds among sites to enhance genetic variability, providing artificial shelters, and enhancing prey populations. (One individual suggested that the Forest Service could start raising and releasing mice.) Apart from being very expensive and of questionable efficacy, these techniques miss a fundamental point. First and foremost, the spotted owl is a flagship for a much larger ecological entity, the entire old-growth forest ecosystem. The owls are certainly not the most conspicuous or dominant components of the ecosystem; the trees themselves fill this role. They may not even be the most vulnerable component of the ecosystem; some little-known insect or lichen may have this honor. They are simply the most vulnerable species that we know about, and as such the Pacific Northwest's old-growth issue revolves around them. If they were to disappear, the issue would not evaporate; it would simply take a new form. As the advisory panel (Dawson et al. 1987) wrote:

> ... the case of the Spotted Owl in northwest California and the Pacific Northwest involves more than the welfare of a single vertebrate species. The value of this bird as an indicator of the condition and extent of old-growth Douglas-fir forest may well facilitate protection of some basic functional relationships of this plant assemblage that are vital to the overall ecology of the Pacific Northwest.

THE BLACK-TAILED DEER

There are 36 species of deer in the world, but only one of them can climb trees,[1] and as a result, deer are very dependent on the vegetation near the ground for both food and cover. Moreover, because many kinds of deer are primarily browsers (and thus are especially fond of shrubs and small trees), deer are often associated with early successional forests where there is an abundance of woody browse. This means that it is an easy metamorphosis, although not an inevitable one, from a cutting site to good deer habitat. This relationship between deer and forest cutting has seemed so simple and fundamental that it came as a considerable surprise when deer biologists studying black-tailed deer along the coasts of British Columbia and southeast Alaska discovered that they do not conform with tradition; they often prefer old growth.

Once thought to be a separate species, the black-tailed deer is now recognized as two subspecies of the mule deer: The Columbian black-tailed deer ranges from central California to Vancouver Island; from there north to Alaska's Kodiak Island is the home of the Sitka black-tailed deer. The forests they inhabit are dominated by conifers such as Douglas-fir, western hemlock, and Sitka spruce and tend to be relatively cool, wet, and productive (Waring and Franklin 1979). Despite the moderating influence of the Pacific Ocean, at the northern end of its range, the blacktail encounters fairly severe winters. Temperatures are not exceptionally low, but snow commonly accumulates to

---

[1]Himalayan musk deer near Mt. Everest have recently been observed foraging in trees during winter. Their long, splayed toes, an adaptation to deep snow, allow them to climb rhododendron trees, which have fairly horizontal limbs (Katel pers. comm.).

depths that present the deer with serious problems, basically lack of food and mobility. It was when researchers became concerned about how blacktails cope with harsh winters that they discovered that old-growth forests are an integral part of the deer's survival strategy (Wallmo and Schoen 1980, Bunnell and Jones 1984).

To understand the relationship of blacktails to forest succession, consider three age classes of forest, early successional (less than about 30 years old), late successional (roughly 30 to 150 years), and old-growth (Figure 5.13). In an early succesional stand, there is a low stratum of shrubs and small trees, much of which is within the deer's reach. By late succession, the trees have formed a high, dense canopy. A climax forest has canopy gaps where large trees have died, and these gaps support a patch of low vegetation.

Differences in the forest canopy have a profound influence over snow depth (Harestad and Bunnell 1981); in late successional and old-growth forests the canopy is dense enough to intercept a large portion of the snow, and thus snow depth is correspondingly reduced. Southeast Alaskan forests with a canopy over 80% closed had an average snow depth of 36 cm, compared to 132 cm in open sites (Bloom 1978). This is a very important difference, because once the snow is over two-thirds of a deer's chest height (about 35 cm), it significantly inhibits the deer's movements, and this is at a time of year when the deer can ill-afford the additional energy demands of travel in snow (Parker et

Figure 5.13   Three stands in southeast Alaska; the first, recently clearcut; the second, a late successional stand with little understory; and the third, an old-growth stand. (Photographs by John Schoen)

Figure 5.13 *(Cont.)*

al. 1984). Furthermore, snow compounds the energy deficit problem by covering up the deer's food.

Although snow depth may explain why deer would prefer older forests over those too young to have developed a canopy, it does not explain why deer prefer old-growth stands to late successional ones. The answer is quite simple. In a late successional stand the canopy is so dense—there are so few gaps through which light and rain can reach the ground—that there is only a very small biomass of herbs and shrubs, often less than 100 kilograms per hectare (Alaback 1982). In contrast, an old-growth stand might have 1,000–5,000 kg/ha in the understory because of all the gaps. In short, old-growth stands offer the best winter habitat for blacktails because they have a sufficiently dense canopy to reduce snow depths, and yet enough gaps to grow an abundance of deer fodder.

The basic story also has some additional angles that are quite interesting: (a) The deer's preference for old growth is not necessarily limited to winter. In one study, the ratio of the number of deer using old growth versus the number using young growth was 7:1 in the winter and remained 5.3:1 in the summer (Wallmo and Schoen 1980). The lower summer use was ascribed to logging residues—slash piles—inhibiting the deer's movements. (b) Although the absolute amount of understory biomass in old-growth stands is not as great as in early successional forests, it may be better forage for the deer in two senses. First, it is taller and less likely to be covered by snow (Harestad et al. 1982); to take a specific example, one of the deer's preferred foods, blueberry bushes, are much more common in old-growth forests than in young stands and are tall enough to be available in deep snows (Bunnell and Jones 1984). Second, browse in old-growth forests is somewhat more nutritious than in young stands; concentrations of inorganic nutrients are about 10% higher on average (Bunnell and Jones 1984). (c) Old growth provides another rather unexpected, but highly palatable, source of food for the deer, arboreal lichens. In one study, an average of 86 kg/ha of lichens fell to the ground in 180 days, and 37-53% was consumed by the deer (Stevenson and Rochelle 1984); immature stands produce very few lichens. (d) Not all old growth is of equal value to the deer. An analysis of snow depths in different Alaskan stands concluded that the best deer habitat was provided by the stands with the highest volume of timber (Kirchoff and Schoen 1987); unfortunately, these stands are rare (less than 2% of the harvestable forest in the Tongass National Forest) and very desirable for timber harvest too.

Keeping deer well-fed through the winter is secondary to the underlying question: How do old-growth forests affect an area's carrying capacity for blacktails? This is difficult to predict with great certainty—one study estimated deer populations would be reduced at least 33-90% if all the old growth were harvested (Bunnell and Jones 1984)—but it is very clear that concern for black-tailed deer can be directly translated into a concern for the maintenance of old growth. On the other hand, it has been proposed that the linkage between black-tailed deer and old growth could be avoided through careful timber management (Nyberg et al. 1987). If what is needed is a dense canopy to minimize snow depths closely juxtaposed to openings in which forage can grow, could not careful interspersion of small cuts among late successional stands provide the appropriate habitat? Maybe. However, the arboreal lichens that deer seem to need would be absent and, more importantly, the habitat needs of many other species, the lichens themselves for example, would go unmet.

Some biologists have speculated that the story of the black-tailed deer may not be unique; perhaps it is not an exception to the general pattern effected by the exceptional nature of the forests of the Pacific Northwest (Schoen et al. 1981, Pac et al. 1984). Perhaps the relationship between blacktails and old growth only seems unique because in other areas virtually all the old-growth stands were destroyed before anyone had a chance to study them and their deer populations. Such speculations are unlikely to cause deer managers in other regions to start advocating the creation of large areas of old forest, but it should give them pause to reconsider their habitat management practices. It is entirely possible that deer habitat quality in many regions would be improved if old forests occupied a reasonable portion of the landscape.

SUMMARY

Because various plant and animal species are associated with different stages of succession, balancing the age structure of a forest accomplishes two objectives: achieving a sustained yield of forest products and providing diverse wildlife habitat. Thus, viewed from the time-scale of succession and the spatial scale of landscapes, forest management can readily enhance biological diversity. There is a very critical exception to this generalization, because forests become economically mature, ready to be cut, long before they are biologically mature, and thus old forests are uncommon in most regions. (There is no universal definition for what constitutes old or old-growth stands except that they are relatively old and relatively undisturbed; specific parameters for age and disturbance need to be developed for each type of forest.) It is very important to protect and maintain a reasonable area of old forest to maintain populations of the species that are associated with this habitat and the community as a whole. In regions where there are very few old, virgin forests remaining, it may be possible to grow old stands in the future and thus maintain habitat for all the species of a forested landscape, not just those associated with younger forests. However, this possibility does not diminish the value of existing old stands.

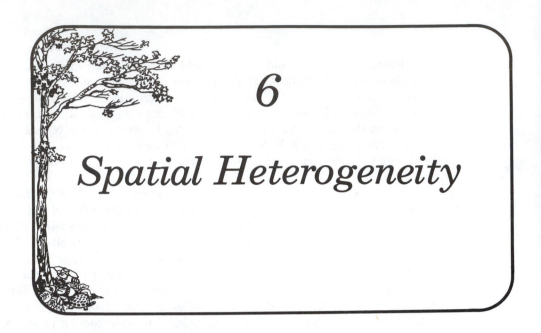

6

Spatial Heterogeneity

All organisms need space—a place to live—but the amount of area needed by different species varies enormously. The thousands of square kilometers occupied by some wolf packs is about 12 orders of magnitude larger than the few square centimeters inhabited by many sedentary invertebrates, such as the web-spinning spiders. And many more species can be added to the lower end of the home range size continuum if one is broad-minded enough to think of plants as having home ranges. Even if we restrict our thinking to solitary vertebrates, the span of home range sizes from salamanders to large carnivores covers eight orders of magnitude. Obviously, big things need more room than little things—more formally, size is a primary determinant of home range area—and elephants and redwoods will always use more space than salamanders and violets. Lifestyle has an even more profound effect on spatial requirements. Because plants are sedentary (ignoring their wonderful repertoire for seed dispersal including cross-country trips of tumbleweeds), their space requirements are determined almost entirely by size. Caterpillars, cattle, and other herbivores have to move around to graze, but if their food is abundant, they can limit their activities to a relatively small area. Predators usually have to cover quite large areas to find enough herbivores to eat, although a few have evolved ways to let the food come to them. Witness the ant-lion, which constructs a sand funnel and waits at the bottom for an ant to blunder in. With wildlife occurring in such a vast assortment of sizes and lifestyles, a forest that is going to support a diversity of wildlife must be spatially diverse. The importance of spatial heterogeneity might best be understood by considering the generic question, ''When does a mixture of black and white become gray?''

In Figure 6.1, the two blocks represent the same tract of forest as seen by two hypothetical species. The left block represents a species that uses a fairly small home range, for example, a warbler; the right represents a species that requires a relatively large home range, e.g., a hawk. In both diagrams, the individual black-and-white squares represent one hectare of forest (black can be 60-year-old stands and white 5 years old); on the left only 4 ha are shown, while on the right there are 400 ha depicted. In the left diagram a warbler might find either of the two black squares a suitable place to establish a breeding territory. In contrast, a hawk, seeing the world as depicted on the right, may view this whole forest as unsuitable habitat if it needs open land in large contiguous tracts. In other words, while the warbler can see black and white and select the black that it needs, the hawk may not see black and white, only gray, and find no suitable habitat.

Organisms also vary in the extent to which they need a diverse environment, and we often speak of habitat specialists and habitat generalists. Habi-

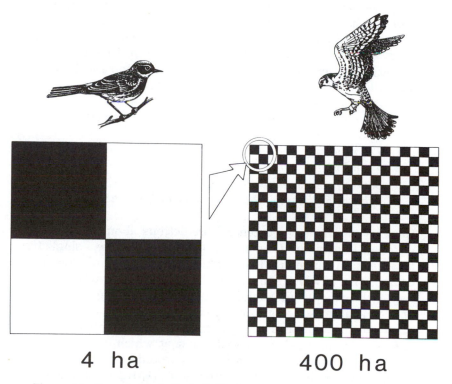

4 ha   400 ha

Figure 6.1 In each block the individual cells represent 1 ha of 60-year-old forest (black) or 5-year-old forest (white). The left figure represents the scale of perception of a yellow-rumped warbler; the right represents an American kestrel's scale. (Redrawn from Hunter 1986)

tat specialists may be subdivided into two groups: those that need a fairly uniform environment, and those that must have a diverse environment. The northern spotted owl's requirement for old conifer forests make it a well-known example of a habitat specialist that prefers a relatively uniform forest (Forsman et al. 1984). The American woodcock is a habitat specialist that requires a diverse habitat: It needs forest openings in which to roost and display, alder swamps in which to forage, and 15- to 30-year-old deciduous stands for nesting and brood-rearing (Sepik et al. 1981). Large mammals are often good examples of habitat generalists—consider bobcats, raccoons, coyotes, and white-tailed deer that inhabit spruce-fir forests, bottomland swamps, semi-arid scrub, and many other environments—but so are smaller species such as American robins and American toads. Species with special requirements for a uniform habitat are usually small creatures with small home ranges; monarch butterflies and milkweed patches are but one example of an insect irrevocably tied to a habitat type defined by a single kind of plant (Strong et al. 1984). A researcher studying red oak acorns in an Illinois forest found nearly 100 species of insects and other creatures living inside these motes of habitat (Winston 1956). As with all generalizations, there are many exceptions. Giant pandas are tied to the cool, moist forests of southeastern China where their favorite species of bamboo grow (Schaller et al. 1985), but *Anthocoris nemorum,* more prosaically the "very small speckled brown bug," is found in deciduous forests, hedgerows, orchards, gardens and many other habitats (Elton 1966). To return to the analogy of Figure 6.1, some animals need black, some need white, some need black *and* white, and some can use either black or white.

If we use the following two ideas as premises:

1. *The diversity of a forest depends on the scale at which it is viewed by individual species;*
2. *Some species can live in a variety of habitats, some require a diverse habitat, and some require a fairly uniform habitat;*

then we can make three general predictions about the biological diversity of different forests. First, on a very small scale, where a single tree could constitute a habitat, diversity is probably greatest in an old forest of mixed species composition in which some of the largest trees are starting to die, thus breaking up the canopy and allowing small groups of younger trees to prosper. Second, on a larger scale, one measured in hundreds of hectares, a mosaic of small stands of different ages and species compositions would have the richest diversity. Finally, on the largest scale, that at which industrial and governmental policy makers operate, forest diversity would be greatest if the landscape were covered by stands of many different sizes (1, 10, and 100 ha), ages, and species compositions.

These predictions are based on intuition and extrapolation from specific research, not on broad-based documentation; only a few studies have considered the relationship between habitat patchiness and species diversity, and they worked only at a scale appropriate to the group of organisms being studied (Wiens 1974, 1976, Roth 1976, August 1983). Nevertheless, it is reasonable to use the predictions to guide forest wildlife management, until they are corroborated or refuted.

Before delving further into spatial heterogeneity, we need to briefly digress with a review of harvesting systems, because spatial heterogeneity of managed forests is determined largely by how they are cut.

A Short Primer on Natural Disturbance and Harvesting Systems

There is inherent spatial heterogeneity in many forests because of soils, topography, and, most importantly, patterns of disturbance. The toppling of a single tree can create a gap in the forest canopy large enough to modify the microclimate and effect a series of small changes in the forest ecosystem (Runkle 1982). Large-scale disturbance—fires that incinerate great sweeps of land—constitute the other end of the disturbance spectrum. Different forests have characteristic patterns of natural disturbance. Many coniferous forests seem to be primarily affected by fires, disturbances that come relatively infrequently but cover fairly large areas (Runkle 1985). Temperate deciduous and tropical forests are disturbed largely by the deaths of individual trees, a fairly frequent event. Interestingly, if you measure the rate of natural disturbance in large areas of forest over extensive periods, the results are quite consistent among a variety of forest types; the average disturbance rate is about 1% per year and generally ranges from 0.5–2% (Zackrisson 1977, Runkle 1985). Put another way, on average about 0.5–2% of a forest will be disturbed in any given year, or any given spot in the forest will be disturbed about once every 50–200 years.

Forests with small-scale, frequent disturbances are generally composed of shade-tolerant trees of many different ages. The characteristic portrait is of small trees waiting in the understory, suppressed until a large tree dies and leaves a vacant spot. Such forests are usually referred to as uneven-aged, although some people call them all-aged and reserve "uneven-aged" for forests with two or three distinct age classes. In contrast, in forests characterized by large-scale, infrequent disturbances, the trees will usually be of just one age class, and thus this type of forest is called even-aged. Either shade-tolerant or shade-intolerant trees grow in these forests, depending on how far succession has proceeded since the last disturbance.

The distinction between even-aged stands and uneven-aged stands becomes rather fuzzy, because virtually all uneven-aged stands could be thought of as composed of many very small, even-aged stands. "Even when a single large tree dies it is replaced not by one new tree but by many that appear nearly simultaneously" (Smith 1986), and this group of small trees could be considered an even-aged stand. Thus, to draw a line between even-aged and uneven-aged stands, one has to turn to questions of scale again—how small can a stand be? Smith (1986) proposed that a group of trees growing in a forest opening which has a diameter greater than twice the height of the surrounding trees should be considered an even-aged stand, because the temperature in the center of the opening would not differ substantially from temperatures in a larger opening. By this criterion, a stand surrounded by 20 m trees would only have to be larger than 0.13 ha (for 60-foot trees this would be about a third of an acre). This is rather small for practical application—most foresters seldom recognize a stand of less than ten acres (4 ha)—and thus we are still left with no definitive distinction.

To return to our polarized perspective, these two situations—uneven-aged forests with small-scale, frequent disturbance and even-aged forests with large-scale, infrequent disturbance—correspond with two harvesting systems designed to imitate natural disturbance patterns. *Clearcutting* is the best known even-aged management scheme, and *selection cutting* is more or less synonymous with uneven-aged management.

In selection cutting, a small portion of the trees are removed at frequent intervals; 5–10 years would be typical. Individual trees are chosen for removal because they have grown to maturity, or because they are of poor quality and by weeding them out the growth of the remaining trees can be improved. (Sometimes selection cutting involves removing only the most profitable trees and leaving just poor-quality timber behind; this is called high-grading and is widely frowned on. A somewhat more acceptable practice is diameter-limit cutting, in which all the trees above a certain size are cut, but again this technique ignores differences in quality.) Because selection cutting produces only a very small disturbance to the forest canopy, roughly 10–100 m^2 (0.001–0.01 ha) per tree, it creates spatial heterogeneity on a very fine scale. A forest managed by selection cutting wil provide a forest habitat continuously and indefinitely, because the felling of individual trees imitates the natural fall of old individuals. Moreover, selection cutting does not open the canopy enough to allow shade-intolerant trees to become established, and thus the continued presence of shade-tolerant trees is favored.

Clearcutting refers to harvesting almost all the trees on a swath of land, and thus it can potentially produce very coarse scale spatial heterogeneity, sometimes measured in hundreds of hectares. The minimum size that a cut can be and still constitute a clearcut is arguable; here it is defined as one hectare. (Incidentally, when most people see a clearcut, they greatly overestimate its size, often by as much as a factor of ten.) Clearcuts initiate succession and thus over time produce an array of different-aged stands with their associated wildlife. The patterns vary a great deal. Clearcuts in which many seedlings were established before the cut, or in which regeneration comes largely from seeds lying dormant in the soil or from root sprouts, might experience little change in forest composition. In contrast, in a clearcut without advanced regeneration, the pattern of succession will be profoundly affected by the availability of different seeds. This, in turn, depends on the proximity of trees that can act as a seed source and the relative mobility of different seeds (Canham and Marks 1985). As a generalization, shade-intolerant trees usually have very mobile, often wind-dispersed seeds, and shade-tolerant trees have heavier seeds with large energy reserves that allow them to grow in shady conditions. Thus, clearcuts tend to favor shade-intolerant species. This effect is further enhanced by the facts that shade-tolerant trees are often outgrown by shade-intolerants in an open environment, and many shade-tolerant trees cannot tolerate the hot, dry conditions in a clearcut (Grime 1979). Foresters often cannot afford to wait for natural regeneration and, therefore, large clearcuts are commonly planted with seedlings of preferred commercial species.

There are four other harvesting systems to mention here. If all the trees are removed from a site but the area is too small to be a clearcut, it is called a *patch cut;* their minimum size is about 0.1 hectares. Patchcuts merge into *group selection* cutting in which clumps of trees to be removed are identified. *Shelterwood* cutting is a harvesting system in which most of the trees are removed in a short period, but enough, anywhere from 25% to 75% of the largest and most vigorous trees, are left behind as a seed source and to provide some shelter to sensitive seedlings. They comprise a sufficient canopy to keep daytime temperatures moderate and reduce desiccation. Later, when the young trees are large enough to tolerate exposure, the residual trees are removed in another cut. Shelterwood systems may imitate the natural disturbance of a mild wind or ice storm in which only strong individuals are left standing (Runkle 1985). If providing an adequate supply of seeds is the only problem and microclimate is not an issue, then a partial cut that only leaves a few scattered, wind-firm trees is appropriate; this is called the *seed-tree* method. Except for group selection cuts and very small patchcuts, these harvest systems constitute even-aged management and can be effectively substituted for clearcutting in the scenarios developed below.

Referring to these methods of removing trees from a site as harvesting systems is rather misleading. If all you care about is extracting trees, it does not much matter how you do it. However, if you are concerned about the forest's future, planning for reproduction should be behind every choice of a harvesting system, because cutting is a key component of a larger management process, the silvicultural system. In succeeding chapters, particularly Chapter 12, "Intensive Silviculture," silvicultural systems are considered in greater depth and in additional contexts, but the present question is how to make forests spatially heterogeneous.

MANAGING FOR SPATIAL HETEROGENEITY

Selection cutting is clearly the management system of choice for maintaining diversity on a small scale, because the habitat discontinuity it creates is quite small, about the size of a single large tree. Furthermore, it assures that there will be trees of many different ages within each stand. Patch cuts or small clearcuts can be used to create a mosaic of stands of different ages and species composition, and thus they are appropriate for producing a diverse forest from a mid-scale perspective. When thinking about forest diversity on a landscape scale, and whenever possible, this is the scale at which to operate; the repertoire should include selection cutting, patch cutting, and clearcutting. The use of large clearcuts is often criticized, but if maintaining wildlife diversity is a goal, clearcuts have an immediate role as habitat for early successional species, and eventually they will become large, relatively uniform stands.

If forest harvesting is to be used to create spatial heterogeneity, then the relative importance of different cutting types has to be determined. In other words, if there is to be an array of different cut sizes, what should be the end points of the array, and how much weight should be given to different points in the array?

A Model for Spatial Heterogeneity

In his seminal work, *The Fragmented Forest,* Larry Harris (1984) addressed a problem in forest spatial heterogeneity when he tried to devise an ideal frequency distribution for different size fragments of old-growth forest (Figure 6.2). The actual numbers do not concern us; the key point is that there

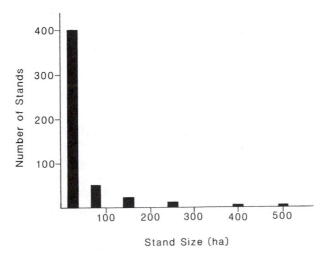

Figure 6.2 Harris's (1984) proposed distribution for the number of old-growth Douglas fir stands showing a predominance of small stands.

are a great many more small stands than large ones. Harris did not have empirical evidence to support the desirability of this type of distribution; it was based on informed speculation along two lines of reasoning.

The first factor concerns the relative size of organisms. Large organisms are less common than small organisms in two senses (May 1986). First, there are fewer species of large organisms; for example, there are only seven species of bears in the world but several hundred species of mice. Second, the absolute populations of large animals are lower; the total world population of all bears probably does not great exceed one million, but the world population of mice must reach well into the billions. Because small species have smaller home ranges than large species and there are more small species, Harris felt the distribution of forest stands should be weighted in their favor. Harris's distribution is weighted in two senses. If one multiplies each stand size by the number of stands of that size, not only are small stands more numerous, but they also collectively cover a larger area than large stands (Figure 6.3).

The second idea Harris considered in developing this distribution involved the precedents set by natural landscapes. For example, if one examines the patterns of stream lengths and lake and watershed sizes, they tend to follow this same distribution pattern with small units more numerous than large ones (Figure 6.4) (Horton 1945). In other words, myriad tiny rivulets lace the landscape, but there are very few continent-draining giants like the Nile, Indus, and Rhine. Less well documented, but of more relevance to trees, are the patterns of the physical environments in which forests develop; Figure 6.5 indicates that the size distribution of soil units may follow a similar pattern to lakes.

There is another set of natural phenomena that could provide a very relevant model for imitation, forest disturbance. For example, the effect of tree

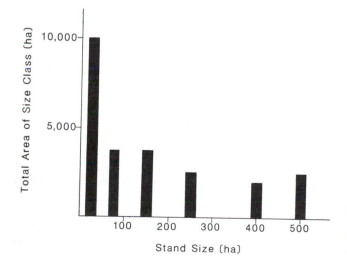

Figure 6.3 Harris's distribution transformed to show the total area of each size class of old-growth stand; small stands are still predominant.

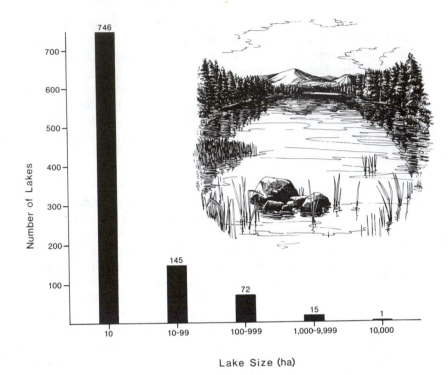

Figure 6.4 Size distribution of lakes and ponds in Piscataquis County, Maine, show that there are more small ponds than large lakes. (Based on lake surveys of the Maine Dept. of Inland Fisheries and Wildlife and U.S. Geological Survey maps.)

falls on the continuity of the forest canopy varies enormously. When a moderate sized tree falls, it leaves a gap no larger than its crown; a bigger tree may take a few neighbors down with it and create a fairly significant hole; a tornado or hurricane can level great sweeps of forest (Foster 1988). Once again, the same pattern emerges: There are more small tree falls than big ones (Figure 6.6; Brokaw 1982, 1985). Similarly, small forest fires are currently much more common than large ones (Figure 6.7; USDA Forest Service 1986*d*). Unfortunately, these data are undoubtedly affected by people's penchant for putting out fires before they become very large, and relatively little is known about the size distribution of fires in the absence of human influence, in part because an occasional large fire can obliterate evidence of previous smaller fires. Some information suggests that presettlement fire regimes varied considerably among different forest types and different terrains. In the flat, unbroken expanses of some boreal regions, large fires (over 400 ha) were quite regular, and fires exceeding 100,000 ha an occasional event (Foster 1983, Heinselman 1981*b*); much smaller fires were more typical in steep, broken terrain (Kilgore

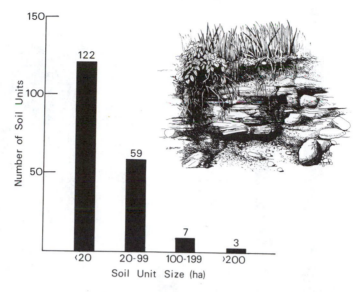

Figure 6.5 Size distribution of soil units from near Homer, Alaska. (Based on maps in Hinton 1971)

1981) and in places where extensive wetlands inhibited the fires' spread (Foster 1983).

It is unclear if there is some important, fundamental design principle at work here, or is it just a trivial observation that there are more small tree fall gaps than large ones, and more grains of sand in the world than boulders? Mathematicians have considered how to measure patterns of size in nature (Mandelbrot 1983), and biologists have explored various ramifications of size such as how organisms overcome gravity to get water to the top of a redwood or to keep an elephant from collapsing (McMahon and Bonner 1983), but no one has answered the question, "Why are there more small things than big things?"

Fortunately, we do not have to completely understand the patterns of nature in order to imitate them, and this pattern of small being more numerous than large has been used in Figure 6.8 to construct a hypothetical distribution of cut sizes for a 5,000 ha tract. At the small-scale end of the continuum, there are 100,000 cuts of 0.01 ha; in other words 1,000 ha of selection cutting. Next there are 10,000 patch cuts of 0.1 ha, again comprising 1,000 ha collectively, and finally three sizes of clearcuts, 1, 10, and 100 ha. Beyond the general rule that there should be more small units than large ones, the shape of the distribution becomes rather arbitrary. In this example, each size class is ten times larger than the one before and has a tenth as many stands. (A ten-fold increase was selected, because intuitively it seems likely that each size class will repre-

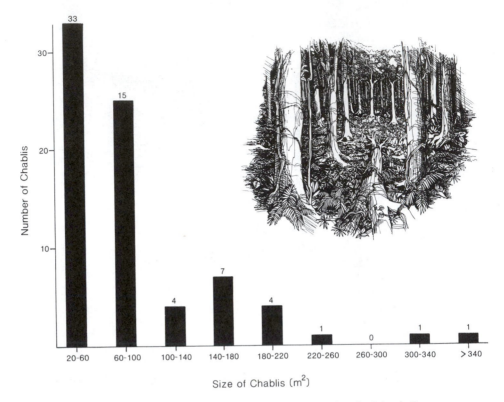

Figure 6.6 Size distribution of chablis on Barro Colorado Island, Panama (Brokaw 1982). (Chablis is a French word used by some forest ecologists; it refers to the fall of a tree, the damage it causes, and the fallen tree itself.)

sent a different type of habitat.) Because the increases in size classes are matched by proportionate decreases in the number of stands, the total area of each class is 1,000 ha and the area-distribution curve is flat; compare Figure 6.3 to Figure 6.9.

Again, the reason Harris felt there should be a larger total area of small units is because there are more small species in the world than large ones, more mice than bears; so why has a flat distribution curve been proposed here? There are two reasons. First, large lakes, fires, chablis, and soil units seem to cover a larger total area than small lakes, fires, chablis, and soil units, respectively. (This statement is based only on the data presented in Figures 6.4, 6.5, 6.6, and 6.7 after organizing them into size classes that increase by increments of ten; it should be tested extensively.) Second, large species that need large habitat units are much more likely to be threatened with extinction than small species, and thus they deserve special attention.

This pattern can be summarized in a general rule of thumb: *Harvest for-*

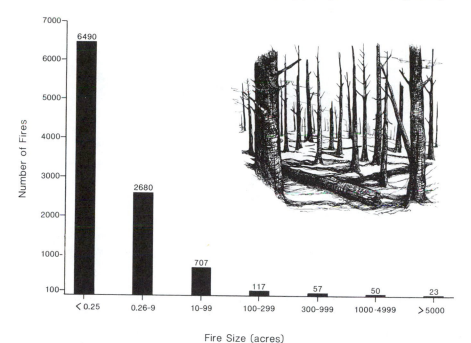

Figure 6.7 Size distribution of forest fires in the United States in 1985 (USDA Forest Service 1986*d*).

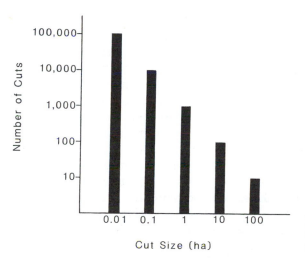

Figure 6.8 A hypothetical distribution of cut sizes designed to maximize the spatial heterogeneity of a forest.

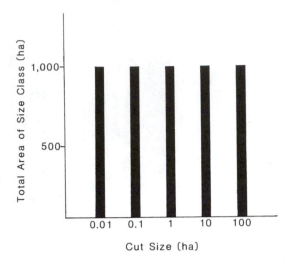

Figure 6.9 Figure 6.8 transformed to show that the total area allocated to each size class is equal.

ests at a range of different scales, and allocate approximately equal areas to different points along the continuum from small scale to large. The exact details of the distribution used will have to be shaped to local conditions, but in most areas it is clear that the small-scale end of the continuum should be defined by selection cutting. It is less obvious where the other end of the continuum should be; in other words, how big should the biggest clearcuts be?

How Big Should Clearcuts Be?

Size of the management unit

The first approach to answering this question is very simple: the more land in a management unit, the larger clearcuts can be. In Table 6.1, this approach is laid out systematically. Starting with a minute piece of land, 0.1 ha or the equivalent of a small suburban house lot, selection cutting is obviously the only viable approach if there is going to be some age diversity on the land. With 2 ha it would be reasonable to manage half the property using selection cuts and maintain ten 0.1 ha patch cuts on the other half. On a 30 ha parcel, ten 1 ha clearcuts could be accommodated along with 10 ha of patch cuts and 10 ha of selection cutting. Very large clearcuts, 100 ha or more, become possible when the tract is several thousand hectares or larger.

Note that in each array there are at least ten units of each size class. This is to assure that several age classes can be represented in each size class. For example, in a 100 ha forest cut on a 100-year rotation, there would be enough room to have at least 10 ha stands that are 5, 15, 25, 35, 45, 55, 65, 75, 85, and 95 years old and still retain all the selection cut, patch cut, and 1 ha clear-

TABLE 6.1   **Hypothetical Distribution of Cut Sizes on Different Sizes of Forest Tracts**

0.01 ha × 10 = 0.1 ha

0.01 ha × 100 = 1 ha
0.1 ha × 10 = 1 ha
         2 ha

0.01 ha × 1,000 = 10 ha
0.1 ha ×   100 = 10 ha
1   ha ×   10 = 10 ha
            30 ha

0.10 ha × 10,000 = 100 ha
0.1 ha × 1,000 = 100 ha
1   ha ×   100 = 100 ha
10   ha ×   10 = 100 ha
            400 ha

0.01 ha × 100,000 = 1,000 ha
0.1 ha × 10,000 = 1,000 ha
1   ha ×   1,000 = 1,000 ha
10   ha ×   100 = 1,000 ha
100 ha ×   10 = 1,000 ha
            5,000 ha

cut areas. Depending on what is happening on other forests in the region, it might be desirable to have at least three or four representatives of each age class, e.g., four 5-year-old 10 ha stands, four 15-year-old, etc.

As mentioned earlier, if forest landscapes are to be diversified, it is desirable to coordinate forest management over extremely large areas. Does this mean that if a management unit is 100,000 ha, then 1,000 ha clearcuts are in order? To answer this, a second approach must be employed, one that involves considering the maximum space requirements of certain animals.

Habitat requirements of animals

At the beginning of this chapter, it was pointed out that some animals, notably large carnivores, have huge home ranges, areas measured in square kilometers, not hectares. Should forest managers then have similarly huge stands of uniform forest? No; virtually all species with large home ranges are either habitat generalists or are very capable of routinely crossing most types of forest habitat to reach the types that they prefer. A lynx might find ideal conditions in 10,000 ha of continuous young forest teaming with snowshoe hares, but it could certainly thrive in a quiltwork of smaller stands. If large carnivores were not fairly catholic about their habitat requirements, they probably would not survive. Small carnivores are more likely to be the species that need the largest areas of uniform habitat. For example, if a pair of American kestrels needs between 25 and 150 ha of open land in which to hunt for rodents, it might be reasonable to assume that some clearcuts should also be at least 25 ha, perhaps as much as 100 ha, to support kestrels plus a range of species with more limited requirements for continuous habitat.

Data to support the idea that some species require large clearcuts is very scarce; research in Australia has shown that birds such as the Nankeen kestrel, brown thornbill, and brush bronzewing are found in large clearcuts but not in cuts less than 10 ha (Recher et al. 1980, 1987). Of course, it is important to remember that today's 50 ha clearcut will eventually (barring conversion to agriculture or other uses) become habitat for a species that needs 50 ha of forest.

It is often desirable to think beyond the space requirements of an individual or social group and consider the area required by a whole breeding population. This perspective may be especially important for relatively immobile species in which genetic isolation can become a problem. To construct a hypothetical example, if a certain species has 10 ha territories and if it takes perhaps 1,000 breeding individuals to have a viable population (Frankel and Soulé 1981, Soulé 1987), then up to 10,000 ha of habitat might be needed to sustain a population. This may be an argument for maintaining very large tracts in natural ecosystems, but it is not likely to be a reason for using silviculture to create extremely large, uniform stands. To maintain gene flow it is only necessary that a few individuals occasionally tolerate the conditions of a "foreign" habitat long enough to traverse it and reach a suitable area on the other side (Endler 1977, Dobzhansky et al. 1979, Greenwood and Harvey 1982, Slatkin 1985). For example, an eastern box turtle may not be able to prosper and breed in a 10-year-old pine stand, but it could probably cross one to reach more suitable habitat. This issue and related questions about fragmentation and habitat corridors is addressed in Chapter 8, "Islands and Fragments."

Aesthetics

Most people dislike clearcuts because they do not like the way they look. It is not open space that people mind. Indeed, in heavily forested regions open vistas are often a welcome relief, and some writers would trace this aesthetic preference back to a time when our humanoid precursors moved out of the forests of Africa and onto the savannah (Orians 1980). It is primarily the slash that is left behind that people find offensive. It looks terrible and makes traversing a clearcut unpleasant. Clearcuts with long straight edges also distress people because such regular interfaces do not seem natural. Although it might be tempting to dismiss these perceptives as sentimental whims, they are real and should be recognized. Placing clearcuts where few people will see them is an obvious solution that can resolve many conflicts before they arise.

Soil and nutrient loss

Knocking down trees and dragging them away with big machines creates a situation fraught with potential for soil erosion: no trees to buffer the rain's impact and a soil that has been scraped and left exposed. Fortunately, techniques to mitigate the calamity have been worked out and are widely employed. (Lifting logs by helicopters or aerial cable networks are among the more spectacular methods.) Nevertheless, the risk still exists, and there will always be some less conscientious loggers whose activities precipitate serious soil erosion.

Ironically, removing the slash that people find unattractive can exacerbate another issue, nutrient depletion (Bormann et al. 1968, Sollins and McCorison 1981). Branches, twigs, and leaves contain a disproportionately large part of the nutrients in a tree. If they are left scattered about a site, eventually they will decompose and their nutrients will be available again. However, some modern harvesting techniques involve carrying whole trees to the roadside and delimbing them there, or alternatively, chopping the whole tree into chips for fuel or pulp. Whether the slash is left at the roadside or taken away to be burned, the effect is much the same; a sizable supply of nutrients has been removed from the forest. It is not generally known if this depletion is critical or is instead balanced by the influx of new nutrients from bedrock weathering or atmospheric depositions (Sollins et al. 1980). It may be economically feasible to replace nutrients through chemical fertilization, but these cannot replace the organic matter that is important to soil structure and as a habitat for small creatures. Actually, the possibility of soil and nutrient loss is not directly a function of clearcut size, but the larger the cut, the greater the potential impact.

Natural disturbances as models

An evaluation of clearcuts should also consider how readily a forest will become established again on the site. In forests that regularly experience large-scale disturbances, notably conifer forests, this should not be a problem. Chances are good that there will be tree seeds or seedlings present that will flourish in the open environment. Initially, they might be outcompeted by herbs and shrubs, but soon a forest community will prevail. In forests where small-scale disturbance is the norm, reestablishing the forest soon after a large clearcut may be a huge task, perhaps almost impossible. For example, clearcutting a tropical rain forest will generally have such profound effects on the microclimate and soil that it will take a very long period, perhaps as much as 1,000 years, before a mature forest ecosystem exists again (Opler et al. 1977). Even in temperate latitudes, seedlings of some shade-tolerant trees cannot survive in the open environment of a clearcut, and thus the cycle from mature forest to clearcut and back to mature forest would be very long. Furthermore, many shade-tolerant trees have fairly heavy seeds that are not easily transported into the middle of a large clearcut.

The problem of a harsh change in environments can be ameliorated to some extent by the shelterwood method, and both the shelterwood and seed-tree methods can be used to assure an adequate supply of seeds. This said, it would be prudent not to allow clearcuts to exceed the size of natural disturbances even if that means no clearcutting at all in some types of forest. By turning this statement around, one could argue that in boreal regions where fires measuring thousands of hectares occur naturally, equally vast clearcuts are acceptable. However, this is probably not ample justification for huge clearcuts because of public sentiment issues and because a fire and a clearcut are not the same thing: They have rather different effects on the soil and vegetation.

The idea of using natural disturbance regimes as models should be tempered by the fact that these regimes change through time. For example, during the past 750 years fires in northwestern Minnesota have occurred in cycles of different intervals, and the lengths of these intervals have been linked to changes in temperature and precipitation (Clark 1988).

Of clearcuts and compromises

Despite its limitations, clearcutting is an important and legitimate part of forestry's repertoire of techniques, and there are even circumstances in which it can significantly diversify a forest from a wildlife perspective because of its marked effect on spatial heterogeneity. However, it is likely to always remain somewhat controversial because of its visual impact and its alleged negative effects on wildlife. Of course, its effects on many individual species of wildlife can be distinctly negative, at least in the short term, and if these species are public favorites, then controversy is inevitable. A good example of this can be found in the story of Maine's deer and moose told below.

Too often, when a controversy over clearcutting arises, wildlife advocates decry large clearcuts as biological wastelands and timber advocates decry selection cutting as uneconomical. The typical outcome is a compromise reached by converging on small clearcuts, in other words, medium-scale management (Figure 6.10). Biological diversity would usually be better served if a broad-based compromise was reached in which there was management at a number of different scales and the economic and ecological effects of selection cuts and large clearcuts were allowed to balance one another.

THE DEER AND THE MOOSE

On September 22, 1980, television crews from each of the major U.S. networks were in the north Maine woods to cover the beginning of what was to become a major controversy, the resumption of hunting moose, Maine's "sacred cow," after a 35-year hiatus. This experimental open season was followed by three years of public debate orchestrated by organizations called SAM (The Sports-

Figure 6.10 In this area of the Ocala National Forest, Florida, many moderate-sized clearcuts (roughly 15 ha, 40 acres) were used. Because most of them are very similar in size, the spatial heterogeneity of this landscape is lower than it might be (Photo by USDA Forest Service).

men's Alliance of Maine) and SMOOSA (Save Maine's Only Official State Animal) that spent $500,000 to win votes in a public referendum (Lautenschlager and Bowyer 1985). Wildlife biologists watched from close range, distraught that a biological issue was being argued in an emotional public forum and that so much money was being wasted filling the coffers of the communications industry. Some effort by the biologists managed to keep the debate close to the facts. Most participants soon realized that there essentially was no biological issue; the moose population was large enough to easily sustain a limited hunt, but not so large that it needed to be culled to avoid an overpopulation of moose damaging the environment. Everything boiled down to personal values and a question of sportsmanship. One side contended that Maine moose had become so tame that shooting one required as much skill as knocking off a cow in a paddock, while the other side maintained that the actual killing of a moose was only part of the total sporting experience and that the moose would become wary after a couple years of hunting. The latter argument prevailed, and moose hunting returned to Maine.

Behind this story of human values lies another one, a story of ecological change and its consequences. The moose share Maine's forests with a second member of the deer family, the white-tailed deer. (Until this century a third species, the caribou, was also found in moderate numbers.) The word "shares" is a bit misleading for two reasons. First, Maine represents an interface of the

two species' geographic ranges, with moose found in the circumpolar band of boreal spruce-fir forest and the white-tail living in the temperate and tropical Americas. Thus, to a certain extent—there are many exceptions—deer and moose have partitioned the state, with moose living in areas where local climate and vegetation are of a more northerly character, and deer predominating elsewhere (Tefler 1970, Kearney and Gilbert 1976). Second, in areas where deer and moose coexist, competition would be a better descriptor of their interactions than sharing. The long-legged moose has the upper hand (or leg!) where deep snows accumulate, because late winter starvation is a major limiting factor for northern white-tails, and moose can move around to forage in a meter of snow (Tefler 1970). Their greater bulk also makes them energetically more efficient; (as animals get larger, heat-producing muscle increases proportionately more than heat-losing surface area). Conversely, in areas with a more moderate climate, the deer can thrive and the moose are kept at bay by a diminutive ally of the deer, a parasite called the meningeal worm. The meningeal worm lives in the brain of many deer without causing them harm, but if a moose or caribou enters an area with a sizable deer population, it is likely to become infested as well, and erratic behavior, paralysis, and death are likely to follow (Gilbert 1974).

Over the eons a balance has evolved, the white-tail unable to penetrate far into moose range because of climate, and the moose unable to venture south because of the white-tails' parasite. This is quite an oversimplification, because there are other considerations such as relative vulnerability to predators and direct competition for food. Let's consider this interaction further by taking an historical perspective, and in doing so we return to the main theme of this chapter, spatial heterogeneity.

Archaeological evidence indicates that both deer and moose were important in the diets of aboriginal Mainers; 4,200 years ago deer were dominant, but there was a gradual shift, presumably because the climate was cooling, and by 500 years ago moose were a more important component of the diet. The advent of European colonists 350 years ago brought many changes; the ones most critical to the balance between moose and deer began in the 1820s. At this time there was a major expansion of logging into the vast, uninhabited portions of northern Maine, and this created a mosaic of young stands of secondary hardwoods interspersed with residual stands of mature softwoods, thus markedly improving habitat for the white-tail. With ample winter food juxtaposed with shelter from inclement weather, deer became more common and widespread in the northern half of the state. The elimination of wolves and overhunting of moose almost certainly facilitated this increase. By the middle of the twentieth century, the deer population was probably as high as it's ever been, while the moose population was near its nadir. Since then the trend has reversed itself and the deer herd has waned while moose have waxed. There are a number of reasons for this recent change; one of them involves a major change in timber harvest technology.

During the 1970s, large machines came to the Maine woods, machines that could move across the landscape snipping off trees with enormous shears, stacking them in a trailer behind, then carrying 20 or more to the nearest roadside. These machines, often called feller-bunchers, were expensive, typically well into six figures, and to pay for themselves they often had to be run day and night. Little time could be spared to move them from site to site, so operational areas became larger; i.e., clearcuts became more common and they got bigger. Before 1970, encountering a clearcut was something to be remarked upon; by 1975 clearcuts were the norm and ones that extended over a hundred hectares were far from rare. A few dozen machines and an insect, the spruce budworm, whose defoliation necessitated huge salvage operations, had changed the landscape of northern Maine and once again shifted the balance between moose and deer.

Simply put, expansive clearcuts are great for moose—they will browse almost across the length and breadth of them—but deer are usually too shy to venture far from the nearest forest and thus can use only the periphery of a clearcut (Hunt 1976). This difference may be due to deer being vulnerable to a wider range of predators. The net effect of this difference is easy to estimate. If one assumes (to keep things simple) that all the clearcuts were 100 ha and square, and that white-tails primarily forage within 100 meters of a forest edge, (the actual distance is quite variable [Sweeney et al. 1984, Williamson and Hirth 1985]), only 20% of the browse would be readily accessible to the deer and the moose would have little competition on 80 ha.

As long as deer hunters significantly outnumber moose hunters, (in 1988, 1,000 moose hunters were selected by lottery; anyone can hunt deer and almost 200,000 chose to do so) people will complain about clearcuts—moose pastures as some would call them. This need not be so. With a balanced approach toward harvesting—one that creates a spatially heterogeneous forest landscape—the habitat requirements of both species can be met in the same region.

SUMMARY

Because there is such a great variety in the home range sizes of various organisms, spanning 8 to 12 orders of magnitude, it is important that forests be managed at a variety of scales. This will involve making silvicultural decisions on a landscape basis, not just stand by stand. Ideally, the management regime should range from the very fine-scale management represented by selection cutting to the coarse-scale management effected by sizable clearcuts. The maximum size of clearcuts should be determined by considering issues such as size of the management unit, the home range requirements of large animals, aesthetics, nutrient loss, and natural disturbance regimes. If wildlife managers condemn clearcuts as bad for wildlife, and foresters write off selection cutting as uneconomic, public policy and law may arrive at a compromise strategy that will promote the extensive use of small clearcuts. A forest managed in this fashion will lack an important component of structural diversity, spatial heterogeneity.

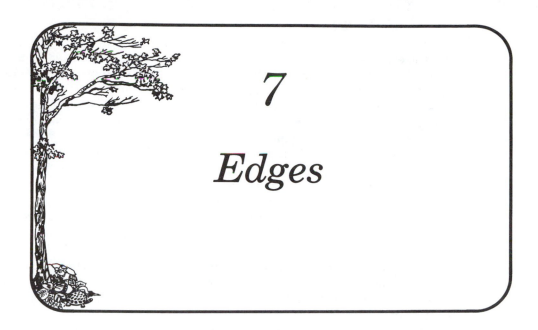

7

Edges

Game is a phenomenon of edges. It occurs where the types of food and cover which it needs come together, i.e., where their edges meet. Every grouse hunter knows this when he selects the edge of a woods, with its grape tangles, haw-bushes, and little grassy bays, as the likely place to look for birds. The quail hunter follows the common edge between the brushy draw and the weedy corn, the snipe hunter the edge between the marsh and the pasture, the deer hunter the edge between the oaks of the south slope and the pine thicket of the north slope, the rabbit hunter the grassy edge of the thicket.

Aldo Leopold (1933) wrote these words in his classic *Game Management,* and "create more edge" has been one of the cardinal rules of wildlife habitat management ever since. Leopold's statement is really just a confirmation of what any astute observer of nature knows intuitively: The diversity and abundance of wildlife, especially game animals, is often greatest near an edge. Unfortunately, this idea—often called the *edge effect*—has long been accepted as a premise for wildlife habitat management with little critical examination. In recent years, the edge effect has been studied to a limited extent and, as we shall see later in this chapter and in Chapter 8, "create more edge" can no longer be accepted as a universally valid rule of wildlife habitat management.

Edges are simply the places where two ecosystems come together. They are never a perfectly sharp line; there is always a transition zone from one set of environmental conditions to another, from one set of plants and animals to another. These transition zones are called *ecotones.* Some edges are so abrupt that the ecotone is hard to recognize; a lake and a forest separated by a shore-line cliff would be a good example. In other cases the ecotone is a very gradual transition; e.g., one forest type grading into another on the side of a mountain. At some point, a very broad ecotone would be considered a separate ecosystem and thus, by definition, ecotones are usually quite long and narrow compared to the ecosystems that comprise them.

Sometimes the transition is based on an inherent feature of a site: A cat-tail marsh grades into an oak forest because a rise in topography reduces soil moisture; a white fir stand gives way to a red fir stand where the altitude creates a cooler microclimate; the edge of an outwash plain of glacial sand marks a shift from pitch pine and scrub oak to white pine and red oak (Wiens et al. 1985). These are called *inherent edges* (Thomas et al. 1979b); they are long-term, relatively stable features of the landscape. They can change, how-ever; volcanoes, floods, and landslides alter the landscape very rapidly; many processes such as erosion produce gradual modifications. They can also be changed by human activities but, again, they are relatively permanent charac-teristics of the landscape.

In contrast, *induced edges* are short-term phenomena created by changes in the vegetation (Thomas et al. 1979b). They exist wherever two different successional stages abut, and thus can be created by logging, fires, wind storms, and other disturbances. Any significant manipulation of the vegeta-tion—planting and seeding, fences that exclude grazing animals—can create an induced edge. Some induced edges are maintained for a moderately long period; for example, certain woodland-pasture edges in Britain have persisted for several centuries (Rackham 1980) (Figure 7.1). However, many induced edges, perhaps most, are relatively short-lived phenomena; they last only a few years or decades before succession or some other change moderates the difference between the ecosystems sufficiently that a significant edge no longer exists. Obviously, foresters have enormous scope for creating induced edges by prescribing various harvesting operations, but before considering the possibilities, we will examine the potential consequences for wildlife.

THE EDGE EFFECT

Edges first caught the attention of hunters and wildlife managers because they are favored by many game species, but over the years the term *edge effect* has been extended to embrace the general increase in wildlife species richness and abundance that often characterizes edges (Giles 1978). It is not immediately apparent why so many game animals show a predilection for edges, but one can speculate that it may be related to the fact that game species tend to be large and shy. Large, shy animals need relatively more food and cover and may be attracted to places where two ecosystems, one offering more food, the other more cover, come together. The conjunction of two different successional

Figure 7.1 Older edges, such as those maintained by agriculture (A), usually have a better developed understory of early successional plants than do younger edges, such as those created by clearcutting (B).

stages is often a good example, the younger community being a better place to forage and the older a better place to hide. Of course, all animals need food and cover, and there are many nongame species that are also associated with edges because they need ready access to two or more ecosystems. For example, European sparrowhawks nest in forest stands but hunt in both forests and open land; and European badgers require two or more types of habitat in their home range to have an adequate array of food available at all times of year.

In addition to wide-ranging animals that move between two ecosystems, there is a second group of plants and animals associated with edges because they find their preferred habitat in the unique conditions of the ecotone itself. Some species are rarely found away from an edge, and one special set of edges—the shores between upland and aquatic ecosystems—supports quite a few unique species; shores are examined in depth in Chapter 9. However, most of these edge species regularly occur in other sites too; they are just more common in edge environments.

Thirdly, there are many species living in an ecotone that are only secondarily edge species. The bulk of their populations are found in either of the ecosystems that form the edge, but they can also tolerate the conditions in the ecotone, and some individuals are found there. Some individuals may even spill over a short distance into the other ecosystem.

These three sets of species—animals found along edges because they need access to two habitats, plants and animals specifically tied to ecotones, and plants and animals that extend into ecotones from the adjacent ecosystems—are represented diagrammatically in Figure 7.2. These are not exclusive groups. They are just a convenient means to understand the different ways various species use edges. Plants and animals adapt to local conditions

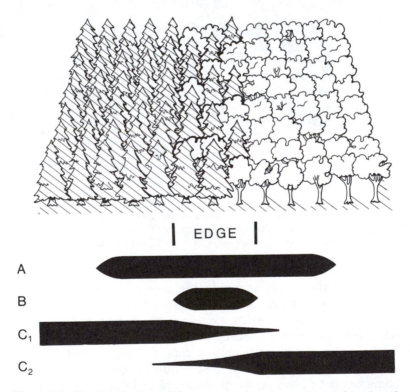

Figure 7.2   The distributions of three groups of species are depicted in a broad and a narrow edge. Group A consists of species associated with edges because they need access to both ecosystems; the B group species require an ecotone habitat; and C species are primarily associated with either of the adjacent ecosystems. Note that Group B species are likely to be absent from an abrupt edge.

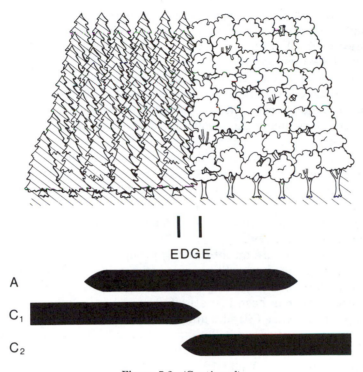

EDGE

A

C₁

C₂

Figure 7.2 (Continued)

throughout their geographic ranges, and this can affect their use of edges. Furthermore, the mobility of animals allows even individuals to modify their use of edges at different times of year or at different stages of their lives (Yahner 1988).

There are two important points here: (a) A large number of species use edges for a variety of reasons; and (b) some species (groups A and B in Figure 7.2) find their preferred habitat in ecotones and probably reach their maximum population densities there. Unfortunately, these generalizations are based as much on simple common sense as on careful research. In particular, the use of edges by wide-ranging animals is based on little data. There have been a few studies—for example, Hanley (1983) used radio-telemetry to document that elk and black-tailed deer in the Cascade Range spend disproportionately more time near clearcut-forest edges—but most information is anecdotal. This is not to say that the anecdotes are inaccurate, and later in this chapter we will discuss an anthropological study that verified anecdotes about deer, wild pigs, and edges in the Philippines.

The idea that some species fare particularly well in the special conditions of an edge (group B) is better documented. For example, the forest at the edge of an open area will experience more wind and sun than a forest interior, thus creating a band with a special microclimate in which a different assemblage of plants can grow. Specifically, the relatively warm, dry, sunny conditions of a forest edge allow species usually associated with early succession to grow under a forest canopy (Wales 1972, Whitney and Runkle 1981), thus creating a combination of late and early successional species (Ranney et al. 1981). There are discernible differences among edges of different orientations; ecotones facing the equator are usually wider than those facing the poles (Wales 1972), as are ecotones facing the prevailing wind direction (Forman and Godron 1986). But microclimate is not the whole story; a study of the fate of pignut hickory nuts discovered that reproduction was more successful near edges in large part because fewer nuts and seedlings were eaten by squirrels and other herbivores there (Sork 1983). Interestingly, plants may exhibit an edge effect in more than just distribution and abundance; plant size, structure, and reproductive strategy (vegetative versus sexual) can all be influenced by their location in an ecotone (Falinska 1979).

Songbirds are an interesting group to examine for edge effects because they are mobile, but unlike many game animals, they often have home ranges too small to readily encompass two ecosystems. The distribution of birds near forest edges has been investigated in several studies spanning a geographic range from Texas (Strelke and Dickson 1980) to Sweden (Hansson 1983), and a range of edge types including clearcuts (O'Meara et al. 1985, Elliott 1987), powerlines (Kroodsma 1982, 1984, 1987; Small and Hunter 1989), maritime shores (Helle and Helle 1982, Small and Hunter 1989), and edges between forest types (Laudenslayer and Balda 1976). Unfortunately, no absolute "truth" can be distilled from these studies, although three generalizations can be made for which exceptions are rare. First, the forest next to an edge usually has a relatively high density of birds; the density of birds in openland near forests is neither consistently high nor low. Second, there is no consistent relationship between bird species richness and proximity to edge: Some studies show an increase near the edge; some a decrease; most show no effect. Third, there are few bird species that can be considered edge specialists, and those that are tend to be very common species. For example, Elliott's (1987) study on Maine clearcut edges (the only analysis based on sophisticated statistics) identified only the American robin, the most ubiquitous bird in the state, as an edge species.

Two observations about these studies should be noted. First, although they can be divided into two groups (those that studied clearcut edges less than eight years old, and those that studied inherent edges or induced edges old enough to have developed an understory of early successional plants), this distinction does not explain the discrepancies in results. Second, most of these studies have focused on breeding birds, and this presents a methodological problem. When the location of a bird is recorded and analyzed on a scale so fine (25 or 50 m is often used) that it is much smaller than the size of a bird's territory, the bird's apparent distribution will be biased toward those places where it is most likely to be detected by a censuser—generally a song perch. Song perch locations may not be an accurate reflection of the bird's overall use of its habitat. It is possible that the apparent increase in density of birds near edges only reveals a preference of birds to sing near edges (Helle 1984), perhaps because territory intruders are most likely to be encountered there. On the other hand, a Finnish study (Helle and Muona 1985) discovered that several types of invertebrate were relatively abundant near forest-clearcut edges, so perhaps birds find food more abundant there. Unfortunately, their collections were only identified to a coarse taxonomic level (e.g., beetle versus fly versus ant) so that it was not possible to generalize about how invertebrate diversity may be affected by edges.

The Other Side of the Edge Effect

Some wildlife biologists have pointed out that even if edges are zones of high diversity and density, this does not necessarily make them ideal wildlife habitat (Noss 1983). A study in Michigan discovered that songbirds nesting near edges suffered much higher rates of nest predation than those in a forest interior and were also more likely to be parasitized by brown-headed cowbirds, a species that lays its eggs in other birds' nests (Gates and Gysel 1978). In a similar Wisconsin study, 67% of the nests within 100 m of an edge were parasitized by cowbirds and the rate only dropped substantially, to 18%, beyond 300 m from the edge (Brittingham and Temple 1983). Another North American study found an edge-related increase in predation on artificial nests of quail eggs that extended 300–600 m into the forest (Wilcove et al. 1986), while a similar Swedish study found predation was high—50–200 m into the forest (Andren and Angelstam 1988). Turtles burying their clutches close to an edge are also more likely to lose them to a predator (Temple 1987). Even African buffalo are more likely to be killed by lions near an edge (Prins and Iason 1989). If one thinks of herbivorous animals as plant predators, the issue takes another dimension. For example, the positive effects of edges on white-tailed deer in northern Wisconsin have led to reductions in the regeneration of many trees, shrubs, and herbaceous plants (Alverson et al. 1988), and given the mobility of deer this effect can extend for kilometers from the nearest edge. These effects have led some writers to refer to edges as "ecological traps" (Gates and Gysel 1978).

Another problem with edges concerns a group of species that do not appear in Figure 7.2; these are species that seem to shun edges in favor of forest interiors. There are many such species; for example, there are various trees—sugar maple, American beech, and a host of tropical rain forest species—that do not thrive near edges (Ranney et al. 1981, Hubbell and Foster 1986). For these forest interior trees, dessication from sun and wind, competition from shade-intolerant plants (in this context they are often called weeds), and windthrow make edges an inhospitable place. We will return to forest interior species and other problems with edges in Chapter 8, "Islands and Fragments," but in the balance of this chapter we will ignore potential difficulties and consider how forest management can create more edges on the landscape.

CREATING EDGES THROUGH FOREST MANAGEMENT

Edges and Geometry

Every time a harvest operation changes the size or shape of a stand, the amount of edge is changed too in accordance with two simple rules of geometry. First, because a circle is the surface with the least amount of edge per unit area, the more a stand's shape diverges from circular the more edge it will have per unit area (Figure 7.3). Squares have only marginally longer edges, and the edge to area ratio of rectangles becomes substantially larger only when the long axis of the rectangle is much greater than the short axis. Harvest areas with convoluted edges have by far the greatest edge to area ratio, especially if they are relatively long and narrow. Furthermore, irregular forest edges have an important secondary benefit in that they appear more natural. Most people do not find long, straight edges bisecting the landscape to be very aesthetic.

Second, the size of a stand or cutting operation affects how much edge it has. Obviously a large stand has a longer edge than a small stand but, on a

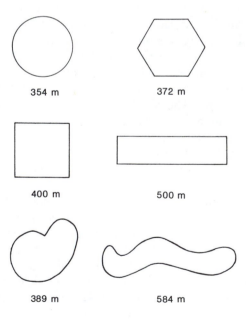

354 m

372 m

400 m

500 m

389 m

584 m

Figure 7.3 Six different shapes with equal area but very different edge length show that the more a shape diverges from circular the longer its edge will be. Using a limnological equation designed to compare the edge length of a lake to that of a circle— Index = Edge Length/($2 \sqrt{\text{Area} \times \pi}$)— one can also calculate indices for these shapes of 1, 1.05, 1.13, 1.41, 1.09, and 1.65, respectively (Patton, 1975)

relative basis, the large stand actually has less edge because of another simple geometric rule: Linear distances increase as an additive function, but areas increase as a square (Figure 7.4). (If you double the length of a square from 3 cm to 6 cm, its area increases from 9 cm² to 36 cm², but its total edge increases only from 12 to 24 cm.) Thus, dissecting a forest into many small stands will increase the relative amount of edge.

The image of arranging squares, rectangles, or convoluted shapes of different sizes on the land has a certain appeal to those who enjoy grand scale manipulations. However, it is important to remember that the lay of the land, patterns of soil, topography, access roads, etc., will determine much of the arrangement. This said, there is still enormous scope for forest managers to increase the amount of edge in a landscape. (On the other hand, because it is much quicker to cut a stand than to wait for it to grow, the scope for *decreasing* the amount of edge is more limited.)

Coverts, Interspersion, and Contrast

Various elaborations of the edge effect theme have been proposed over the years, but unfortunately few have spawned any research to judge their effectiveness. One idea came from the father of the edge effect concept, Aldo Leopold (1933); knowing that optimal bobwhite quail habitat includes forest, brushland, grassland, and cultivation, he induced that if the places where two ecosystems come together are good game habitat, then the places where three or more habitats adjoin must be even better (Figure 7.5). In these places, an animal with limited mobility is more likely to find all its requirements within a convenient distance. These habitat focal points have been termed *coverts*, and elaborate schemes for maximizing the number of coverts and length of edge have been proposed (Conlin and Giles 1973).

Ideas about optimal configurations of edges and the ecosystems that comprise them go beyond the concept of bringing three or four ecosystems together in one spot. Words such as interspersion, dispersion, and juxtaposition are used to describe the arrangement and proximity of ecosystems on the

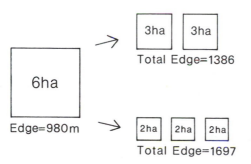

Figure 7.4 Large areas have relatively shorter edge lengths than small areas.

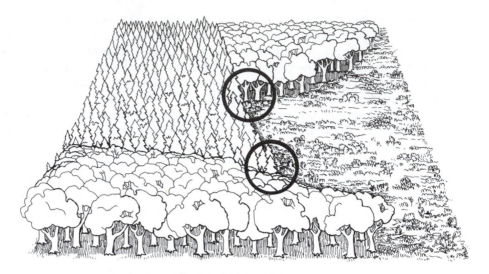

Figure 7.5   Places where three or more types of ecosystem come together are called coverts.

landscape. It is difficult to define these terms in a precise and meaningful way, although Giles (1978) has attempted it by proposing some quantitative indices. It is much easier to understand the concepts visually; Figure 7.6 shows some landscapes with and without very much interspersion, dispersion, or juxtaposition.

Because it is the difference between two ecosystems that creates edges and ecotones, it is generally thought that the edge effect will be greatest when

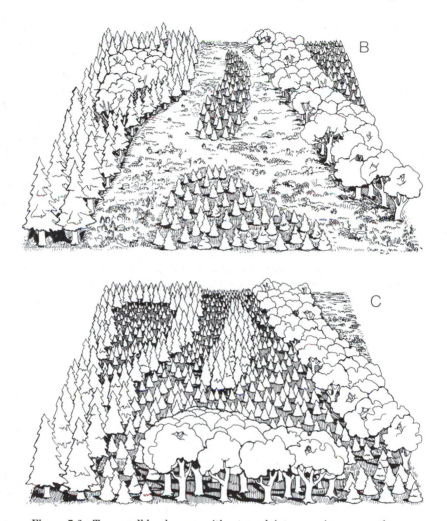

Figure 7.6 Two small landscapes without much interspersion among the eco-systems (A), and with more interspersion (B), are depicted. Note that the total length of edge is much longer in B, implying that this might be a good index of interspersion. In many cases this will be true, but in C a landscape is depicted that has an edge length equal to B but less interspersion because its diversity of edge types is lower. (Note that it is possible to have interspersion without converts.)

two adjacent ecosystems are very different from one another. "The greater the contrast, the more likely the adjoining habitats are to be very different in structure and in the wildlife species they support. This tends to increase the species richness of the ecotone" (Thomas et al. 1979b). This feature is usually called *contrast*, although drama and (unfortunately for the sake of clarity) juxtaposition have also been used (Harris 1980, Harris and Skoog 1980).

Maximizing the contrast of edges takes us back to a topic covered in Chapter 4, "Age Structure": What is the best way to distribute cuts? If a new cut is located next to an 80-year-old stand, there will be much more contrast than if it abuts a 5-year-old stand. Of course, if the 80-year-old stand is then harvested the following year, there will be a new cut next to a 1-year-old stand. In the long term, adjacent stands will have a maximum age difference (and presumably edge contrast) if stands are cut when the adjoining stand is half-way through the rotation period (Harris and McElveen 1981, Mealey et al. 1982).

These ideas about edge geometry, dispersion, and contrast are based more on intuition and common sense than on extensive research. A Florida study did find a larger number of bird species associated with the high contrast edges between cypress domes and a clearcut than with the low contrast edge between cypress domes and 7 m tall pine plantations (Harris and McElveen 1981). Another project concluded that the bird species richness of suburban woodlots in Ontario correlated more closely with the woodlots' perimeter length than with their area, suggesting that woodlot shape is an important feature (Gotfryd and Hansell 1986).

PEOPLE AND PIGS IN THE PHILIPPINES

On the northeastern coast of Luzon, most northern of the major islands of the Philippines, is an isolated area called Palanan Bay. The area is dominated by tropical rain forests and inhabited by two groups of people, the Palanan and the Agta. About 10,000 Palanan farmers create and maintain agricultural land for corn, rice, and root crops such as yams and manioc, and about 800 Agta live by hunting and fishing. (Interestingly, the Agta do not conform to the conventional image of a hunter-gatherer society in which men hunt and women gather plant foods and snare an occasional rabbit; many Agta women are active and accomplished big game hunters.) There is a great deal of economic interdependence between the groups revolving around the exchange of carbohydrates and protein; the Agta give game and fish to the Palanan and receive crops in return. Less conspicuous, but also very important, is an ecological interdependence between the groups. Anthropologist Jean Peterson (1978, 1981) discovered that Agta choose to live near the Palanan not only to be able to trade for crops, but also because the edges of the farmers' fields are excellent hunting areas.

Agta hunters pursue a number of quarry—fish, birds, lizards, snakes, monkeys, etc.—but most of their effort is directed at pigs and deer (Javan pigs and sambar, to be precise), two species that the Agta closely associate with edges or *digdig*. The hunters believe that pigs and deer are concentrated near edges because the animals' preferred food plants are most abundant there;

(they recognize a few dozen kinds of plant upon which pigs and/or deer forage). They are also well aware that the pigs and deer frequently raid the Palanan crops—primarily for fruit and tubers in the case of pigs, and corn and rice in the case of deer. In turn, the Palanan are well aware of the damage caused by game animals and report 80–100% crop losses in new fields near the forest edge, especially if the field is bordered on two sides by forest. Crop losses drop to 10–15% in established fields that are farther from the forest and receive some protection from guards or scarecrows.

There is a seasonal component to the story as well. One of the pigs' primary natural foods, a variety of forest fruits, is very scarce between March and July, the same season when their nutritional demand is high because piglets are born in March and nursed through April and May. Consequently, this is the period when pigs are most concentrated around the field edges where they can raid crops. However, this is not a major hunting time, because the incentive to hunt is low; the game is lean, and fish and crops are abundant. Deer are not as dependent on fruit as pigs are and their distribution pattern does not change as much seasonally.

The Agta also recognize a second type of edge—the interface between mountain slopes and flat floodplains—as an important area for pigs. At the base of these steep slopes, fruit rolling downhill accumulates, creating another edge with a concentrated food supply.

The preceding story is based on what the Agta and Palanan people told Peterson and thus is largely anecdotal. However, Peterson was able to quantify some components of the story. For example, by recording the distance from kill locations to the nearest edge, she determined that a disproportionate percentage of the game was killed near edges; and by doing vegetation analyses in different areas, she found that forage plants preferred by deer and pig were common near edges. Thus, the balance between Agta and Palanan, between wild animals and crops, and between forest and croplands provide a good example of how important edges can be. Interestingly, in their conversations with her, the Agta expressed a serious concern that the abundance of game associated with edges will be a short-lived phenomenon. They realize that if the current trend of deforestation and conversion to agriculture continues, there will soon come a time when there will not be enough forest to support the game on which their lives and culture are dependent.

SUMMARY

The edges or ecotones where two ecosystems come together often support a high diversity of wildlife species: Some are animals that need access to two ecosystems; some animals and plants require the special environment of an edge; and many are primarily associated with one of the two ecosystems that comprise the edge. By determining the size, shape, and arrangement of forest stands, forest managers have considerable control over the amount of edge in a forested landscape. Small stands have relatively more edge than large stands; irregularly shaped stands have more edge than stands that approximate a circle; and interspersing stands on the landscape creates more edge. Unfortunately, there are some significant negative sides to creating more edge. There are some species that avoid edges, and some other species that readily use edges but are subject to high rates of predation and brood parasitism there. These problems and others are explored in Chapter 8, "Islands and Fragments."

8

Islands and Fragments

Islands are powerful metaphors. They bring images of independent, self-sustaining refuges where the turmoil of the world is a distant noise, where life is buffered by an expanse of water. But they also carry images of tiny, isolated fragments that are too small to be resilient and thus are buffeted by disturbances. Unfortunately, the latter image is a more accurate portrayal of the "islands" of forest habitat that persist on the land when the intervening forest has been removed. A three hectare copse surrounded by pastures and croplands can maintain only a vestige of the original flora and fauna associated with a region's forests. Even bisecting a forest with a narrow road or powerline right-of-way may diminish its value as habitat for some species.

These are important issues among conservationists, who often ask questions such as, How long can a tiger population survive in a 50 km^2 enclave encircled by rice paddies? and seek the answers in a body of ideas and observations known as *island biogeography.* Island biogeography concerns the distribution of plants and animals on islands—a topic that intrigued Darwin, Wallace, and generations of biologists since. It was first formalized with a quantitative theory and brought to prominence by a 1967 monograph by Robert MacArthur and E. O. Wilson. Its application to conservation problems was largely stimulated by a 1975 paper by Jared Diamond, in which he likened nature reserves to islands in a sea of human developments and used the theory to make recommendations about the design of reserves. Another milestone was reached in 1984 with Larry Harris's book, *The Fragmented Forest,* an analysis of how forest harvesting can affect biological diversity by leaving stands as isolated habitat islands. Before considering habitat islands and forest fragmentation, we need to review some of the basic ideas of island biogeography. In the box, the quantitative foundation of MacArthur and Wilson's theory is briefly described; a less formal presentation that incorporates some ancillary ideas follows.

ISLAND BIOGEOGRAPHY THEORY

The basic idea of MacArthur and Wilson's equilibrium theory of island bio-geography is that the number of species on an island represents a balance between extinction and immigration. The rate of immigration is determined largely by how isolated an island is; the more isolated an island is, the lower its immigration rate. (In Figure 8.1 the curve for remote islands [*far*] is lower than the curve for islands near the mainland [*near*]). Extinction rates are a function of island size; populations on large islands tend to be larger and thus less vulnerable to extinction. (The extinction curve for *large* islands is lower than the curve for *small* islands.)

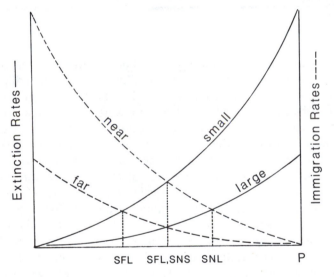

Number of Species on Island   **Figure 8.1**

    For any given island, there is an extinction rate and an immigration rate that will balance one another and keep the number of species relatively constant. (In this example, the numbers of species for four equilibria are represented: S_{FS}, number of species on a far, small island; S_{FL}, far, large island; S_{NS}, near, small island; S_{NL}, near, large island; P is the total number of species that could potentially immigrate to the island from a nearby landmass.)

    The specific relevance of this theory to conservation questions is easily challenged (Zimmerman and Bierregaard 1986), but it does give us a well-defined framework from which to proceed (Haila 1986).

ISLAND BIOGEOGRAPHY

Island Size

A fundamental tenet of island biogeography is that islands have fewer species than the nearby mainland, and small islands have fewer species than large islands. This relationship is not unique to islands. If you were to go into any ecosystem and identify species in a series of progressively larger plots, the number of species recorded would increase in a fairly predictable pattern. This is called the species-area relationship (Figure 8.2), and it exists for two primary reasons (Connor and McCoy 1979, Boecklen 1986). First, as area increases, so does the diversity of the physical environment (Lack 1976, Williamson 1981). The South Island of New Zealand has glacial-carved fjords, dry, wind-blown plateaus, mountains shrouded in snow, and rain-soaked valleys. In contrast, a tiny coral atoll could not begin to have such a wealth of ecological opportunities. Even when comparing a 10 m² area to a 50 m² area, the larger area is likely to have a greater range of microhabitats—a fallen log, a pile of stones—than the smaller area.

A second reason for the species-area relationship is that in a large area there are more likely to be uncommon species that live at low population densities as well as common species (Haila 1988). To take a silly example, the odds of finding a bear while censusing a 10 m² plot are rather small. Here lies a fundamental difference between different sized plots on the mainland and different sized islands. Although at any one time the chances of a bear being in a given 10 m² area might be very small, such a tiny area could easily be part of a bear's home range and routinely used. In contrast, a small island would almost never be used, because it would be too isolated for a bear to visit readily

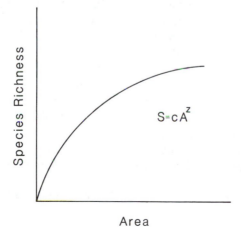

$S = cA^z$

Figure 8.2 The number of species on an island or in a sample plot increases as area increases, although it eventually levels off. The steepness or slope of the curve varies considerably, e.g., it is usually steeper on islands than on sample plots. (The curve is described by the formula $S = cA^z$, where S = number of species, A = area, and c and z are constants.)

and too small to accommodate the entire home range of a bear. In other words, at a very minimum an island usually has to be large enough to encompass the home range of at least one individual of a species, if that species is going to be represented in the island's biota (Rusterholz and Howe 1979). In practice, most species are represented by populations or not at all. The effect is not limited to wide-ranging animals, for plants and sedentary animals can also have low-density populations.

There is a third reason why there are fewer species on small islands: Small islands will, on average, support small populations, and small populations are more likely to go extinct. The likelihood of small populations going extinct is increased by problems such as genetic inbreeding, random fluctuations in the sex and age structure of the population, changes in the environment, and cata-strophic events (Soule′ 1987). For example, if a disease were to kill 95% of a large deer population, one that numbered in the thousands, the population could probably recover in 5 to 10 years. However, if the population was an is-land herd of 100 animals, the remaining five individuals might be four bucks and a doe too old to breed any longer; the result—local extinction.

Island Isolation

Isolation is the second major factor affecting biotic diversity on islands; the more isolated an island is, the fewer species it will support. There are two pri-mary ways a piece of land comes to be an island isolated in the sea. Some islands are called *continental* or *land bridge islands* because they were once part of a continental land mass and became isolated only after sea levels rose. Immediately after separation, continental islands have a fairly complete biota, but the number of species gradually dwindles because the small, isolated popu-lations characteristic of islands are prone to extinction. Local populations on the mainland seldom become extinct—or at least they do not remain extinct for very long—because there are usually new individuals immigrating into the population (Brown and Kodric-Brown 1977). In contrast, curtailed immigra-tion on islands makes local extinction a more common and long-lasting event. (Restricted immigration may present a second problem for island popula-tions—inbreeding—but this may also present an opportunity for new species to evolve.)

There are also islands that rise out of the sea, slowly erected by reef-building corals, or dramatically deposited by a volcano; these are called *oceanic islands.* They begin with no terrestrial plants or animals, but over time they are gradually colonized by various species, and an ecosystem, albeit a depau-perate one, develops. The number of species on an island stabilizes when the rate at which new species colonize equals the rate at which established ones go extinct.

Immigration and colonization are greatly influenced by the degree of isolation; islands that are very isolated are reached relatively seldom. Distance from the mainland is the simplest measure of isolation, but isolation is much more complex because water is not an equally effective barrier for all species. Birds accustomed to migrating across the Caribbean would not be as daunted by a stretch of open water as a ruffed grouse would; a mouse is more likely to find itself transported to a distant isle inside a hollow log than is a moose; spiders wafted aloft on strands of silk are more commonly among the first animals to arrive at a newly-created island than are slugs; and a coconut is more likely to wash up on a tropical beach and germinate than is an apple. Other factors that affect an island's effective isolation include its position with respect to air and ocean currents, its proximity to other islands that might function as stepping stones, and its relationship to migration routes.

Symbiosis on Islands

Island size and isolation may be primary determinants of diversity, but their effect is accentuated by the fact that many species have developed fairly tight, often symbiotic, relationships with one another. A small island will have relatively few animal species, in part because it has relatively few plant species on which the animals can forage for foliage, fruit, pollen, other animals, etc. (Power 1972). To some extent, the converse will be true; low animal diversity may beget low plant diversity. For example, the relative scarcity of island plants with large, showy flowers has sometimes been attributed to a shortage of butterflies and long-tongued bees (Carlquist 1974). On the whole, however, plants tend to be less dependent on animals than vice versa; e.g., many plants can persist without their specialized pollinators by using an array of generalist pollinators (Woodell 1979, Feinsinger et al. 1982). In short, the net impact of these relationships on island diversity is a snowballing effect or positive feedback loop, depending on whether you prefer simple imagery or the language of cybernetics.

ISLANDS IN THE LANDSCAPE

Attempts to apply island biogeography theory to a variety of distribution questions have significantly extended the concept of what constitutes an island. It may be difficult for some people to think that an oak or a mouse might constitute an island for caterpillars or mites (Janzen 1973, Dritschilo et al. 1975), but it is an easy conceptual leap to recognize the insular qualities of caves, mountain-tops, and the isolated patches of forest scattered across many human-dominated landscapes. Such landscapes characterize much of Europe and eastern North America and, as burgeoning populations in tropical developing nations precipitate the conversion of forest lands into agricultural ecosystems, more and more of the world's forests will exist as small, residual tracts encircled by a sea of croplands (Figure 8.3).

The similarities between a forest fragment and an island, particularly a continental island, are quite obvious. In both cases, a small unit of land with its various plants and animals has become separated from other, similar ecosystems by the intrusion of a barrier, either water or open land. In both cases, we can predict that many species have become locally extinct because the variety and size of habitats are too limited. The documentation of these phenomena, especially over a reasonable time span, is quite limited, except for correlations between forest patch size and bird species richness, which have been described in many parts of the world (e.g., Bond 1957, Moore and Hooper 1975, Forman et al. 1976, Galli et al. 1976, Whitcomb et al. 1981, Helle 1984, Lynch and Whigham 1984, Loyn 1985, Blake and Karr 1987). The patterns for other organisms, and the processes by which the patterns develop, are not well known (Hoehne 1981, Levenson 1981, Matthiae and Stearns 1981, Murphy and Wilcox 1986).

There are also some important differences between islands and forest fragments. First, water and open land represent two very different types of barriers. To make some gross and obvious generalizations: Terrestrial animals will be isolated most by water; aquatic animals by land; and for flying animals and wind-dispersed plants it will make relatively little difference. Plants that disperse by clinging to animals or residing briefly in their digestive tracts—burdocks, blueberries, and the like—will be limited by whatever limits the animals' mobility. The upshot is that immigration to islands is probably significantly more limited than immigration to forest fragments.

In thinking about immigration and isolation, the perspective "isolated from what?" is important. Island biogeography theory implicitly assumes that a mainland with a large pool of potential immigrants exists some distance away from any given island. With forest fragmentation, the "mainland" is often being inexorably degraded, and an entire region may be devoid of very large tracts of forest (Harris 1984). In all of Great Britain (228,300 km²), there are only a handful of forests greater than a thousand hectares, and even these are criss-crossed with roads; in fact, there are only four forested National Nature Reserves over 250 ha (Streeter 1974). Consequently, it is critical that the issue of forest fragmentation be addressed from a broad-sighted, regional perspective.

Time is the basis of another difference between continental islands and forest fragments. Most continental islands became isolated during post-glacial sea level changes several thousand years ago. Although from a geological time scale this was yesterday, it is much longer than the decades or centuries that forest patches have been isolated by human activities on the landscape. It is quite likely that the process of gradual extinction has not run its full course in most forest fragments. In Great Britain, where the forest has been fragmented for some 2,000 years, there is a much stronger correlation between

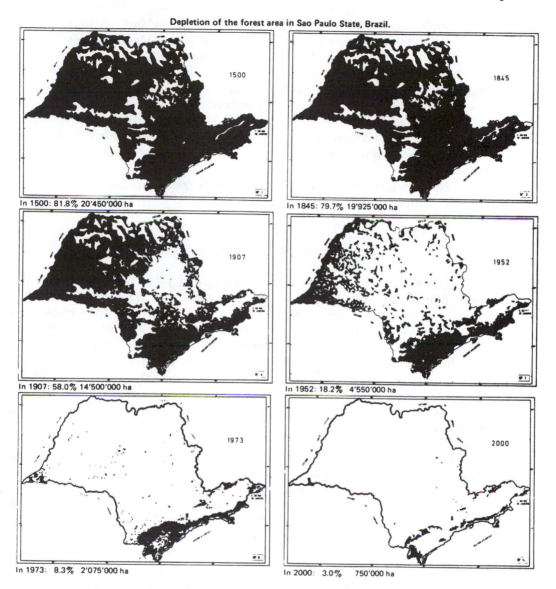

Figure 8.3 Progressive fragmentation and reduction of forest land in São Paulo State, Brazil 1500–2000 A.D. (Reproduced from Oedekoven 1980).

forest patch size and bird species richness than in Maryland, where the forests have been extensively cleared for less than 300 years (McLellan et al. 1986).

Finally, the scope and intensity of interactions between members of a forest fragment community and members of the surrounding terrestrial community is likely to be much greater than the degree of interaction between island and aquatic communities. In other words, predators, parasites, and competitors from adjacent communities are more likely to impinge on a fragment than on an island (Wilcove et al. 1986). For example, suburban cats are very likely to go hunting in a forest fragment, but sharks never come ashore to hunt on islands.

The differences between islands and forest fragments are well-illustrated by a particularly mobile group, birds.

Birds and Forest Fragments

The relationship between bird species diversity and forest fragment size seems simple and predictable, but only if the analysis is limited to a narrow range of forest sizes (Forman et al. 1976, Galli et al. 1976). With a more thorough evaluation, the picture becomes much more complex and has some surprising elements. Some North American studies have shown that much of the bird diversity–forest area relationship hinges on the fact that certain long-distance migratory species are usually found only in fairly large forests (Whitcomb et al. 1981, Ambuel and Temple 1983, Askins et al. 1987). Birds that winter in Latin America, such as the ovenbird and Kentucky warbler, are much more likely to be absent from a small tract than are birds that are year-round residents or short-distance migrants, such as white-breasted nuthatches and pine warblers. This seems rather counterintuitive; birds that find their way to Latin America and back should be *less* sensitive to forest isolation than their sedentary associates, not more so (O'Connor 1986).

The explanation may lie with some other characteristics of long-distance migrants. Consider the following: (a) Long-distance migratory birds tend to produce only one small clutch of eggs per year, and this is often laid in an open nest close to the ground (Whitcomb et al. 1981); (b) these habits make them relatively vulnerable to nest predators such as crows, grackles, skunks, and raccoons and to the brown-headed cowbird, a brood parasite, that lays its eggs in other birds' nests (Gates and Gysel 1978); (c) nest predators and parasites are known to be more effective near edges and in small fragments that are composed primarily of edge habitat (Gates and Gysel 1978, Brittingham and Temple 1983, Wilcove 1985); and (d) the late arrival of long-distance migrants gives them less opportunity for renesting.

There are other possible explanations for the paucity of bird species in small tracts (Whitcomb et al. 1981).

Maybe competition with edge-habitat species is important in small forests (Askins and Philbrick 1987).

If destruction of the winter habitat of long-distance migrants—e.g., tropical deforestation—has reduced their populations, this might be first evidenced by their disappearance from marginal habitat, i.e., forest fragments (Ambuel and Temple 1982).

Perhaps some birds that require forest interiors have a psychological need for larger tracts of forest than they actually occupy.

In a similar vein, there is some evidence that certain forest songbirds exhibit a type of loose coloniality and would thus prefer a forest tract large enough to sustain a colony, not just a pair of birds.

Interestingly, none of these explanations relies on MacArthur and Wilson's (1967) dynamic equilibrium theory, because the colonization half of the model (the part based on isolation and restricted immigration rates) seems irrelevant to highly mobile birds. It is now widely recognized that the MacArthur-Wilson model is an incomplete explanation for species distributions in forest fragments (Connor and McCoy 1979; Haila 1983, 1986, Wilcove et al. 1986). Indeed, these alternative explanations represent an implicit warning against one of the traps set by the simplicity of most models: assuming all species to be equal (Lovejoy and Oren 1981). If all species had an equal probability of immigrating or becoming extinct, each forest fragment would have a random, unpredictable set of species. This is clearly not the case. Some birds can thrive in small forest fragments—these tend to be common, widespread species such as gray catbirds, American robins, and common grackles—while others, notably several long-distance migrants, are seldom found in small forests. The result is a fairly nonrandom, predictable pattern of certain species being added to the avifauna as forest size increases. Whatever model offers the best explanation for this pattern, it is clear that large forest fragments support more diversity than small ones. That seems simple enough, but when this and related observations were proposed as premises for setting conservation policies, a major controversy ensued.

Nature Reserves as Islands

In 1975, Jared Diamond used island biogeography theory and his observations on the distribution of birds on the islands surrounding New Guinea to propose six design features that could help maintain species diversity in nature reserves surrounded by a human-dominated landscape (Figure 8.4):

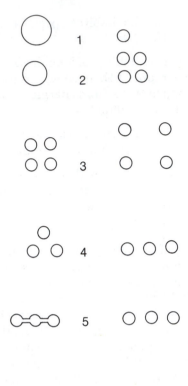

Figure 8.4 Schematic representations of design principles for nature reserves. In each pair, the design on the left will probably have a lower extinction rate and higher species diversity. (Redrawn from Diamond, 1975)

1. A large reserve will hold more species than a small reserve because of the aforementioned species-area relationships.
2. For the same reason, a single large reserve is preferable to several small reserves of equal total area, assuming they all represent the same habitat type.
3. If it is necessary to have multiple small reserves, they should be grouped as closely as possible to minimize isolation.
4. Arranging small reserves in a cluster, as opposed to a linear fashion, will also facilitate movement among the reserves.
5. Connecting the reserves with corridors will make dispersal easier for many species.
6. By making reserves as circular as possible, dispersal within the reserve will be enhanced and the negative effects of edges will be minimized.

These ideas and similar expressions (Wilson and Willis 1975, Diamond and May 1981) were quickly and widely accepted even though a number of the points have been challenged (Ambuel and Temple 1983, Blouin and Connor 1985) and one—a single large reserve is better than several small ones of equal total area—has generated an extraordinary degree of controversy (Diamond 1976, Simberloff and Abele 1976a, 1976b, 1982, Terborgh 1976, Whitcomb et al. 1976, Boecklen and Gotelli 1984, Simberloff 1986). Theoretical models, common sense, and a modicum of data have been brought to bear in the arguments, along with some heated rhetoric. Defenders of Diamond's model have sometimes reacted as though the first design principle—large reserves are better than small reserves—was under attack and have accused the opposition of advocating the dismembering of nature reserves (Simberloff and Abele 1984, Willis 1984). This is not the case; Daniel Simberloff, Lawrence Abele, and others have only expressed doubts about the second principle. To them there is no universal answer to the question: If you have a finite amount of money, should you buy one large nature reserve or several small ones of equal total area? In terms of island biogeography theory, the answer to this question depends on whether the abilities of different species to colonize and survive in a small, isolated area are equal, or whether they vary along a gradient in a fixed and predictable fashion (Connor and Simberloff 1984, Gilpin and Diamond 1984). In the box, this question is explored in detail; suffice it to say here that there is no consensus on the correct answer.

SINGLE LARGE RESERVE OR SEVERAL SMALL

The answer to the single large or several small debate (often abbreviated SLOSS) lies somewhere between these two alternatives, but to illustrate the difference between them, Table 8.1 depicts the two extreme cases. Diamond's approach would be supported if Scenario 1 (in Table 8.1) described the real world. Each successively larger reserve contains all the species of the smaller reserves plus additional species which have more stringent minimum area requirements. There is a predictable gradient among the species. At one end of the gradient, daisies are found in all the reserves; at the other end hawks need so much land that they can survive in only the 240 ha reserve. In this situation, if you were given $1,200,000 to save forests from being turned into parking lots and land cost $5,000 per hectare, you should buy the 240 ha reserve and thus maintain 224 species. For the same amount of money, you could buy reserves D, E, and F, but you would protect only 199 species.

Scenario 2 describes a situation that would definitely favor the Simberloff approach. Again, large reserves have more species, but each reserve has a unique set of species, a more or less random selection from the species pool. There is no pattern of adding new species with more stringent area requirements, because all the species have an equal ability to survive in limited areas or because the uniqueness of each set masks any

TABLE 8.1   A Hypothetical Series of Seven Progressively Larger Reserves*

SCENARIO 1

Patch Size (ha)	Number of Species	Number of New Species	Accum. Number of Species	Representative Species
A (10)	119	—	119	Daisy, etc.
B (10)	119	0	119	Daisy, etc.
C (20)	137	22	137	Daisy, sparrow, etc.
D (40)	159	16	159	Daisy, sparrow, snake, etc.
E (70)	175	24	175	Daisy, sparrow, snake, robin, etc.
F (130)	199	25	199	Daisy, sparrow, snake, robin, squirrel, etc.
G (240)	224	25	224	Daisy, sparrow, snake, robin, squirrel, hawk, etc.

SCENARIO 2

Patch Size (ha)	Number of Species	Number of New Species	Accum. Number of Species	Representative Species
A (10)	119	—	119	Daisy, etc.
B (10)	119	119	238	Sparrow, etc.
C (20)	137	137	375	Ivy, grackle, etc.
D (40)	159	159	534	Trillium, blackbird, tortoise, etc.
E (70)	175	175	709	Lily, toad, rabbit, shrew, etc.
F (130)	199	199	908	Holly, snake, warbler, mouse, pine, etc.
G (240)	224	224	1,132	Robin, lizard, frog, squirrel, fox, hawk, etc.

*The area of each reserve is shown (Column 1), as well as the total number of species in each reserve (Col. 2), the number of new species added to the series total by each reserve (Col. 3), and the accumulative number of species in the series (Col. 4). The last column gives a hypothetical sample of the species found in each reserve. In Scenario 1, each reserve has all the species of the smaller reserves plus some new species. In Scenario 2 each reserve has a unique set of species. (Each reserve has the same area as the total of the three preceding reserves. Species numbers were calculated from $S = cA^z$ with $c = 75$, and $z = 0.2$; this might roughly approximate the number of vascular plant and vertebrate animal species in a temperate forest.)

tendency for some species to occur only in large areas. Here, the best approach would be to buy reserves A, B, C, D, and E; they would harbor 709 species and cost just $750,000. The G reserve would still cost $1.2 million and only have 224 species.

In summary, the fundamental difference between the two scenarios depends on whether the abilities of different species to colonize and survive in a small, isolated area are equal, or whether they vary along a gradient in a fixed and predictable fashion. Obviously, neither of these scenarios describes the real world, but which is more accurate? The primary consideration may be habitat homogeneity. Diamond's assertion that one large reserve is superior to several small ones explicitly assumes that all the reserves have the same habitat. (Obviously, if each small reserve was a unique environment, it would have a unique biota.) But is it realistic to assume that different reserves can represent the same habitat? Critics claim that at a microhabitat scale there will always be differences among reserves and that these will be reflected in the biota.

Conceptually, it would be easy to determine which scenario is more accurate; one would measure total species richness for a variety of different size islands and determine what portion of their biota is unique or shared with other islands. In practice, most studies have been limited to one or two taxonomic groups. The critics of Diamond's view have taken their best supporting evidence from studies of plants and invertebrates, species that usually have very small area requirements and are quite sensitive to microhabitats (Simerloff and Abele 1976*a*, Higgs and Usher 1980, Jarvinen 1982, Simberloff and Gotelli 1984, Nilsson et al. 1988, Quinn and Harrison 1988). Diamond's defenders have been largely ornithologists and thus in tune with species that may be less sensitive to microhabitats and more sensitive to area than plants and invertebrates. Overall, the limited field evidence does not seem to clearly favor either side (Simberloff and Abele 1982), especially because the data are confounded with questions of habitat homogeneity and how long a patch must be isolated before the extinction rate stabilizes.

In the absence of clear guidance from island biogeography theory, there are many other considerations that bear on the argument of single large versus several small reserves. Firstly, there is the fact that large, wide-ranging animals such as wolves, Sumatran rhinos, and harpy eagles must have very large reserves; small reserves will simply not suffice unless they are closely connected to one another. Indeed, to protect these species over a long period, reserves must be large enough to hold a fairly large population unless there is some movement of individuals among reserves. It has been estimated that to avoid the likelihood of extinction, a minimum viable population is likely to require hundreds of individuals, perhaps thousands, if it is to evolve and survive in the face of changing environments (Soule' 1987). (The minimum viable population concept is discussed in Chapter 13, "Special Species.") Under these criteria, many large mammals will require reserves measured in the thousands of square kilometers. The special needs of large mammals and birds are of particular importance because (a) they play key roles in many ecosystems; (b) they are typically the species that society values most, whether for utilitarian or aesthetic reasons; and (c) they are often in dire need of the attention of conservationists. In short, it would be naive to treat all species as equal, and accommodating the needs of species that need special attention will tend to favor large reserves over multiple small ones.

Secondly, the prospects of catastrophes such as fires, severe storms, diseases, or the introduction of exotic predators, parasites, or competitors tend to argue against "having all your eggs in one basket" or a single large reserve. The heath hen might not be extinct today if there had been another population beside the one eliminated from Martha's Vineyard by fire, predation, and disease. Alternatively, it could be argued that a single large population would be much better at recovering from a catastrophe than several small populations should they all be afflicted. Furthermore, because cycles of disturbance and succession are essential to the long-term survival of many species, reserves should ideally be large enough to provide for natural disturbance regimes (Pickett and Thompson 1978). For example, a truly adequate reserve could sustain a wildfire, perhaps 10,000 ha or more, without being decimated.

Thirdly, sometimes two species do not coexist because they are competitors; in this case two small reserves could, together, support both species, but one large reserve could not (Gilpin and Diamond 1982).

Finally, there are practical considerations. It is almost always harder to find large tracts of relatively undisturbed land than small parcels, and it may be more difficult to turn them into reserves if this will have a negative impact on the local economy. On the other hand, the management costs for a single large reserve will presumably be lower than for multiple small reserves. The capabilities of different organizations are important too; perhaps governments and large conservation groups should focus on large reserves on the assumption that small groups and individuals will protect small reserves (McLellan et al. 1986).

One solution to the single large versus several small dilemma is to take an intermediate course that involves creating reserves of a variety of sizes. Only very large and very small reserves would be excluded from the array. Very small reserves would be excluded, because their biota is typically composed of common, widely distributed species, most of which are edge habitat species that will thrive on the periphery of larger reserves. Very large reserves should be avoided if (a) they far exceed the size required by any population inhabiting the region; and (b) protecting them directly conflicts with the establishment of a number of smaller reserves. To take a far-fetched example, imagine that the State of Maryland had no reserves at present but had decided to acquire 2,000 km² of land to establish a new reserve system. It would make little sense to purchase a single tract of this size, given that it is very unlikely that a viable population of wolves and mountain lions could be reestablished there. It would be far wiser to distribute 20–200 smaller reserves from the Appalachians to Chesapeake Bay and the Delmarva peninsula to represent Maryland's whole array of environments. A single reserve of 2,000 km² might make sense in Idaho where mountain lions, wolves, and grizzly bears can be maintained; indeed an analysis of the mammal fauna of western North American parks found only one 20,736 km^2 complex of four reserves (Kootenay, Banff, Jasper, and Yoho in Canada) has not experienced a loss of mammal species (Newmark 1987). In practice, it is unlikely that a potential reserve would be rejected because it is significantly larger than necessary; few potential reserves of such enormity exist.

Again, it is obvious that there is no simple course to follow. In each situation the merits of alternative courses must be evaluated with full consideration of the many issues raised here. This said, all other things being equal, it is probably best to lean toward large reserves simply because as the inexorable process of fragmentation continues, they will become rarer and rarer, and the process is effectively irreversible.

Corridors

In landscapes that have seen extensive deforestation, the remaining stands are dotted across the land, separate and isolated. If instead, these forest patches were strung together in a network of patches linked by long, narrow bands of forest, would this mitigate the forest fragmentation problem? Could habitat bridges minimize isolation? Unfortunately, there is little information to use (MacClintock et al. 1977, Wegner and Merriam 1979, Henderson et al. 1985, Lovejoy et al. 1986, Szacki 1987), but some answers are intuitively obvious.

The answers begin with a predictable caveat: The effectiveness of corridors will depend on the type of organism, the type of movement, and the type of corridor. It is hard to imagine that migratory birds would require corridors to find a suitable patch in which to settle. However, for forest animals with more limited cruising ranges, a ribbon of trees wide enough to give them a sense of security surely facilitates their movement. Such movement may only be the one-time passage of a young garter snake dispersing from its birthplace, or it may be the daily to-and-fro of a gray fox or blue jay patrolling its territory. If the corridor is wide enough to be a reasonable facsimile of a forest, even sedentary plants might move along it very slowly—reproducing vegetatively—at a rate measured in meters per generation, rather than the meters per minute characteristic of animals and plant seeds carried by animals or wind.

Narrow strips of trees are a regular feature of many landscapes, and shelterbelts, hedgerows, and road and trail rights-of-way undoubtedly expedite the travels of some species (Forman and Godron 1986). Unfortunately, most are too narrow to provide a forest environment and thus may restrict the movement of many species, especially plants that need to live, grow, and reproduce along the way (Forman and Baudry 1984). In trying to determine how wide corridors should be, it is easy to say the wider, the better. On the other hand, narrow corridors would not be utterly worthless. Eastern chipmunks move among woodlots along fencerows consisting of nothing but a barbed-wire fence and a narrow strip of herbaceous vegetation (Henderson et al. 1985). Certainly one of the best ways to provide corridors would be to maintain forest cover along all the waterways that reach across the land (Recher et al. 1987). This approach has much to recommend it, in part because shorelands have so many other ecological values, and thus it is easy to justify maintaining fairly wide bands of forest vegetation near shores; the topic of shorelands is considered in Chapter 9. One can also think of corridors on a grand scale, to link up parks and other public lands to form very large reserves, even to allow species to shift their geographic ranges in response to long-term climate change (Hunter et al. 1988).

Not surprisingly, some of the controversy surrounding the "single large or several small" issue has spilled over into questions about the desirability of corridors. Some would argue that they are essential means for ameliorating the various ills associated with fragmentation (Harris 1985, Noss and Harris 1986, Noss 1987), while others think they may be too expensive and too likely to allow diseases, exotic predators, and other forms of disturbance to spread (Simberloff and Cox 1987). Perhaps the most telling argument in favor of corridors is that natural landscapes are far more connected than those heavily shaped by humans (Noss 1987).

FOREST MANAGEMENT AND FOREST FRAGMENTATION

Forestry and Nature Reserves

Although the acquisition of nature reserves and connecting corridors is not in the mainstream of forestry, the preceding discussion of nature reserves as islands may be of considerable importance to foresters, because as often as not the reserves in question will be tracts of forest. Consequently, foresters need to collaborate with other natural resource managers and regional planners to guide economic development and the establishment and management of reserves. The forester's role is also critical because seldom would it be practical, or even desirable, for all forest patches to be inviolate reserves where active forest management is excluded. Through sensitive, low-impact techniques, timber can be extracted from many of these patches without compromising their value as wildlife habitat (Whitcomb et al. 1977). Admittedly, the economic return from timber harvest may be minuscule compared to what a landowner could make from selling out to a developer, but there may be an important psychological satisfaction in seeing the land produce some income, even if it covers only the property taxes.

Fragmentation in a Forested Landscape

Most of the studies documenting the negative effects of forest fragmentation have been undertaken in agricultural regions where forests have been isolated from one another *and* there has been a large decrease in the region's total area of forest (Askins and Philbrick 1987). Unfortunately, this situation is probably more the rule than the exception because, according to one estimate, the total area of forest has decreased from about one-quarter of the earth's land surface in 1960 to less than one-fifth, and it will not stabilize until the year 2020, at about one-seventh the land area (CEQ 1980, WRI and IIED 1986). However, when one thinks of forest management, the image is usually of large expanses of forests, not an agricultural landscape with scattered groves. How relevant is island biogeography to regions where forests still dominate the land?

Empirical evidence is rather limited. One group undertook a three-year study of amphibians, reptiles, birds, and mammals in the Douglas-fir forests of northwestern California where about half the forests had been clearcut during the preceding 30 years, thus fragmenting what was formerly almost continuous forest (Rosenburg and Raphael 1986). They found that some edge and early successional species were more common in patches associated with clearcuts, and a few species, such as the spotted owl and fisher, seemed to avoid edges and areas where the cutting had been particularly heavy. Their overall conclusion was that the composition of the vertebrate animal communities was very similar in forest patches that varied widely in size and isolation. This study does not necessarily disprove the idea that forest fragmentation could have a significant effect on wildlife in heavily forested regions. The cutting was relatively recent, much of it within the duration of a few generations for many species, and even within the lifetimes of some individual animals. In a hundred years the effects of fragmentation may be far more pronounced. Some analyses of long-term changes in the forested regions of northern Finland concluded that fragmentation was responsible for declining populations of several bird species, especially nonmigratory species that nest in cavities (Helle 1985*b*, 1986).

Other studies have identified some more specific problems that could potentially stem from fragmentation in a forested region. Predation of artificial nests of quail eggs was higher in small forest fragments than in large fragments, even though many of the fragments were only isolated by narrow powerlines, roads, and streams, and the study was conducted in Maine where forests cover 90% of the landscape (Small and Hunter 1988). A German study on the movements of small mammals and ground beetles (carabids) found very few incidences of animals crossing roads, thus suggesting that a forest carved up by roads might experience some consequences of isolation (Mader 1984). One of the roads small mammals were reluctant to cross was just a 3 m wide lane closed to most traffic. In Australia, even an overgrown, unused track inhibited small mammal movement (Barnett et al. 1978).

Indirect evidence suggesting that fragmentation can threaten wildlife in a forested region was assembled by Larry Harris (1984) in *The Fragmented Forest: Island Biogeography and the Preservation of Biotic Diversity*. Specifically, Harris examined the likely consequences of cutting most of the remaining tracts of old-growth Douglas-fir in Oregon and Washington and leaving only small, residual old-growth stands surrounded by younger forests. Obviously, populations associated with old-growth would decline because of the reduction in their habitat, but Harris argued that the declines would be much worse than this, to the point of regional extinction for some species, because the destruction of old-growth habitat would be exacerbated by isolation effects. Fortunately, Harris was predicting a possible future, not reporting the past, and his suggestions for mitigating the impact may influence land managers to take positive steps that will belie his predictions.

Fragments and Edges

Throughout discussions of island biogeography, one basic message, avoid further fragmentation of the forest, is very clear. Even in the heat of the "single large or several small" controversy, no one ever proposed carving up existing large tracts to make them into several smaller ones. But is not "avoid fragmentation" in direct conflict with the "create more edge" maxim discussed in Chapter 7? Yes, there is a clear conflict here. For some species, edge habitats are optimal or even essential (Gotfryd and Hansell 1986), but other species need large tracts of interior forest and may find edges to be ecological traps in which competition, parasitism, and predation are problematic (Gates and Gysel 1978, Brittingham and Temple 1983).

Balancing these conflicts will, to a certain extent, involve balancing people's preferences for different species. If the choice is between an edge species, such as the song sparrow that can be found around every brushy corner, versus a forest interior species whose long-term fate is of some concern, such as the Kentucky warbler, then the choice seems straightforward: The warbler should be favored. If the choice is between popular game species that favor edges, such as white-tailed deer and bobwhite quail, versus common, widespread songbirds that prefer forest interiors, e.g., scarlet tanagers and ovenbirds, again the choice is fairly clear: The deer and quail will probably win. Unfortunately, these choices divide people who would like to see common edge animals, notably the game species, become more common, from people who favor the uncommon interior species. In a tropical forest the choice might be between common, early successional tree species associated with edges versus rarer shade-tolerant species of the forest interior (Hubbell and Foster 1986). How each individual decision is resolved is not critical, but across a whole landscape care should be taken to see that the choices balance out. This again is the "diversity of diversity" argument of Chapter 2. In regions where human manipulation of the landscape has left forests reduced and isolated, there already is an imbalance, and avoiding further fragmentation should take precedence over creating more edge. Where forests still substantially dominate the land, a management program that optimizes spatial heterogeneity by managing stands at a variety of scales can produce both edge habitat and large stands (see Chapter 6, "Spatial Heterogeneity"). At typical scales of forest management, an abundance of edge habitat will result from patch cutting (middle-scale management), and large stands can be created or maintained in two ways: (a) selection cutting on large tracts (fine scale); and (b) clearcutting large stands (coarse scale). (However, we should recall the concept of "when does black and white become gray" (Figure 6.1) and realize that a forest that seems fragmented and full of edges to a salamander may seem homogeneous to a moose.) Using a model of the interrelationships of patch size, edge length, patch configuration, and extent of cutting, Franklin and Forman (1987) suggested a somewhat different solution: clustering small clearcuts and thus concentrating the edge effect in limited areas.

One could have large stands *and* plenty of edge by simply creating large stands with extremely irregular shapes. Unfortunately, even this would probably not accommodate the needs of forest interior species very well (Figure 8.5). In a paper called "The Eternal External Threat," Dan Janzen (1986) pointed out that the longer the edge around a patch of forest, the greater the likelihood of intrusion by poachers, pesticides, exotic plants and animals (Usher 1988), fire, and other agents of disruption. In forested landscapes, the edge effects created by harvesting may largely disappear in a few years if there is rapid regrowth (Ranney et al. 1981), and this effect could be accelerated by having a buffer zone of partial cutting between a clearcut and the residual forest (Ratti and Reese 1988).

On a larger scale, buffering forest fragments from the external world is clearly a wise idea; an idea for which the term *multiple-use modules* has been coined (Harris 1984, Noss and Harris 1986). A multiple-use module (MUM) might consist of a core of undisturbed forest, surrounded first by an inner buffer zone in which dispersed recreation and selection logging was allowed, and then by an outer buffer zone where a variety of forestry activities and limited agriculture could be permitted. The idea of surrounding reserves by low-intensity land use (e.g., forestry and grazing rather than crops and housing) characterizes much of the public land in western North America and is widely used in the establishment of new national parks, especially in developing countries where economic exigencies make it critical to integrate reserves into the local economy.

Thinking in terms of MUMs may be especially timely in the Amazon Basin, a region where the forest still stretches for thousands of kilometers, but drastic change has begun.

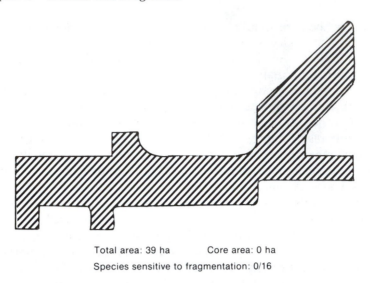

Total area: 39 ha   Core area: 0 ha
Species sensitive to fragmentation: 0/16

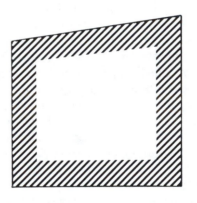

Total area: 47 ha   Core area: 20 ha
Species sensitive to fragmentation: 6/16

Figure 8.5 Two forest fragments with similar total areas, but in A there is no core area (over 100 m from an edge) and no bird species sensitive to fragmentation occur there. In B there is a 20 ha core, and six sensitive birds occur. (Reproduced from Temple, 1986)

FOREST FRAGMENTS IN AMAZONIA

Throughout much of the Amazon Basin, large tracts of tropical lowland forest are being cut and burnt to produce pastures in which cattle can be raised for export. Tropical deforestation has precipitated widespread concern among the world's conservationists, because these forests contain an extraordinary diver-

(Photograph by R. O. Bierregaard.)

sity of unique species and may play a critical role in the global processes that determine carbon dioxide concentrations and thus the earth's climate. The story of tropical deforestation has been well-told before (Myers 1980, 1984, Caufield 1984); here the focus is on some events that began in 1979, 80 kilometers north of Manaus, the major inland port on the Amazon. Two organizations, the World Wildlife Fund (WWF) and Brazil's Instituto Nacional de Pesquisas da Amazonia (INPA), arranged with local authorities and landowners to orchestrate the pattern of forest-clearing in a manner that would eventually create an elaborate experimental design. As deforestation proceeded, 25 tracts of forests would be left isolated from one another and the region's continuous forest; there would be eight 1 ha fragments, nine 10 ha, five 100 ha, and two 1,000 ha, and a single tract of 10,000 ha.

These tracts form the basis of the Biological Dynamics of Forest Fragments Project (initially called the Minimum Critical Size of Ecosystems Study), a research endeavor directed by Thomas Lovejoy, Richard Bierregaard, and Herbert Schubart, and executed by scores of fieldworkers.[1] The researchers began by establishing plots in the midst of continuous forest and carefully inventorying the flora and fauna. Over the ensuing years, the forest around these plots would be cut, leaving them as isolated fragments, and by monitoring the changes that followed isolation, the researchers would be able

[1]Most of the information presented here was obtained from three summary papers, Lovejoy et al. 1983, 1984, and 1986, but detailed reports on various groups of organisms are being published.

to document the effects of forest fragmentation with unprecedented rigor. Virtually all investigations of island biogeography begin decades, if not millennia, after the isolation took place, leaving scientists to assume what the original biota was like. Another major advantage of the design was inherent in the huge range of tract sizes (1 to 10,000 ha), a range spanning four orders of magnitude. Furthermore, with replicates of all the smaller size classes, it would be possible to determine how similarly the biota of various fragments would respond to isolation. Would each tract end up with quite a different set of species, or would they largely converge on a similar set? Finally, by continuing the project for 20 years, changes that might not be manifested immediately could be detected. The magnitude of the project is enormous whether you consider the area, the duration, or the variety of organisms; combining all these considerations, it has to be one of the largest ecological experiments ever undertaken. It will be many years before the entire story can be told, but already some interesting preliminary observations have been made on the few small patches that have been isolated.

One of the most conspicuous consequences of isolation has been seen among the trees. Because trees are long-lived, immobile, and use relatively small areas, they would seem unlikely candidates to abandon a small patch, and indeed, thus far tree species have not been going extinct in the patches. However, in the 1 ha and 10 ha patches, they have been dying at almost twice the normal rate. Because mortality is especially high among the trees near the windward edges of patches, it is assumed that a change in the microclimate, i.e., the prevalence of dry, hot winds, is primarily responsible. The death of a few trees may not immediately threaten their populations, but it certainly can have significant influences on the rest of the community. The gaps created by the tree falls have been occupied with successional plant species, and it has been speculated that, within a decade, secondary vegetation will largely replace primary forest in the 1 ha and 10 ha patches. And needless to say, this will have profound effects on the many animals that have evolved as herbivores, pollinators, seed-dispersers, etc. of the plants.

The larger mammals have responded predictably; species such as jaguars, margays, pacas, mountain lions, and white-lipped peccaries ceased using small patches after isolation. (There is an interesting anecdote about linkage among species here: Following the disappearance of peccaries from one reserve, the wallow they maintained dried up, eliminating the only breeding pool for four species of frog.) Among monkeys and marmosets, the results are also quite predictable; the species that live in troops that have large home ranges are absent, even from a 100 ha tract, but species with more moderate area requirements persist. Even some smaller mammals, rodents and marsupials, have begun to disappear from the smaller patches.

Some of the initial results from bird censusing seemed rather surprising; there was a marked *increase* in the number of birds soon after isolation. This was attributed to a short-term influx of individuals displaced by the surrounding habitat destruction. After a few months, the numbers declined to a point lower than in nonisolated forest. There also appeared to be a diminution in bird species richness, although this conclusion was based on analyses of the rates at which new species appeared among birds captured in mist-nets, not complete counts of species. Using mist-nets to census birds (standard operating procedure in tropical forests) only provides a reasonable assessment of the birds of the forest understory. These birds, with their predilection for dark, humid places, are probably much more likely to be adversely affected by being surrounded by a hot, dry swath of cut-over land than are the species that dwell high up in the forest canopy.

One group of birds provides a particularly interesting example of how species can become extinct in an area too small to meet their needs; these are the birds that follow army ants as they swarm across the forest floor. Birds such as the rufous-throated antbird and the white-chinned woodcreeper use army ants as beaters by foraging on the insects stirred up out of the forest litter by the ant hordes. For some birds, the relationship is obligate; others can alternate between following ants and more traditional foraging styles. An army ant colony uses about 30 ha of forest in its peregrinations, so an isolated tract much smaller than this will probably have neither army ants nor ant-following birds. In one case, all the ant-followers left a patch even though one army ant colony lingered there, for ant-followers cannot survive following a single colony because the ants have a 35-day cycle during which they are swarming (and thus useful to the birds) only part of the time. Ant-followers even disappeared from a 100 ha patch, suggesting they may need three or more ant colonies to follow.

Two groups of insects were selected for study, butterflies and euglossine bees. The euglossine bees were censused by recording the rates at which they visited three chemical baits. Comparing visitation rates before and after isolation, there was a distinct decline in 1, 10, and 100 ha patches. This preliminary result could be of some consequence; euglossine bees are the only pollinators for many orchids and are important pollinators for over 30 families of plants. Rather different changes were noted by the researchers studying butterflies. Soon after isolation, there was a decrease in the number of butterfly species, probably because burning accompanied the forest clearing, but later, species richness increased again, eventually surpassing the preisolation levels. The invasion of light-loving species explains the increase. After isolation there was more light in the patch, extensive second growth around the patch, and pockets of second growth in the gaps created by tree falls. It has been estimated that edge habitat butterflies will venture 200–300 m into the forest, creating a broad zone of overlap with the forest interior species. The outcome of this interplay between the two groups is not known, but it may involve competition for nectar sources and egg-laying sites.

In a world with ample funds for ecological research, this study would be extended in time for several decades, perhaps in perpetuity. It would be expanded in space to include replicates in all the world's forest biomes. Finally, it would be broadened conceptually to examine fragmentation of forest landscape in which residual stands remain part of a forest mosaic, not isolated parcels in a barren landscape. This last scenario would provide information on how to achieve a balance between maintaining biological diversity and styles of timber management that may cause only modest degrees of fragmentation.

SUMMARY

Small patches of forest isolated in a landscape of open land share certain similarities with islands in the ocean; in particular, they are likely to have an impoverished biota. There are many possible reasons for this phenomenon—low habitat diversity, high rates of local extinction, absence of species that live at low population densities, deleterious effects associated with edges—but the upshot is the same: Wherever forests have been reduced to scattered tracts, further fragmentation should be avoided. Furthermore, these forest fragments should generally be subjected to less disruptive forms of forest management such as selection cutting. In regions where forests still dominate the landscape, there is also some evidence that fragmentation may have negative effects on some species of wildlife, but the issue is less critical. Managing extensively forested landscapes at a variety of scales and with a careful distribution of stands should provide suitable habitat for both species that need large areas of unbroken forest and species that need forest edges. Because some of the most vulnerable species—notably certain large mammals and birds—require such extensive tracts of habitat, a conservative approach to maintaining biological diversity should emphasize avoiding fragmentation.

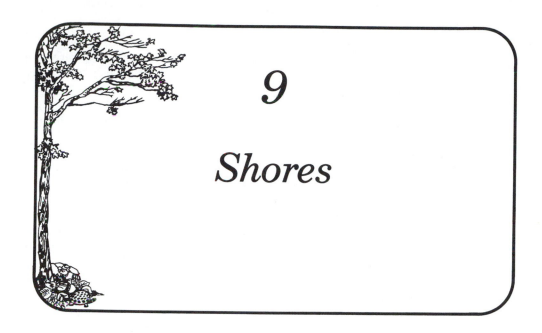

9

Shores

When we refer to earth as the watery planet, we recognize both a substance that is critical to supporting all forms of life and an environment that was the sole habitat for organisms for most of our history, about 3 billion years. Even though terrestrial creatures now dominate the planet, the interface between aquatic and terrestrial realms remains of great importance, because shores are often home to a unique biota and some special ecological processes. Shorelines are equally important to many human endeavors because they have been traditional foci for activities such as fishing, agriculture, and shipping. The inevitable consequence is that we have extensively altered some of our most valuable ecosystems and must be especially careful to wisely manage those that remain in some semblance of a natural state.

There are many kinds of shores, ranging from broad expanses of sand buffeted by ocean waves to the narrow fringe of moss-covered rocks that lines a tiny woodland brook, and the scope of human activities to be found here is similarly immense. This chapter focuses on just one part of the whole picture: the effects of timber management on the streams that dissect forested landscapes, specifically timber management in shoreline or *riparian* forests. There are three points here that require elaboration.

First, although it may be obvious that urban development along marine and estuarine shores is beyond the scope of this book, it is not so clear that the riparian ecosystems of arid and semi-arid landscapes should be excluded, especially because it is here that interest in riparian ecosystems first blossomed (Johnson and Lowe 1985). In deserts and grasslands a ribbon of trees often marks a watercourse and forms the basis of a rich and distinctive community, but unfortunately in many areas these communities are rare or severely degraded because of various encroachments such as excessive livestock grazing. Rather than present a superficial summary here, I encourage interested readers to consult the volumes that have been compiled on the ecology and management of riparian ecosystems in arid environments (e.g., Johnson et al. 1985).

Second, forest streams come in many sizes, growing from spring-fed trickles to substantial rivers as they move downhill and converge with one another to drain larger and larger watersheds, and along this gradient the ecological characteristics of a stream ecosystem change in a gradual continuum (Vannote et al. 1980). Ecologists often use a classification scheme to separate this continuum into meaningful units. In this scheme, the smallest streams are called first-order (see Figure 9.1); streams formed from the convergence of two or more first-order streams are called second-order; when two second-order streams join they form a third-order stream, and so on. The Mississippi River is a tenth-order stream, the only one in the United States (Leopold et al. 1964). Because small streams are far more abundant than large ones and collectively have longer shorelines, we will concentrate primarily on first-, second-, and third-order streams.

Finally, we need to ask, just what is a riparian zone? Or, more specifically, what sort of parameters—hydrologic, climatic, biological—define the bounds of this terrestrial-aquatic interface? At the smallest scale the riparian zone would be the immediate water's edge where some specialized plants and animals form a distinct community (Swanson et al. 1982). At a larger scale it would be the area periodically inundated by high water; i.e., the banks and floodplain of a stream. Finally, on the largest scale it would be the band of

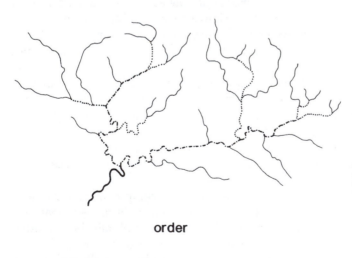

order

——— 1   ·········· 2   ·–·–·– 3   ——— 4

Figure 9.1 Stream patterns from central Spain showing small first-order streams converging to form second-order streams, third-order streams formed by two second-order streams, and finally a fourth-order stream formed from the third-order streams. (scale 1 cm: 5 km).

forest that has a significant influence on the stream ecosystem, or conversely is significantly influenced by the stream. This latter definition, although still quite nebulous, is the one we will use predominately.

THE IMPORTANCE OF RIPARIAN ECOSYSTEMS

The Aquatic Perspective

Most streams are not especially productive ecosystems, simply because photosynthesis is limited by a scarcity of plants. Currents sweep plankton downstream too rapidly for a large community of phytoplankton to develop; rooted plants have difficulty persisting in the scoured substrates that line the shallows; and often little more than a crust of algae and mosses can flourish (Hynes 1970). Because of this paucity of organic matter production, many riverine ecosystems depend heavily on organic matter imported from adjacent ecosystems. In a small stream lined by trees, the largest portion of organic matter, as much as 99%, will enter the food web from external sources (Fisher and Likens 1973, Triska et al. 1982).

The major source is leaves, twigs, and other debris that become both food and cover for a variety of invertebrates collectively called shredders. These organisms perform the first conspicuous step in the process of decomposition by shredding detritus, although a closer perspective would actually see decomposition beginning with the microbes that are sustained by dissolved organic matter leached from the leaves and other terrestrial sources. Microbe or mayfly larva, leaf or soluble compound, the important point is that streams are substantially subsidized by surrounding forests. Indeed, from bacteria to trout and otters, most of the stream's organisms ultimately rely on imported energy to a large degree. For example, it has been calculated that in the forest streams of coastal Oregon, well over 50% of the energy obtained by young coho salmon is derived from terrestrial production (Chapman 1966). The connection between fish and forest is much more direct in the Amazon basin where hundreds of species of fish consume the fruits and seeds of trees (Goulding 1980). Anything that interrupts this flow of organic matter will have significant consequences, and thus the condition of riparian forest is very important to stream ecosystems. To put it another way, it only requires a modest stretching of ecological linkages to think of a can of salmon as a forest product.

Not all the terrestrial organic matter that enters a stream is quickly decomposed as a source of energy. Large logs often fall into streams where they can remain for many decades before rotting away or being swept downstream (Triska et al. 1982, Sedell and Swanson 1984). Besides becoming food for decomposers, such logs are an important source of habitat diversity, serving as a substrate for many small species and cover for larger ones (Meehan et al. 1977, Anderson et al. 1978). Two studies in Alaska discovered that juvenile salmon and trout were significantly more abundant in streams with large amounts of woody debris (Bryant 1983, Dolloff 1986). By acting as dams, logs can divide what might otherwise be a uniformly flowing stream into one marked by alternating pools and riffles, thus further enhancing habitat diversity. The pools are particularly important, for it is here that organic matter collects and stays long enough for the decomposers to work on it (Bilby and Likens 1980.)

Logs lying across streams and along banks also tend to stabilize the dynamic equilibrium between sediment deposition and erosion (Heede 1985). These processes shape a stream's channel as the stream moves back and forth and up and down, cutting into banks and dropping piles of sediment (Richards 1982). Shoreline vegetation is also important; a bank interwoven with a mesh of roots is far less likely to erode than a barren one (Smith 1976).

Organic matter dragged into streams and arranged into dams by beavers is yet another way that materials originating in the forest end up shaping stream morphology (Parker et al. 1985). The ability of beavers to manipulate their environment is truly extraordinary; if they were confined to a remote region of Africa, rather than distributed throughout much of the northern hemisphere, they might be thought the most remarkable mammal next to humans. Furthermore, the ecological impact of beavers is not confined to their engineering activities. It has been estimated that in an area of healthy beaver populations, they may import 3,000 kg of woody material per kilometer of stream; about 75% will be left to decompose; about 25% will be consumed and then enter the aquatic system as feces (Waring and Schlesinger 1985).

Streams are natural receptacles for both terrestrial plant production and the by-products of soil erosion within their watersheds. Unfortunately, this latter process has been greatly enhanced by human activity to a point where many water bodies have been severely degraded by turbid water and sediments that coat the bottom (Dunne and Leopold 1978). For example, gravel beds are useless as salmon and trout spawning areas if they become so clogged by fine sediments that water cannot percolate through them (Phillips et al. 1975). Forestry operations, especially road building, can produce excessive erosion, but if riparian forests are left intact, they can help to buffer streams from this problem. Not only will erosion next to the stream be minimized, but sediments transported from distant erosion sites can be filtered from surface runoff before reaching the stream.

The organisms that live in a forest stream have usually adapted to a fairly narrow range of water temperatures, and this benign thermal environment is provided, in part, by the forest canopy that shades such streams (Barton et al. 1985). Shade is also important in keeping streams from becoming too warm, because warm water holds less dissolved oxygen than cold water.

Because streams represent a distinct ecosystem with hundreds of species found in no other type of ecosystem, a forest landscape laced with healthy stream ecosystems will support a much greater variety of wildlife than a forest landscape drained by impoverished ditches. Moreover, as previously explained, if a landscape is to be laced with healthy stream ecosystems, it must have healthy riparian ecosystems. Therefore, one can conclude that riparian forests are key components in maintaining the wildlife diversity of forest landscapes.

The Terrestrial Perspective

For terrestrial plants and animals, riparian zones represent a place of abundant water. In a tropical rain forest awash with over two meters of rainfall annually, this may be of modest consequence, and along some streams the availability of water may be only an ephemeral phenomenon. Still, in every riparian ecosystem there are some species for which the simple presence of water makes it their preferred, or even sole, habitat. Some of the associations are very obvious; with most amphibians living on land and returning to water to breed, it is hardly surprising that many of them spend much of their lives in riparian ecosystems (Brinson et al. 1981, Vickers et al. 1985). Many of the reptiles, birds, and mammals that are commonly thought of as aquatic species could just as easily be considered riparian. Beavers, kingfishers, otters, and ospreys might come to mind first for a person residing in the northern hemisphere, but duck-billed platypuses, anacondas, African fish eagles, and hundreds of other species fall into this category too. The mere presence of drinking water will make riparian areas preferred habitats for some animals. For example, anecdotal evidence suggests that in the winter, white-tailed deer will seek open water so that they do not have to tax their limited energy reserves by eating and melting snow.

Many species of plants also have strong associations with riparian zones (Nilsson et al. 1988). Some require the presence of open water and grow along the water's margin. However, overall, the riparian zone's humid microclimate and moist soil, created by both open water and ground water near the surface, are more important to plant growth than open water per se (Swanson et al. 1982). The humidity of riparian ecosystems also tends to make them favorable habitat for many amphibians and some small mammals (Miller and Getz 1977).

Larger streams often flood, and then water ceases to be an essential resource and becomes an agent of stress, especially for plants which cannot seek high ground as most animals can (Ruffer 1961, Batzli 1977). Some plants persist in floodplains because they can tolerate an occasional period of standing in water or even being inundated, especially if it comes outside the growing season (Kozlowski 1984). Others survive because they can quickly recolonize a site after a surge of water (and perhaps ice) has removed the previous community leaving a denuded bank. In either case, some species are far better at adapting to water level fluctuations than others, and thus the stressful aspects of a riparian environment also contribute to the uniqueness of the biota.

The stressful effects of water can be seen at a glance in the distinct units of vegetation many riparian zones contain, particularly those along larger streams. Beginning at the water's edge and progressing into the forest, successive bands of vegetation are often defined by very small changes in topography (Metzler and Damman 1985, Menges 1986). Just a one-meter rise in elevation may see a complete shift from the strip of herbaceous plants that covered the bank after ice scoured it during the last major flood, to full-sized trees that can tolerate an occasional soaking. In some respects, one can think of these strips of vegetation as part of a sequence of forest succession with water acting as the disturbance agent that keeps setting back the successional clock (Lindsey 1961).

The vegetation diversity has a strong vertical component as well; from the water's surface to the top of the canopy, there are often several distinctive layers of vegetation (Figure 9.2). This is especially true along larger streams where the stream's breadth creates a break in the forest canopy allowing light to reach the ground (Swanson et al. 1982).

Not surprisingly, all this vegetation diversity is reflected in a great variety of animals. In the Blue Mountains of eastern Oregon and Washington, 285 of the 378 species of terrestrial vertebrates are either directly dependent on riparian areas or prefer them to other habitats (Thomas et al. 1979a). This pattern is repeated in study after study, which show that a large portion of a region's vertebrate fauna is often associated with riparian zones (Brinson et al. 1981), and it may well be true for invertebrates too. A partial survey of

Figure 9.2   Riparian vegetation is often characterized by distinctive horizontal zones and vertical strata.

insects and spiders in a Czechoslovak elm-ash floodplain forest found 1,261 species, and it was estimated that 10–20% of Czechoslovakia's whole arthropod fauna occurred in this one forest (Kristek 1985).

Riparian diversity is also enhanced by habitat differences along the length of a stream, differences that are exploited by a variety of species. Moving down a watershed, changes in the flora are very apparent, especially as streams become large enough to have a distinct floodplain. Less conspicuous changes occur too. Along a tiny headwater stream, the characteristic predator might be a water shrew, whereas larger animals, such as minks and otters, would be more common downstream where larger fish can be found (Harris 1984). A reverse pattern is found in some salamanders. Spring salamanders are confined to headwater brooks, while smaller species, dusky and two-lined, also inhabit larger, warmer streams (Markowsky and Hunter in prep.) The small species probably can tolerate a wider range of conditions, because relative to their body size they have a large surface area over which to obtain adequate oxygen. The much larger spring salamander's low surface to volume ratio means it cannot survive in warm, poorly oxygenated streams.

Edges, successional patterns, vertical layering, special microhabitats defined by physical features—riparian ecosystems have many attributes that enhance their biological diversity. Minnows, frogs, and squirrels, algae, alders, and oaks; they all come together along shorelines.

Besides being diverse and unique, riparian forests have yet another feature that makes them very important: They are often exceptionally fertile and productive. Riparian areas that lie in floodplains usually prove to be nutrient-rich, because whenever a stream escapes its banks it leaves a deposit of sediments behind (Brinson et al. 1980). Researchers used tree rings to reconstruct the growth of bald cypress trees in an Illinois riverside swamp and found that growth was strongly influenced by the frequency and magnitude of floods; i.e., periods when large amounts of phosphorus were deposited in the forest (Mitsch et al. 1979). This is not just a short-term phenomenon; over time, a rich alluvial soil is created, the kind of soil that makes river valleys popular with farmers. The water that flows through a riparian zone also expedites nutrient recycling and thus plant growth by moving oxygen through the soil and removing carbon dioxide and metabolic waste products; this contrasts sharply with the anaerobic conditions that often prevail in wetlands with stagnant water (Brinson et al. 1981, Grunda 1985).

Plants benefit directly from this enrichment, and ultimately plant consumers and other animals also take advantage of the increased production of nutrient-rich forage. The nutrient-rich environment may explain why, in a study of riparian bird communities in Iowa, floodplain forests supported an average of 506 breeding pairs per 40 ha compared to 339 for upland forests, even though both habitat types had a similar number of species (Stauffer and Best 1980). Many studies have shown a greater density of breeding birds in riparian forests than in nearby upland areas; indeed, densities of up to 900 pairs per 40 ha have been reported (Brinson et al. 1981). Floods that enrich the soil and keep vegetation in an early successional stage may also explain why riparian areas are favorite moose habitat in many regions. An aerial survey of Siberian moose found 45% of them were in riparian forests (Kistchinski 1974).

The luxuriant vegetation of a riparian zone is not just so much forage for the animals to consume. It also represents cover—concealment from or for predators, and shelter from inclement weather. Flocks of migrating songbirds probably use riparian forests disproportionately—perhaps because food is abundant there, perhaps because they seek strips of dense cover—although this has been documented only in arid regions (Stevens et al. 1977). In Maine, dense stands of conifers are common along waterways, and it is in such stands that over 85% of white-tailed deer wintering areas are located, in large part because the dense cover produces a benign microclimate (Banasiak 1961).

The linear nature of riparian forests and their dense vegetation make them an obvious choice as corridors to connect isolated forests and avoid the island effect (see Chapter 8), (Recher et al. 1987). The elk of western North America often use riparian corridors to travel up and down the mountainsides between their winter and summer grounds (Thomas et al. 1979a, 1979d). Radio-tracking studies of coyotes, bobcats, red foxes, pine martens, and fishers in Maine suggest that these carnivores travel along streams more than one would expect by chance (Dibello unpubl.).

The Rarity of Riparian Ecosystems

The importance of riparian ecosystems is most striking in terms of the habitat preferences of one particular terrestrial animal: humans. As manifested by the distribution of cities, roads, farms, campgrounds, etc., people are drawn to shorelines for a variety of recreational and practical purposes. Even in the absence of a defined purpose, aesthetic appeal brings people to the water's edge (Black et al. 1985, Hoover et al. 1985). No doubt the rarity or riparian ecosystems heightens their value in human estimation.

Riparian forests have never been exceedingly common; in the continental United States it has been roughly estimated that about 6% of the land is subject to periodic flooding and thus could be considered, at least potentially, as riparian ecosystems. Over 70% of this area has been converted into urban and agricultural land, inundated by reservoirs, etc., and thus less than 2% of the land area remains in some semblance of a natural riparian ecosystem (Brinson et al. 1981). Specific examples are even more dramatic. An estimated 2,239,000 ha of elm-ash forest once associated with rivers in the midwestern United States have been reduced to 279,000 ha (Klopatek et al. 1979). Bottomland hardwood forests in southeast Missouri were reduced from nearly a million hectares to some 40,000 between 1780 and 1975 (Korte and Fredrickson 1977.)

MANAGING RIPARIAN ECOSYSTEMS

Riparian forests are inarguably very important—many would assert that they are the most valuable and sensitive parts of the landscape (Thomas et al. 1979*a*)—and thus their management must be undertaken very carefully. One conservative approach is to simply eliminate all timber management practices from riparian zones and let them become old stands. This laissez-faire strategy has much to recommend it: It protects riparian values well; it creates old forest stands which are usually in short supply (see Chapter 5); and it is simple.

On the other hand, the cost of a no-cut policy can be considerable. Consider the Siskiyou National Forest in southwestern Oregon; there are two billion board feet (4.7 million m^3) of very valuable Douglas-fir and Port Orford cedar on commercial forest land within 100 feet of streams (Anderson 1985). And yet those same streams support populations of salmon and trout that produce a fisheries worth several million dollars per year.

It may be feasible to protect riparian values and still extract timber. Indeed, it is possible that careful silviculture could do a better job of protecting some specific riparian values than a hands-off policy (Brouha and Parsons 1986). For example, managers in Idaho used a model to predict the rates at which large trees would fall into streams lined by western hemlock-grand fir forests and thus provide cover for fish (Rainville et al. 1986). They concluded that harvesting 4–5% of the forest per decade (a 200–250 year rotation) would result in more large logs falling into a stream than a policy of no cutting. Management of riparian deer wintering areas provides another example. White-tailed deer need dense cover as shelter, but will fare better if the canopy is broken in a few small patches where they can find saplings on which to browse (Crawford 1984). Thus, a riparian forest subject to some patch cutting may serve them better than an unmanaged forest. If shading a stream were a primary objective, an intensively managed, uneven-aged stand might be superior to an old, unmanaged stand, because the latter would probably be more vulnerable to disturbances such as tree falls that would break the canopy.

The issue of maintaining the forest canopy to ameliorate high temperatures illustrates the potential complexity of the intensive management approach. While some biologists have determined that this is critical to the survival of trout (Barton et al. 1985), other researchers have shown that when the canopy is opened and more sunlight reaches the stream, there will be more photosynthesis, greater productivity, and more trout (Murphy et al. 1981). Both studies are correct; the question is how much can you allow photosynthesis to be enhanced before excessive water temperature becomes a problem, and obviously the answer will vary from region to region and even stream to stream. The complexity grows when you start adding on other issues such as mitigating soil erosion and providing travel corridors for terrestrial animals.

The problems are not insoluble but they do undermine the value of general prescriptions and argue for case-by-case consideration. The sensitivity and value of riparian forests should make a basic guiding principle obvious: Treat riparian forests with moderation. With this principle in mind, one can consider the various management practices described throughout this book and use those with the least impact whenever managing riparian forests. For example, this would mean favoring small-scale management (Chapter 6), long rotations (Chapter 5), and a natural assemblage of tree species (Chapter 4).

The shortcoming of the intensive management approach is that it assumes (a) a substantial understanding of the riparian ecosystem being considered; (b) that there will be considerable dialogue and cooperation between wildlife managers and timber managers; and (c) logging operators will follow the management plan very carefully. These assumptions are quite reasonable for forests managed by an organization that has a multiple-use mandate and a large, diverse staff of managers and researchers such as the USDA Forest Service. In regions of the world where riparian ecosystems have not been studied, where forest managers cannot find a wildlife manager to consult, and where loggers operate with little or no supervision, it is probably preferable to adopt the simple no-cut policy. Indeed, many governments have decided to protect riparian zones by designating a buffer strip of a certain width within which timber harvest is restricted.

How Wide Should Buffer Strips Be?

Buffer strips ranging from 8 m to 400 m wide have been recommended (Brinson et al. 1981). (These are measured from the edge of the normal stream channel and refer to one bank only; both banks combined would be twice as wide.) Ideally, the optimum for any given stream would be determined by stream width, topography, soil type, hydrologic regime, climate, and, most importantly, the goals of the management policy (Budd et al. 1987). Quite narrow buffer zones might suffice if you were interested only in maintaining some features of the riparian environment. For example, researchers in Ontario determined that maximum water temperature in streams was the main factor limiting trout populations and estimated that a continuous strip of riparian forest just 10 m wide would provide adequate shading (Barton et al. 1985). However, if riparian forests are to be habitat for many terrestrial animals, they must be significantly wider. Gray and fox squirrels in eastern Texas were never seen in buffer strips less than 20 m wide and only rarely in strips less than 50 m wide; they were fairly common in buffers 73 m and 93 m wide (Dickson and Huntley 1985). Among 40 bird species inhabiting riparian forests in Iowa, nine species occurred only in riparian strips wider than 90 m (Stauffer and Best 1980). Buffer strips must also be fairly wide to have some stability (Steinblums et al. 1984); they are of little value if completely blown down, although some blow-down is natural and desirable, as discussed previously.

It is unfortunate that government regulations must be simple to be practical. For most purposes it would be preferable to define the outer edge of the buffer zone ecologically, and both vertical distance above water level and horizontal distance from the bank would be good correlates of this. For example, a buffer zone could be defined as 100 m wide or the extent of the 10-year floodplain (the area that is flooded, on average, once every 10 years), whichever is wider. In regions where there are many streams crossing the landscape and riparian forests are reasonably common, a narrower buffer zone could be designated for small streams; e.g., first- and second-order streams or those that drain a watershed of less than 100 km^2. Again, specific parameters need to be developed for different regions. A 100 m buffer may be fine for a mountain torrent in the Andes, but very inappropriate for a stream winding through a bottomland in Georgia.

Exactly what practices to allow in a buffer zone also merits considerable thought. A conservative approach would allow no cutting at all, but typically some harvesting, e.g., removal of 25% of the trees' basal area, is permitted. A two-tier approach that created a 10–25 m wide no-cutting zone and a wider zone where limited harvesting could take place might alleviate concern about allowing logging operations immediately next to streams. Ideally, roads would be excluded from riparian zones because they are often a major source of soil erosion; in some harvesting operations as much as 90% of the sediments have come from roads (Anderson et al. 1976). Recently, forest roads have been much improved in many areas, but they still represent a vulnerable point where a mistake in either construction or maintenance can have marked consequences (Hynson et al. 1982). Unfortunately, in mountainous areas the flattest terrain, and thus the easiest place to put a road, is often near a stream. Under such constraints it might be necessary to allow a road within 100 m of a stream, but there should always be at least a 25 m buffer.

The Watershed Perspective

Although streams are most sensitive to changes in their riparian zones, intensive forest management activities anywhere in their watersheds can have significant effects. For example, it is well known that clearcutting a watershed will increase stream flows, primarily because less precipitation will be returned to the atmosphere by evapotranspiration from the trees (Hibbert 1967, Anderson et al. 1976). In regions where water is a scarce commodity, resource managers depend on this relationship and use clearcutting of significant portions of watersheds to increase water yields.

There is danger that an increase in water flow may be expressed in more dramatic floods; the devastating consequences of deforestation in the Himalayas—massive flooding in the Gangetic Plains—is a widely known example. However, the research on this effect is a bit more equivocal. After aspen stands in Minnesota were clearcut, peak flows increased by as much as 250% (Verry et al. 1983), but other studies have found relatively minor effects on peak flows (Hewlett and Helvey 1970). It is likely that exacerbation of flooding is a problem particularly when the watershed's soil is extensively disturbed or snowmelt contributes a large part of the annual flow (Anderson et al. 1976).

One might reasonably speculate that extensive cutting in a watershed could have negative effects such as raising stream temperatures and increasing erosion and sedimentation problems. This may be true, but the few studies that can distinguish the effects of riparian zone cutting from watershed cutting suggest that maintaining adequate riparian buffer strips effectively avoids these problems (Rishel et al. 1982).

Riparian buffer strips probably ameliorate, but not eliminate, one effect of large-scale cutting in a watershed: loss of nitrogen (Martin and Pierce 1980). This has been a controversial matter ever since Gene Likens, Frank Bormann, and their colleagues (1970) experimentally denuded a watershed in New Hampshire by clearcutting and repeated herbicide applications, and then discovered nitrate nitrogen concentrations in the drainage stream were up to 56 times greater than the concentrations before treatment. Their study was widely cited as an indictment of clearcutting, while critics claimed that it was of limited relevance because the treatment was too severe and the soils atypical. In subsequent studies examining commercial clearcuts on a variety of soils, nitrogen losses have varied from virtually nil to 14 times those of precut conditions (Patric and Aubertin 1977, Martin and Pierce 1980). Increases in nitrogen can cause problems in aquatic ecosystems by temporarily making them excessively productive (i.e., causing eutrophication and related problems); there is also concern that soil fertility in the forest may be critically diminished.

The movement of chemicals from forest soils into streams could also affect one of the most fundamental qualities of a stream, its pH. There is some evidence that the establishment of conifer plantations in Great Britain may have exacerbated the problems associated with acid rain by increasing the deposition of atmospheric contaminants and contributing organic acids to water runoff (Harriman and Morrison 1982, Ormerod et al. 1987).

Multiple Use in Riparian Forests

This chapter has been limited to the role of timber management in riparian forests, but it is important to remember that timber management is just one of many activities that affect riparian forests (Brinson et al. 1981). Many, probably most, riparian forests have already been totally eliminated, replaced by agricultural and urban developments. Those that remain are threatened by dams, dikes, levees, irrigation canals, and other structures designed to make water flow faster or slower, or from places where it is not wanted to places where it is. Making stream channels wider, straighter, and deeper, better known as channelization, leaves no structures in its wake, but often results in the razing of riparian vegetation. It is reasonable to assume that water pollution harms riparian ecosystems, but this is not well-documented. Better known is the ability of riparian areas to cleanse contaminated water by retaining heavy metals and pesticides (Schlesinger 1979). Cattle and other livestock are attracted to riparian areas for the same lush forage that wild herbivores prefer, and thus overgrazing is a serious problem, especially in drier regions (Bryant 1985). Finally, the simple beauty of riparian areas make them among the first areas to experience excessive use by recreationists.

OF SALMON AND TROUT

From the shores of Korea to Monterrey Bay in California, the lands that ring the northern Pacific are home to one of the world's great biological dramas, the spawning runs of Pacific salmon. After years of foraging and growing large at sea, the salmon return to the fast-flowing streams and rivers where they were born, and there they reproduce and die, repeating a cycle that has evolved and continued over many millennia. These streams are cool and clear and rich with oxygen; they are often lined by striking forests of large spruce, fir, and hemlock trees. Inevitably, the quality of the riparian forests and the quality of the streams' water are closely related, and thus the salmon are dependent on the state of the forests that fringe their spawning streams. The fact that these forests, especially in British Columbia, Washington, and Oregon, are some of the most commercially valuable timberlands in the world has an upside and a downside from the salmon's perspective. The upside is that the value of these forests make them less likely to be converted into another type of ecosystem; the downside is that the incentives to harvest the forests intensively make the salmon vulnerable to certain logging practices. And the salmon have another "upside" going for them: They themselves are an extremely important economic, recreational, and ecological resource.

Given this situation and the unavoidable conflicts of interest, it is not surprising that there have been many studies on the effects of timber harvest on salmon and the other fish with which they share their streams, notably trout. Among these studies one stands out because of its intensity and duration; it is the Alsea Watershed Study conducted in the drainage basin of the Alsea River on the central coast of Oregon.[1] Beginning in 1959, over a dozen federal, state, and private organizations collaborated to monitor a wide array of biological, physical, and chemical parameters on three small tributaries of Drift Creek, itself a tributary of the Alsea River. After gathering seven years of baseline data, the watersheds of two of the streams were cut in 1966; the

---

[1] Over 60 papers and 20 theses have been published from this study; the information presented here was extracted principally from the project summary report, Moring and Lantz 1975.

Needle Branch watershed was almost completely clearcut and the remaining slash was burned; and the Deer Creek watershed was cut in three clearcut blocks, representing about a quarter of the area, with uncut buffer strips about 30 m wide left along both sides of the streams. One watershed, Flynn Creek, was left as an uncut control. After the 1966 manipulation, studies continued for another seven years.

Logging the watershed of Needle Branch had a profound effect on the stream. Maximum temperatures increased dramatically, by as much as 12.7°C, and did not fall to prelogging levels for seven years, by which time young vegetation was shading the creek again. Dissolved oxygen levels at the stream surface were very low during the summer of logging, but in subsequent years, after debris had been removed from the stream, they returned to normal. The dissolved oxygen levels in the interstitial water of the gravel beds remained low for much longer, in part because of sediments clogging the interstitial spaces. There was a 205.3% increase in suspended sediments in Needle Branch after logging roads were constructed. Finally, annual stream flows increased an average of 26.9%, and for six years after logging the loss of nitrate nitrogen increased by four-fold. In Deer Creek, the limited extent of the clearcuts and the presence of buffer strips seemed to prevent most impacts of logging. One notable exception came following road construction when the suspended sediments increased 53.5%; during the same period, unlogged Flynn Creek experienced only a 0.1% increase.

How did all these changes affect the fish populations? Two species, cutthroat trout and reticulate sculpin, experienced a direct negative effect in Needle Branch; their populations plummeted and, in the case of cutthroat trout, remain depressed for the duration of the study. By 1973, the cutthroat population was still only 21.2% of the average prelogging population, even though stream temperatures and dissolved oxygen levels had largely returned to normal. The negative effects on coho salmon were more subtle. Their population density seemed unaffected during the duration of the project, but careful study of several thousand fish revealed that juveniles in Needle Branch were relatively small. Moreover, when these salmon returned from the sea as adults they were still small, indicating that they were unable to compensate for their stunted growth as juveniles, and, more importantly, their fecundity was probably reduced.

Even in a comprehensive, relatively long-term study such as this one, it is difficult to interpret the results with absolute confidence. Would the results have been the same on three different streams? Would the cutthroat trout population eventually regain its former numbers? What would have happened if there had been a severe drought during one of the years? Despite these unknowns, a simple take-home message seems obvious: Unrestricted cutting near a stream has a significant negative effect on the stream ecosystem, but this effect can be largely mitigated by buffer strips.

SUMMARY

Riparian ecosystems are often the most valuable component of a forest landscape because they support a large diversity of plants and animals, some of which are found nowhere else, and because they are very productive. Just as importantly, they are essential to the well-being of streams, another type of ecosystem that contributes substantially to the overall biological diversity of a landscape. The simple rarity of riparian ecosystems is another index of their importance; they have never been very common, and over the centuries most have been destroyed by human activities. Management of riparian ecosystems can take two approaches: leaving a buffer strip along a stream in which no timber harvesting will take place, or carefully controlling the intensity of timber harvesting near streams to minimize its impact. A reasonable compromise would be to combine both approaches by identifying a narrow, no-cut strip along the stream, backed by a wider zone of low-intensity harvesting. Ideally, the actual widths of these strips and the nature of timber management to be allowed in the second zone should be determined on a case-by-case basis, but in most regions it will probably be necessary to set a policy on a larger scale.

Part III

The Micro Approach, Managing Forest Stands

Although it is essential to take a landscape approach if a region's entire biota is to be maintained, it would be short-sighted to see *only* the macro view. Imagine flying over a landscape on which a diverse array of stands representing different ages, species compositions, and sizes were arranged in a manner that balanced concerns about forest fragmentation and edge effects. It would present a very pleasing vista to anyone concerned about the fate of the organisms that live in these stands. But how would that same forest look from the ground? What if the forest canopy was so dense that no shrubs and herbs blanketed the forest floor? What if an efficient harvesting system had removed all trees before they died and thus the ground was devoid of any fallen logs or branches? What if overexploitation of certain species had selectively removed them from the scene? These critical elements of a diverse forest only become apparent when one takes a micro view.

The four chapters in Part III provide four micro views of the forest. Chapter 10 describes the miniature worlds, the microcosms, that revolve around dying, dead, and down trees and then outlines management procedures that can assure their continuation, even in heavily managed forests. Chapter 11 returns to the theme of Chapter 6, spatial heterogeneity, but now the plane has turned from horizontal to vertical and the scale has shrunk from hundreds of hectares to the height of a forest. Here the critical issue is the development of vegetation strata and their associated biota. Chapter 12 reverses the perspective of the rest of the book by first reviewing the array of silvicultural techniques that foresters can use to protect, artificially regenerate, and treat stands, and then cataloguing some of their impacts—both positive and negative—on wildlife. Finally, Chapter 13 moves from the broad goal of maintaining biological diversity to the narrow one of maintaining individual species because of their special values, by examining the interface of these two goals.

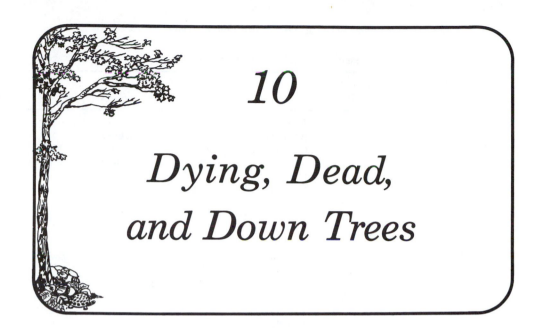

10

Dying, Dead, and Down Trees

To many people there is nothing quite so useless as a dead tree, except possibly a fifth wheel on a cart, and when people are called "deadwood" it is not because they are mature and respected members of an organization. It is hardly surprising that people find dead things unattractive. It is even less surprising that most foresters would perceive a dead tree as an affront to their efforts because, if they have done their job well, trees are harvested before they die and go to waste. No one denies truisms such as "death is part of life"; it is simply a choice between a quick, meaningful death effected by a machine versus a slow, rotting death brought on by disease and other debilitating agents. Unfortunately, there is a basic flaw in this perspective: Dying, dead, and down trees are important components of forest ecosystems and cannot be excised without incurring many negative consequences.

Simply noting the abundance of dead and dying wood in a mature natural forest would lead one to assume that it is important as one of the ecosystem's major reservoirs of organic matter and nutrients. Dead, standing trees, or *snags,* commonly represent 5-10% of the trees in a forest (Table 10.1) and fallen logs alone can amount to an enormous mass. For example, in an old Douglas-fir stand, up to 581 metric tons per hectare (518,000 pounds per acre) of logs can cover the forest floor (Grier and Logan 1977). An extraordinary variety of organisms are associated with dead trees, ranging from conspicuous, well-known creatures such as squirrels and woodpeckers to huge numbers of invertebrates, fungi, lichens, mosses, vascular plants, and microorganisms that exploit this pool of organic matter and nutrients. Focusing just on British animals, Charles Elton (1966) estimated that nearly a thousand species (perhaps one-fifth of Britain's entire forest fauna) rely on dead and dying wood for food

TABLE 10.1   Density of Snags in Different Forest Types

Snags per Hectare	Percent of All Trees	Location	Stand Age	Forest Type	Source
43.1	4.5	Kentucky	100	oak-hickory-maple (U)[1]	5
83.8	8.1	Kentucky	35	oak-hickory-maple (M)[2]	5
39.2	13.7	Connecticut	85	maple-birch-hemlock (U)	6
14.8	8.9	Connecticut	60	maple-birch-hemlock (M)	6
0.5	—	Oregon	10	Douglas-fir (M)	1
18.0	—	Oregon	200	Douglas-fir (U)	1
15.6	7.6	Massachusetts	many	oak-maple (M)	2
19.6	5.7	Massachusetts	many	oak-maple (U)	2
0.7	2.8	Colorado	mature	plains cottonwood	9
26.4	7	California	many	pine-fir	8
61.1	9.8	Arizona	?	mixed conifer	3
4.2	3.0	Florida	all	pine	4
6.8	3.8	Florida	all	pine-hardwoods	4
10.5	3.5	Florida	all	hardwoods	4
2.6	?	Florida	0-30	pine	4
7.2	?	Florida	31-60	pine	4
10.7	?	Florida	61 +	pine	4
7.8	12.7	Maine	all	coniferous trees	7
3.5	9.5	Maine	all	deciduous trees	7

Sources and minimum dbh (diameter at breast height) and heights used to define a snag:

1. Chadwick et al., 1986—15.2 cm/?m
2. Cline et al., 1980—9 cm/4.4 m
3. Franzreb, 1978—7.6 cm/?m
4. McComb et al., 1986a—12.7 cm/1.4 m
5. McComb and Muller, 1983—10 cm/1.8 m
6. McComb and Noble, 1980—10 cm/1.8 m
7. Powell and Dickson, 1984—12.7 cm/1.4 m
8. Raphael and White, 1984—13 cm/1.5 m
9. Sedgwick and Knopf, 1986—?

[1]U = Unmanaged forest
[2]M = Managed forest

or habitat. To understand how all these organisms interact in this special arena, we will follow the process by which a recently fallen tree gradually returns to the soil.

THE DECAY OF TREES

Imagine a recently fallen tree that was alive and vigorous until it was toppled by high winds, for example.[1] (See Figure 10.1, Table 10.2.) At first the tree will

[1]Information for this section comes from four syntheses: Savely 1939, Elton 1966, Maser et al. 1979, and especially Maser and Trappe 1984.

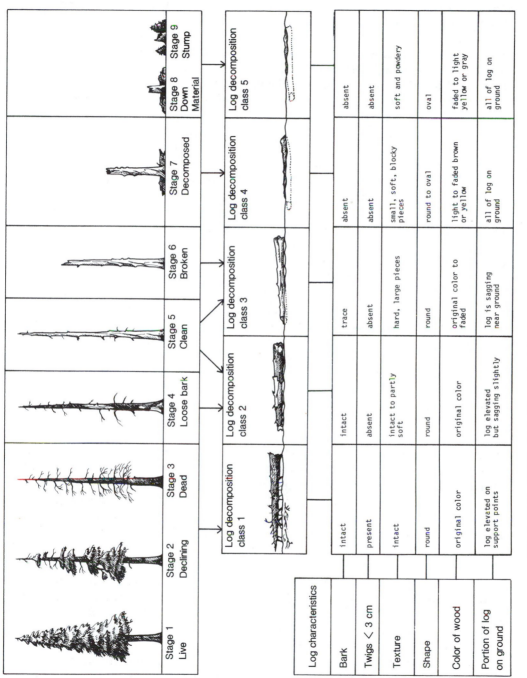

Log characteristics	Log decomposition class 1	Log decomposition class 2	Log decomposition class 3	Log decomposition class 4	Log decomposition class 5
Bark	intact	intact	trace	absent	absent
Twigs < 3 cm	present	absent	absent	absent	absent
Texture	intact	intact to partly soft	hard, large pieces	small, soft, blocky pieces	soft and powdery
Shape	round	round	round	round to oval	oval
Color of wood	original color	original color	original color to faded	light to faded brown or yellow	faded to light yellow or gray
Portion of log on ground	log elevated on support points	log elevated but sagging slightly	log is sagging near ground	all of log on ground	all of log on ground

Figure 10.1 and Table 10.2 The gradual decay of snags and logs can readily be seen in many external changes. Note that when a well-decayed snag falls, it enters an advanced stage of log decomposition. (This figure and table are based on a description of the decay of Douglas-fir snags and logs as adapted by Maser et al. 1979, and Thomas et al. 1979c, from Fogel et al. 1973.)

159

retain much of its original form; the bark remains intact, the wood sound, and branch stubs keep it propped above the ground. In this elevated state it may serve as a platform where male grouse can display their sexuality and squirrels can open pine cones while scanning the forest. It also provides a shelter beneath which small animals can avoid sun, rain, and predators. Furrows in the bark will fill with debris making small beds in which seeds can germinate. More important than everything happening outside the log are the hidden changes within. In the vanguard of these are insects that can penetrate the outer bark to reach the nutritious tissues that lie just beyond. Bark beetles and woodborers work their way through the inner bark (phloem) and the layer of growth cells (cambium), thriving on the protein-rich material and leaving tunnels and deposits of feces in their wake. With the tree's outer defenses breached, fungi and bacteria soon invade—indeed, many spores are carried in by the insects— and flourish in the environment created by the beetles. As the most nutritious material is digested and decomposed, species of fungi that are better able to decompose the less palatable sapwood—where cellulose and lignin predominate—begin to replace the initial invaders. They are also joined by animals capable of subsisting on such a diet; termites, for example.

After the inner bark and outer sapwood have decomposed, the outer bark is just a fragile, clinging shell that eventually sloughs off, exposing a dark, humid world to the desiccating sun. Many species will be lost in this change, but in time the new surface is covered with vegetation—algae, lichens, mosses, liverworts, a few seedlings—and inhabited by a host of invertebrates small enough to view this thin veneer as habitat. New decomposers arrive too, continuing the decay process through the sapwood and into the heartwood. Whether it is the boring of an insect or the inexorable growth of a seedling's root, the effect is to fracture the wood, peeling off layer after layer of sapwood, splitting what remains with a network of fissures. Mites, slugs, nematode worms, pseudoscorpions, collembolans, millipedes, and centipedes can make their way along these cracks, while larger animals—salamanders, mice, and shrews—find burrowing through the crumbly mass a simple task. All this fragmentation creates more and more internal surfaces, substrates for fungi and bacteria, and makes it easier for seedlings growing on the log's surface to tap the nutrients and moisture within. Even the layers of bark and wood that drop off a log are not really lost; they accumulate in a loose pile that also hosts many of the log's biota.

Eventually the log is detectable only as an elongated mound on the forest floor, a mass of crumbly chunks of heartwood covered by litter and forest floor plants. In time it will merge into the humus of the forest floor, its former existence perhaps revealed by a line of young trees that got their start on the log.

It should be stressed that this description is a generalized one; for example, termites and salamanders are not omnipresent in all forests. Even within a single forest there are differences in the flora and fauna associated with logs of different tree species. This is true especially during the earliest stages; as the logs come to resemble piles of organic matter, rather than dead pines or oaks or beeches, their biotas become more similar. Despite some variations, the overall patterns of change are quite consistent. They are analogous to classic models of large-scale ecosystem succession in that the activities of one group of organisms changes the environment in a fashion that leads to the establishment of another set of organisms.

How would this scenario differ if a tree died while still standing and thus became a snag? The critical events are very similar—the invasion of bark beetles, fungi, and bacteria followed by colonization of other organisms, sloughing of the bark, and so on. Typically, the snag becomes shorter and shorter until, at some point, it tumbles and joins the logs on the ground (Figure 10.1). There are some differences. In most places, a standing dead tree will decompose more slowly than a log because, raised above the soil and into the wind, it will be drier than a log and thus a less favorable habitat for many decomposers, particularly fungi. Conversely, in some moist environments, a log will decompose more slowly than a snag because it is so wet that oxygen is limited. External developments will differ more conspicuously; for example, seedlings are less likely to grow on a snag and woodpeckers are far more likely to perforate a snag's bark in search of bark beetles.

Many trees begin to rot from within long before they are dead. Whenever a tree's bark barrier is bridged—a branch is torn off in a windstorm, a skidder scrapes it—there is a good possibility that fungi will invade and initiate decay in the heartwood. Moreover, injuries and general stress such as drought can cause trees to release volatile compounds; these may be smelled by bark-boring beetles which are attracted to stressed trees within 24 hours, thereby accelerating the decay process (Dunn et al. 1986). Eventually the decay may proceed so far that the tree is hollowed out from within; trees in which decay cavities are formed—they are often deciduous species—are sometimes called *den trees* (McComb et al. 1986b). Alternatively, cavities are often created by woodpeckers. It is thought that drumming woodpeckers may hear the special resonance of a tree with a rotten core and drill a hole through the living wood to reach soft wood where they can easily excavate. Certainly, it is quite well-documented that most woodpeckers need to have their work eased by fungi (Conner et al. 1976, Conner and Locke 1982). Whether a cavity forms in a live tree or a snag, whether it is quickly excavated by an animal or formed by inexorable decay, any mass of wood with a sizable hole in it is likely to have special value for wildlife.

THE VALUE OF DEAD AND DOWN TREES

Tree cavities are very important to many animals. They can provide a place to sleep, rest, and rear young that is secure, dry, and thermally buffered (e.g., warm in winter, cool in summer), and thus a striking variety of animals compete to use them. Hundreds of bird species from Carolina chickadees to California condors, from aracaris to barbets to cockatoos, have been known to nest in cavities. Even birds associated with other ecosystems, such as wood ducks, common goldeneyes, Australian shelducks, and smew, venture into forests in search of trees with cavities. The world's 200 odd species of woodpeckers are the best-known cavity-using birds, in large part because they excavate their own cavities. This feat is imitated by only a handful of other species; for example, some kinds of chickadees and nuthatches pluck out a cavity in very soft wood. (Animals that excavate their own cavities are called *primary excavators;* those that use decay cavities or ones abandoned by primary excavators are called *secondary cavity-users.)* Although birds dominate the list of cavity-users, they are not alone. A variety of mammals, ranging in size from bats and shrews to bears, and some amphibians and reptiles, regularly use cavities too. In fact, if we consider cavities in both trees and logs, it is likely that most species of forest-dwelling mammals, reptiles, and amphibians seek shelter in cavities at least occasionally. Spiders, slugs, wasps, and many other invertebrates often occupy cavities too, as does perhaps the most popular species of invertebrate, the honeybee (McComb and Noble 1982).

To demonstrate the importance of snags in a scientifically rigorous fashion, one should census the biota of some study areas; remove all the snags from a subset of the areas; and then recensus the biota. This has been done in two bird studies and, as might be expected, a dramatic decrease in the number of cavity-nesting birds was documented (Scott 1979, Raphael and White 1984). This is particularly important because cavity-nesting species often comprise 20–40% of the birds in a forest and sometimes as much as 66% (Scott et al. 1980).

The value of cavities is also revealed by demand. In some Netherland forests 54–93% of all the cavities were occupied (Van Balen et al. 1982); cavities in an Australian eucalypt forest were 47% occupied, primarily by eight species of parrot (Saunders et al. 1982); and, in South Carolina, southern flying squirrels alone used 36% of the cavities (Carmichael and Guynn 1983). Adding an abundance of artificial cavities—nest boxes—increased the population of pied flycatchers in a Scandinavian forest 16-fold (Lennerstedt 1983). Anecdotes about competition for cavities convey the same message (Steinhart 1981). A Nuttall's woodpecker was watched as it defended its nest cavity against a series of potential usurpers: downy woodpeckers, violet-green swallows, western bluebirds, ashy-throated flycatchers, and house wrens. A single cavity was once found to contain eggs of three species: hooded merganser, common goldeneye, and barred owl. These stories are vaguely amusing, but the consequences of competition may be rather dire. When Puerto Rican parrots were on the brink of extinction (at one point only 12 individuals were known to survive in the wild), competition for nest cavities with pearly-eyed thrashers was the major problem they faced (Horn 1985).

Whenever wildlife biologists think of dead trees, their minds turn first to

wildlife that use cavities, but this is only one part of the story. It is very important not to forget all the thousands of small life forms that participate in the
decomposition of a tree. Legend has it that the distinguished British biologist
J. B. S. Haldane once found himself in the company of a group of theologians
and was asked what one could conclude about the Creator from a study of
his creations; Haldane replied, "He has an inordinate fondness for beetles"
(Hutchinson 1959). The diversity of beetles is truly remarkable, and just the
small subset that makes its living on dead and dying wood outnumbers *all* the
world's species of mammals, birds, reptiles, and amphibians at least two to
one (Parker 1982). And, of course, beetles are only one of the many groups that
flourish in dead wood.

In truth, you do not have to be inordinately fond of millipedes, mites,
mushrooms, and other minute creatures to appreciate the value of dead trees.
Besides the cavity-users, there are raptors that use snags as perches, bats that
roost under flakes of bark, and hundreds of other species that use snags and
logs for a wealth of purposes (Thomas et al. 1979c, Glinski et al. 1983). In the
Blue Mountain forests of eastern Oregon and Washington, almost half the vertebrate species (179 of 378) make some use of logs (Maser et al. 1979), and in
forests west of the Cascade Crest the figure is 150 species (Bartels et al. 1985).
Although it is not known to what extent dead wood is critical to the survival
of these animals, it is reasonable to assume that it contributes significantly to
the structural diversity of the environment and thus to the biotic diversity of
the ecosystem.

One might also argue that the primary importance of logs lies beyond the
welfare of individual species and involves a major ecosystem process, nutrient
recycling, on which all species ultimately depend. The availability of nutrients,
particularly of nitrogen, is potentially a limiting factor in many forests. Thus,
it is a matter of some concern if the nutrients lost through erosion, leaching,
and removal of organic matter are not replaced by nutrients made available
by the weathering of rocks, the deposition of atmospheric dust, or, in the case
of nitrogen, by being extracted from atmospheric gases by nitrogen-fixing bacteria (Bormann and Likens 1979). One part of this equation that is readily
influenced by the timber industry is the removal of organic matter, i.e., the
harvesting of trees. Consequently, people concerned about the fertility of forest soil strongly advocate leaving branches, twigs, and leaves—slash—on a site
after timber harvesting. They usually do not question the removal of the tree
boles because, of course, the boles are the primary commercial product, and
compared to leaves and twigs, boles have relatively low concentrations of nutrients (Wenger 1984). Nevertheless, logs are special, because long after the
leaf and twig litter has decomposed, a log will still be there, providing a moist,
fertile seed bed. The logs are not just a large mass of damp organic matter;
they can act as a sponge accumulating nutrients from a variety of sources. The
rain that falls on a log will bring with it nutrients washed from the leaves in
the canopy overhead or leached from leaf litter covering the log. Fungi may

enrich the wood they cover by immobilizing nutrients, and nitrogen-fixing bacteria may do the same (Maser and Trappe 1984).

In some forests, for example, old-growth Douglas-fir stands, logs are major sites for tree regeneration, and a line of small trees growing on a well-rotted log—a "nurse log"—is a common sight. These nurse logs are evidence of a fascinating relationship among certain trees, fungi, and small mammals that Chris Maser, James Trappe, and others have described (Maser et al. 1978*a*, 1978*b*). One of the reasons seeds find logs a favorable place to germinate is because these are good sites for forming a symbiotic relationship with particular fungi, a *mycorrhizal* relationship, from Greek words meaning "fungus-root" (Harley and Smith 1983). In a mycorrhizal relationship, part of a fungus may penetrate the tiny rootlets of a host plant and facilitate the plant's absorption of nutrients. In turn, the plant helps nurture the fungus with energy from carbohydrates formed during photosynthesis. This is often an obligate relationship; neither party can thrive without the other, and thus it is critical that they find one another. For fungi that produce above-ground fruiting bodies (e.g., mushrooms) this is relatively straightforward, because enormous quantities of spores are produced and scattered by the wind. The chances of some of these spores finding a host plant, or conversely a host plant finding some of these spores, are reasonably good. In contrast, some fungi produce underground fruiting bodies (e.g., truffles); for these species, small mammals such as voles and squirrels may act as dispersal agents by consuming the fruiting bodies and depositing the ingested, but still viable, spores elsewhere. This could be important to forest regeneration on a cleared site. For example, Townsend chipmunks regularly venture into forest openings from the surrounding cover and undoubtedly leave behind spore-bearing feces from which new mycorrhizal partnerships can develop. Logs are the center of all this activity; they are important as cover for the chipmunks and as substrates on which the fungi can grow.

Logs are not the only kind of dead and down woody material; stumps, limbs, and slash also function as sites for decomposition and thus are habitat for many of the small organisms involved in the breakdown of wood. In addition, they can provide needed cover for larger animals; wildlife managers have long encouraged quail and rabbit hunters to make brush piles as a means of enhancing their sport. Occasionally, the slash left after a clearcut can be so deep as to impede the movement of larger animals, deer and the like, but, on the whole, residual woody material is a boon to most wildlife.

To what extent can slash and other small pieces of wood provide the same values as logs? Beetles and centipedes may be able to use a wrist-thick piece of wood, but it would not be large enough to accommodate mice, salamanders, and other large species. Perhaps the most important shortcoming of small pieces of wood is simply that they disappear quickly.

This line of reasoning has led Chris Maser to some rather creative thinking about the role of large logs in the Douglas-fir stands that he knows best

(Maser pers. comm.). He believes that large logs may be so important in nutrient cycling that some very long cutting cycles should be instituted to assure their presence. He notes that only a tree that has grown slowly for 400 years will be large enough, and in particular will have enough heartwood, to last another 200 to 250 years as a log on the forest floor. Does this mean that Douglas-fir should only be grown on a 400-year cycle so that after each harvest some logs can be left to last through the next cycle? Not necessarily. Maser thinks it would probably be adequate to follow a few short cycles, say two or three 120-year cycles, with a 400-year-long cycle that would provide logs for the next two cycles (Figure 10.2). These ideas are inevitably rather speculative; we do not understand forest nutrient dynamics in a time frame of decades, let alone centuries. However, it may be wise to consider the possibility that long-term nutrient depletion in forests will become a problem. To date, depletion of forest nutrients has not risen as a major concern but, on the other hand, there are very few forests, even in Europe, that have been heavily cut more than twice or thrice. Few farmers would think about removing crop after crop from

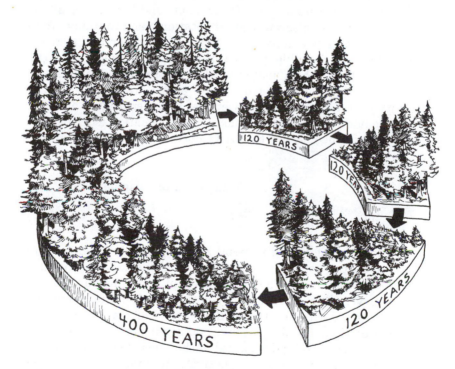

Figure 10.2 After two or three cycles of cutting a Douglas-fir stand at 120-year intervals, a long rotation of 400 years can be included. During this period, large trees can be grown that will form large, persistent logs. (This idea is based on a conversation with Chris Maser and is used with his permission.)

the land without replenishing the nutrients, and after a few more forest crop removals, problems may become evident.

To summarize, there are many specific values associated with dead and down trees: providing a stage for an amorous grouse, a refuge for a squirrel, and so on. Furthermore, it does not take much imagination to see dead trees as miniature ecosystems, systems with a unique flora and fauna interacting in some fascinating processes. Finally, they may well have a critical role in the larger system—the forest—because of their importance to nutrient cycling.

MANAGING DEAD WOOD

Snags and logs may be critical components of a forest ecosystem, but just how many are needed? And what kinds? We could turn to some of the natural, unmanaged forests described in Table 10.1 as models to emulate, but this approach is patently impractical. In an unmanaged stand every tree eventually becomes a log, and most are snags for a while before falling; using natural forests as a model would leave no trees for harvesting. In short, timber extraction means making do with fewer snags and logs, especially when timber extraction involves clearcuts, short rotations, and thinning. To minimize the impact that this will have on wildlife, it is important that during a harvest operation a reasonable number of appropriate snags and logs be left behind. We will address the question of what is a reasonable number by focusing on a more specific question: How many snags are needed to meet the needs of cavity-users? However, first we must consider what is an appropriate snag. A snag is not a snag is not a snag. Its size, location, species, and state of decay will all influence its use by cavity-users.

What Kind of Snags?

Size

The size of a snag is manifestly important; a slender birch pole is a fine site for a titmouse but could never hold a pileated woodpecker, to say nothing of a bear or a bobcat. The converse is not true; small animals are quite capable of using very large snags. Cavity-nesting birds in the Sierra Nevada mountains selected trees with an average diameter of 62 cm, compared to an average diameter of 32 cm for all trees (live and dead), and even tiny birds such as mountain chickadees and red-breasted nuthatches selected trees with average diameters of 56 cm and 70 cm, respectively (Raphael and White 1984). Diameter is probably the main consideration, but height is important too (Table 10.3); (the fact that tall trees tend to be thick trees makes it hard to separate these factors.) Some animals will use a cavity that is practically on the ground, but

TABLE 10.3   The Number of Cavity-using Species in the Blue Mountains of Eastern Oregon and Washington Having Various Minimum Size Requirements for Snags. (Adapted from Thomas et al. 1979c)*

Minimum Nesting Height	Minimum Diameter at Breast Height					
	4 inches 6		10	12	15	20
≥ 6 ft	3	—	8	9	3	7
≥ 15	—	1	7	13	3	1
≥ 30	—	—	—	3	—	4

*Generally, animals can readily use snags much larger than their minimum requirements. For example, although only four species—barred owls, Vaux's swifts, pileated woodpeckers, and fishers—must have snags that are at least 30 feet tall and 20 inches in diameter, all 62 species can probably use snags of this size.

most prefer to be fairly high, probably because they are more secure from ground predators. This said, under some circumstances it may make sense to leave tall stumps (e.g., 2–4 m) after cutting, for example, in a clearcut where full height snags would interfere with aerial access for silvicultural treatments. Perhaps the best work on the morphology of decayed trees and its relevance to animals was undertaken by some Australian researchers who carefully dissected 15 blackbutt eucalyptus trees (Mackowski 1984). The upshot of that study and virtually all others is simply that large snags are more important to keep than small ones. Not only can a greater variety of wildlife use a tall, girthy snag, but large snags are likely to last longer than small snags (Cline et al. 1980).

State of decay

Woodpeckers are known for their chisel-like bill, thick skull, and tough neck, but not all primary excavators are as well-equipped to excavate a cavity in a hard snag, and even many woodpeckers prefer to nest in a well-rotted tree. Consequently, it is important to have both soft and hard snags. Having snags in various stages of decay plus dying trees—perhaps den trees—is also in inevitable by-product of assuring a continuous supply of snags over time; this is a critical issue we will return to later.

Species

A review of tree species selection by cavity-users often reveals some definite patterns of preference and avoidance, but they are by no means absolute (Raphael and White 1984). Northern flickers preferred aspen in Ontario, Douglas-fir in British Columbia, western larch in Montana, and white fire in California (Raphael and White 1984). The simple, conservative approach would be to let snag species composition reflect the species composition of the live trees. Of course, there would be an intrinsic bias in favor of species that remain upright longer; e.g., birch and aspen trees may dominate a stand when alive, but if most of them fall down within five years of death, then another species, red spruce perhaps, could dominate the snags. It may be desirable to accentuate this tendency, to favor slowly decaying species, because they will provide wildlife habitat, both as a snag and a log, for a relatively long time. Alternatively, it is a fairly simple process to inspect a sample of trees and determine which have been used by primary excavators or which have developed decay cavities, and then favor these species in leaving snags. On the whole, deciduous species are more likely to develop decay cavities and become den trees than conifers.

Location

Several considerations bear on the location of snags. First, snags will be used in virtually any environment. Some people assume that there is no point leaving a snag in the midst of a clearcut, but this is not true; open country creatures such as kestrels and bluebirds often occupy isolated snags (Dickson et al. 1983). Another reason for having snags spread across the landscape is the territorial nature of many species. For example, a 30 hectare stand could support 10 pairs of downy woodpeckers if it had an abundance of snags scattered about, but if all the snags were crowded into one corner of the stand, two or three pairs might successfully exclude all other downy woodpeckers. On the other hand, some clumping of snags may be a practical necessity. For example, if tall snags cannot be left in a clearcut because they would hinder aerial access for silvicultural treatments, then it might be necessary to concentrate them in residual borders around the cut or in strategically located clumps.

Having many clusters of snags distributed through the forest may be an ideal situation for some cavity-users. For example, snag clusters are often preferred sites for nesting and may facilitate efficient foraging by woodpeckers and other snag foragers (Raphael and White 1984). Furthermore, clumps of snags can be important if the overall number of snags is low; i.e., if a few snags are scattered across the landscape, there may not be enough within the home ranges of individual cavity-users. In this situation it is definitely preferable that the snags be clumped so that at least some parts of the forest are suitable habitat for cavity-users (Neitro et al. 1985).

How Many Snags?

Deciding what general kinds of snags to have is much easier than deciding how many. Jack Ward Thomas and his coworkers (1979) have proposed a predictive model for snag requirements based on woodpecker demographics and habitat use that has received extensive attention from other researchers; no alternative models have been proposed, just reformulations of the original. The box gives a fairly detailed description of this model along with footnotes that describe some of the variations proposed by others. Here we will review the basis of the model and discuss two of its limitations.

For each species of woodpecker, the model estimates how many snags (above a certain minimum diameter) are required per 100 acres. This is done by multiplying three numbers: (a) the maximum number of pairs that could

A MODEL FOR ESTIMATING SNAG REQUIREMENTS

TABLE 10.4a   Snag Requirements for Some Woodpecker Species of the Blue Mountains of Eastern Oregon and Washington. (Adapted from Thomas et al. 1979c)

	Hairy Woodpecker	Northern Flicker	Blackbacked Woodpecker	Williamson's Sapsucker	Pileated Woodpecker
Approximate territory size	25 ac.	40	75	10	300
Maximum number of pairs per 100 acres[1]	4	2.5	1.3	10	0.3
Number of cavities per pair per year[2]	3	1	3	1	3
Minimum snag size[3]	10 in.	12	12	12	20
Number of snags needed per 100 acres[4]	160	38	59	150	14

[1]Thomas et al. derived these density estimates by dividing 100 acres by what they felt to be a reasonable territory size based on literature and some observations; others have used minimum territory size to estimate maximum population density. Raphael and White (1984) argue that using minima will greatly overestimate actual densities; they advocate measuring densities directly.

[2]Many woodpecker species excavate more than one cavity each year to use for nesting and roosting (Evans and Conner 1979).

[3]This table emphasizes minimum acceptable sizes, but there are many reasons to target the average size of snags used by various species, not the minimum (Conner 1979). A cavity that is squeezed into a snag just big enough to hold it may be crowded, fragile, and poorly insulated.

[4]$S = (N)(C)(16 - 1)$ where S is the snags required per 100 acres, N is the maximum number of nesting pairs per 100 acres, C is the number of snags excavated per year, and $(16 - 1)$ reflects the ratio of unacceptable to acceptable snags. Literature estimates indicate that for every snag used there are from 0.3 to 55 snags that are not used (Neitro et al. 1985). Presumably many unused snags are not suitable for excavation; however, in some forests some snags probably remain unused because there are more than the cavity-users need. Most workers (Harlow and Guynn 1983 are an exception) have felt it essential to include a reserve of 3 to 16 snags (e.g. Evans and Conner 1979, Thomas et al. 1979c, Raphael and White 1984, Neitro et al. 1985). Models based on a reserve of only three snags assume that the snags are suitable and are predicated on careful reseach to determine suitability (Raphael and White 1984).

TABLE 10.4b   **Number of Snags of Different Sizes Required per 100 Acres by All the Woodpecker Species in White Fir and Western Juniper Stands**

	10 Inch	12	20
White fir[1]			
Number of snags for 100% of the maximum population[2]	30	136	14
Number of snags for 50% of the maximum population[3]	15	68	7
Western juniper			
Number of snags for 100% of the maximum population	—	38	—
Number of snags for 50% of the maximum population	—	19	—

[1]White fir stands are inhabited by hairy woodpeckers, northern flickers, black-backed woodpeckers, Williamson's sapsuckers, and pileated woodpeckers; the western juniper stands have only northern flickers.

[2]These estimates assume that each species needs are complementary, not supplementary or additive. For example, the 180 ten-inch snags required by hairy woodpeckers in a white fir stand could, in part, be provided by the 14 twenty-inch snags required by pileated woodpeckers and the 136 twelve-inch snags required by three other species, leaving a need for only 30 ten-inch snags (180 − 136 − 14 = 30). Neitro et al. (1985) assumed that individual species' needs would be additive.

[3]Thomas et al. assumed that half as many woodpeckers would need half as many snags; in other words, that there was a linear relationship between snag and woodpecker density. Raphael and White (1984) have demonstrated that the relationship is curvilinear, but for management purposes they advocate assuming that it is linear (Van Horne 1983).

occupy 100 acres; (b) the average number of snags used annually by each pair; and (c) the number of snags that are not used for each one that is.

These species-by-species estimates are combined for each forest type to estimate how many snags, of what sizes, would be required to sustain maximum populations of all the woodpecker species living in each forest type. For example, in western juniper forests only 38 twelve-inch snags per 100 acres would be required because only northern flickers nest in this forest type. However, 30 ten-inch, 136 twelve-inch, and 14 twenty-inch snags would be needed in white fir forests that are inhabited by five woodpecker species. By interpolation, the model can also estimate how many snags are required to maintain some portion of the maximum population: 90%, 80%, 70%, and so on. These proportional figures highlight the importance of a simple, but easily overlooked, mathematical relationship. If a manager chooses to implement a snag retention program on just half of a forest, and the goal is maintaining 50% of the maximum woodpecker density, then the woodpecker population of the whole forest will only be 25% (0.5 × 0.5) of the maximum.

How broadly can this model be applied? First we must question the underlying premise that providing enough snags for woodpeckers will also sustain the needs of other cavity-users. This would be a reasonable assumption if woodpecker excavation were the only means of cavity formation; then it would be as logical as assuming that if you want houses you should be sure there is an adequate number of carpenters. However, woodpeckers are not the whole story; decay cavities may be quite important too, especially for cavity-users that are larger than woodpeckers, e.g., most ducks, raccoons, bobcats, etc. Decay cavities may also be heavily used in places where woodpeckers are uncommon or absent. To take an extreme case, there are hundreds of cavity-nesting species in Australia, New Guinea, and New Zealand, but *no* woodpeckers. Some kingfishers and parrots can excavate a cavity, but they do not begin to fill the woodpecker niche. It also seems unlikely that the 38 twelve-inch snags per 100 acres required by northern flickers in Thomas et al.'s western juniper stands would meet the requirements of the 11 birds and 10 mammals that also use cavities in these stands. There may also be a constraint in applying the model in mature deciduous forests, because dead limbs borne by live trees, not snags, may be a primary location for cavities in these forests (Sedgwick and Knopf 1986, McPeek et al. 1987).

The second limitation of the Thomas et al. model is a practical one. For most forest ecosystems the current understanding of woodpecker biology is too limited to employ the model, and in many parts of the world financial constraints would make it difficult to conduct the requisite research. In the absence of the necessary information, many forest and wildlife managers have been willing to take more of a "seat-of-the-pants" approach and suggest that five to ten snags per hectare (2–4 per acre) will probably be adequate. Within the United States, biologists studying forest types from nearly every region of the country have arrived at recommendations for snag densities that are remarkably consistent (e.g., Scott 1978, Evans and Conner 1979, Thomas et al. 1979*c*, Harlow and Guynn 1983, Raphael and White 1984, Zarnowitz and Manuwal 1985, McComb et al. 1986*a*). Furthermore, in at least one context, U.S. National Forests in the Pacific Northwest, forest managers are following the biologists' advice (Bull et al. 1986). It is not certain to what extent this concordance represents independent arrivals at an ecological "truth," especially since it is all based on North American data, but until better models are derived, 5–10 large snags per hectare seems like a reasonable target. Using this quota as a rule of thumb may be rather simple and unsophisticated, but it is preferable to deciding that the model is too complex and ending up with no snags at all.

Long-term Snag Management

To understand the long-term nature of snag availability, it is useful to borrow some terms from population dynamics and to ask: Is the "recruitment rate" for snags (i.e., the rate at which snags are formed by trees dying) sufficient to balance the "mortality rate" (i.e., the rate at which snags fall over) and thus maintain the "snag population"? Obviously, it is important that the answer be affirmative. The benefits of leaving an adequate supply of snags after a harvesting operation will be quite short-lived if all the snags fall down within ten years and all the living trees are many years away from being old enough to be susceptible to decay fungi and large enough to be a valuable snag. How can we plan for future generations of snags?

Trees that will comprise a second generation of snags can be identified and set aside to await death. Because one can assume that a portion of them will fall before they become snags, it is advisable to leave more than will ultimately be needed as snags. The same snag selection factors of size, species, and location apply, but there are two additional considerations. Firstly, trees with poor form, and hence of low commercial value, can be selected to minimize economic loss. Similarly, it is sensible to choose trees that already are infected with heart rot because they will be of low value and are in the process of becoming snags; indeed, they may already have cavities. Infected trees can be identified by the presence of conks of heart rot fungi; wounds such as broken branch stubs and fire scars; dead portions of the crown; and woodpecker holes (Conner 1978). Unfortunately, retaining poorform, infected trees is unwise if one is trying to improve the genetic quality of tree populations.

Planning for a third and fourth generation of snags requires understanding patterns of snag recruitment and falling and how they change as a stand ages, and then carefully considering long-term approaches to rotation ages and harvesting techniques (Neitro et al. 1985).[2] Trees are always dying, especially when a stand is young and rapidly growing. At this stage, many trees are outcompeted by their neighbors and die because of this suppression. Unfortunately, these snags are too small to be of great value to wildlife. It is only in an older stand that snags of a variety of sizes and stages of decay can be found, and thus the simplest solution to the need for snags is to have long rotations in which trees can grow big and old. However, because under current economic constraints it is often not economically feasible to have large areas of commercial forest land managed in long rotations, other solutions need to be explored.

In stands that are being managed for small-scale heterogeneity through selection cutting, future snags can be reserved simply by identifying and marking individual trees that will not be cut. In stands being managed on a short-rotation, even-aged basis, the problem is more complex, but it is not impossi-

---

[2]Neitro et al. (1985) have described a detailed method for predicting snag abundance over extended periods and suggested long-term management strategies; only the highlights of their ideas are presented here.

ble. These stands need not be devoid of cavity-users because many species will, under the right circumstances, nest in one stand and forage in a nearby one. This means that stands could be arranged so that those that do not have an adequate supply of snags are intermingled with those that do. Stands that could act as snag reservoirs might be uneven-aged, selection-cut stands such as riparian strips. If these are not available, small long-rotation patches dotted among the short-rotation stands would help sustain cavity-user populations over a fair portion of the landscape. It has been proposed that a 0.1 ha patch of old forest should be reserved for every two hectares of short-rotation forest (1/4 acre per 5 acres) (Zeedyk and Evans 1975, Raphael and White 1984). To what extent this could be stretched to a larger scale (e.g., one hectare per 20 hectares) is not known (Evans and Conner 1979). Perhaps as an alternative to clumps, long-rotation forest could be kept as a narrow strip between two short-rotation stands in forests where windthrow is not a serious problem. Thus located, it would provide snags for both stands and may act as a travel corridor and visual buffer (Neitro et al. 1985). Obviously, careful juxtaposition of managed stands with unharvested old-growth reserves, riparian forests, etc. can also alleviate a shortage of dead wood.

Creating Cavities and Snags

Sometimes forest managers do not wish to allow trees to grow old enough to die; sometimes wildlife managers are willing to let trees die, but do not wish to wait very long. These perspectives have led people to explore artificial ways of creating cavities and snags.

Artificial nest cavities

People have been making structures for cavity-dwelling animals for a very long time. Victorian bird houses may come to mind first, but Egyptian hieroglyphics and similar evidence indicate that people have been making bee hives out of cylinders of bark, hollow logs, and ceramic vessels for many millennia (Crane 1983). Some of the first-known bird houses were also inspired by people's appetites; Swedes satisfied their desire for duck eggs by building nest boxes for goldeneyes (Phillips 1925). Native Americans, such as the Chactaws and Chickasaws, had a different motive for turning gourds into homes for purple martins; according to Audubon (1831), "the bird keeps watch and sallies forth to drive off the Vulture that might otherwise commit depredations on the dear-skins or pieces of venison exposed to the air to be dried."

In more recent times, nest boxes have been successfully used by wildlife managers to enhance populations of beleaguered species, such as the wood duck and eastern bluebird, and by wildlife scientists to enhance our understanding of avian reproduction (Froke 1983). (Having sizable numbers of birds nesting in predictable, accessible places is a great boon to ornithology.) The role of nest boxes in forestry began in the twentieth century when European foresters recognized that by putting up nest boxes they could increase populations of birds to help control forest insect pests.

Undoubtedly, installing nest boxes could address some specific problems—too few squirrels for the local hunters, too many caterpillars for the local forester—but do they have a place in the larger picture of maintaining a diverse forest? Most wildlife biologists would say "no" or "very limited." Their potential role is limited because there are so many species that require cavities, and thus providing the requisite array of different sizes, styles, and locations of nest boxes would be a logistical nightmare. An even more basic problem is that nest boxes can never accommodate all the species that use snags and logs for myriad purposes beyond seeking a cavity. For example, over a three-year span a dead elm in front of my house was inhabited by great-crested flycatchers, northern flickers, hairy woodpeckers, common grackles, and countless invertebrates, fungi, and bacteria too small to readily notice; it was used as a perch by bald eagles and belted kingfishers; and now that it has fallen over, redback salamanders and short-tailed shrews have occupied it. During this same period, the nest box at its base was used only by tree swallows and wasps. Even if we restrict our concern to cavity-nesting wildlife, nest boxes cannot meet all needs (McComb and Noble 1981). Most woodpeckers will not use them; apparently excavating a cavity is a critical part of their courtship ritual. People have tried giving plastic trees—cylinders of polystyrene—to woodpeckers, but thus far only downy woodpeckers have readily made the switch (Grubb et al. 1983, Petit et al. 1985).

Whether one is making artificial cavities out of gourds or petrochemicals, costs are likely to be prohibitive. It is important to remember that constructing and installing nest boxes is only part of the story; cleaning, maintaining, and replacing the boxes is a continuing expense. Some data from a study in ponderosa pine stands are illustrative: Nest boxes made of concrete and wood chips cost $9.30 each to purchase; two person-hours to install; and one person-hour per year to maintain (Brawn and Balda 1983). Only four species made regular use of the boxes, occupation rates rising from 5% to 31% over three years. Certainly, it is preferable to use nest boxes to support a modest set of secondary cavity nesters rather than have no cavity nesters at all; this is common practice in much of Europe. However, one must consider the limited set of benefits and significant costs and compare these to the benefits and costs of an alternative strategy featuring dead trees.

Killing trees

If a current shortage of dead trees is the crux of the problem, an obvious solution is to kill some trees. Various methods have been proposed and tested on a pilot basis; none have been implemented on a large scale or tested over an extended period. Perhaps the simplest way to kill a tree is to girdle it; the disadvantage of this method is that decay begins in the sapwood before the heartwood and thus by the time the tree is rotten enough to excavate, it is fragile and easily knocked over; girdling also often fails to kill a tree (Bull and Partridge 1986). Some herbicides will kill trees without eliminating decay fungi (Conner et al. 1981, McComb and Rumsey 1983*a*), but again, the decay proceeds from the outside in and herbicide-killed trees fall even sooner than girdled ones (Conner et al. 1983*a*, Bull and Partridge 1986). Accelerating the natural decay process has also been attempted; this involved drilling holes in trees and inoculating them with a fungus culture (Conner et al. 1983*b*). Evelyn Bull and Arthur Partridge (1986) compared the effectiveness and costs of these techniques in a controlled experiment along with some other methods that ranged from crude (e.g., blowing the top off a tree with dynamite) to elegant (attracting an abundance of beetles with a sex pheromone). They concluded that the best method was simply to climb a tree and cut off the top, 15–25 m above the ground, and remove lower limbs. Perhaps the most elaborate technique proposed, but not tested, involves injecting a fungicide into the phloem of a tree; this kills the tree and preserves the sapwood, leaving a decay-resistant shell that might increase the longevity of the snag. After this treatment, decay in the heartwood can be initiated by the fungus inoculation technique (Goodell et al. 1986). Here the toxicity of many fungicides could be problematic. Finally, in some circumstances it might be desirable to create den trees, rather than snags, by injuring a tree instead of killing it. Because trees can compartmentalize decay, wounding the bark of a 10 cm diameter tree might eventually produce a 20 cm tree with a 10 cm hollow (DeGraaf and Shigo 1985).

Managing for the Other Values of Dead Wood

If enough suitable snags are maintained to meet the needs of cavity-users, will this also meet the needs of all the other species associated with snags and logs? Will there be enough flakes of loose bark to shelter tree creepers and bats, enough crumbling logs to harbor voles and salamanders, enough high, open perches for raptors, enough slash to hide hares and quail? Will there be enough large logs to fulfill their role in nutrient cycling? Will the right kinds of logs be left in the right places?

The last question can be answered with a modicum of confidence. A well-distributed supply of large logs will provide more habitat than a few big piles of small logs, and managing forests for cavity-users will tend to provide this. Large logs are important because they will hold larger species and because they take longer to disappear.

Questions about, "Will there be enough . . ." are almost impossible to answer with any certainty, given our current understanding of these systems.

This is especially true with regard to nutrient cycling. Perhaps the best approach is to think of this uncertainty as additional justification for supporting as large a number of cavity-users as possible, on as much land as possible. Conversely, it is worth recalling Maser's idea (see Figure 10.2) for alternating three short rotations with one long one. This arrangement would provide both large logs as nutrient reservoirs and large snags for cavity-users.

More immediately, log and slash management raises some specific issues that are distinct from problems of snag management. Various silvicultural treatments such as crushing or relocating woody debris with heavy machinery, and incinerating it through prescribed burning, can profoundly influence the abundance and nature of woody debris. For example, forest managers often burn the slash remaining after a clearcut to reduce the danger of a wildfire, to prepare a seed-bed for desirable trees, and for a variety of other reasons. However, if they are too successful, all that will remain is a few heavily charred logs, and wildlife values will be compromised. Silvicultural treatments that affect woody debris have been reviewed by Pierovich et al. (1975) and Maser et al. (1979) and are discussed in Chapter 12, "Intensive Silviculture."

Costs and Liabilities

Whether snags and logs are important as habitat for mosses and mice today, or as nutrient reservoirs for forests hundreds of years in the future, leaving large amounts of dead wood in the forest presents some potential conflicts with timber management. First, much of this material could be harvested. In areas where wood is sold as biomass fuel, every tree, even those that are very small and very rotten, can be chipped and sold. This material may have a minimal value compared to that of live, healthy timber, but it is a value none the less. A demand for firewood can also tax the supply of snags; in 1979, private firewood cutters removed nearly 100,000 snags in just one district of the Arapaho-Roosevelt National Forest in Colorado (Scott et al. 1980). One hundred snags were reserved as wildlife trees and sign-posted; 97 of these were cut. Of course, it is one thing to leave a tree that is already dead and has little value; it is another thing to allow healthy trees to decay and die. To roughly estimate the reduction in wood production required to meet the snag requirements of woodpeckers, one could just consider the recommendation of leaving a 0.1 ha clump of older forest for each two hectares of short rotation forest (Raphael and White 1984). If no trees were ever harvested from the clumps, this would mean a 5% reduction in wood production. There are much more refined methods for estimating production losses (Menasco 1983, Wick and Canutt 1979). Wick and Canutt's model is based on: (a) the Thomas et al. (1979c) method for estimating snag requirements of woodpeckers; (b) predictions of how long different species and sizes of snags will stand before falling; and (c) stand yield tables (i.e., models of long-term growth, production, and mortality). They used their model to estimate that maintaining 60% of the maximum woodpecker population of a ponderosa pine forest would reduce commercial wood production by 5.8%.

A second conflict concerns safety. Snags have earned epithets such as "widow-maker" by occasionally falling on forest workers' heads. In the United States, this has prompted the Occupational Safety and Health Administration to require that: "Dead, broken, or rotted limbs or trees that are a hazard (widow-makers) shall be felled or otherwise removed before commencing logging operations, building roads, trails or landings in their vicinity." If snags are left standing, loggers are not supposed to work within two tree lengths of them. The extent to which this law is observed varies considerably; some loggers assiduously remove all snags, some largely ignore them, and some remove the snags that seem hazardous and leave those that do not (Arbuckle 1986). Many consider it more dangerous to fell a snag than to leave it alone. It would seem that this is a case where blanket regulations are inappropriate, and loggers should be encouraged to use their own judgment to distinguish between safe and unsafe snags. This is particularly true as more and more harvesting is done by large machines with reinforced housing to protect the operator. Neitro et al. (1985) have described some logging system layouts that should minimize snag hazards. Although soft snags are far more likely to fall than hard snags, the consequences of a soft snag mishap may be less because they are, on average, shorter and lighter than hard snags. If a snag must be cut, leaving a high stump (ca. 2 m) would retain habitat for a few species (Morrison et al. 1983).

Fire hazards are another potential problem; tall, isolated snags are liable to be struck by lightning and dead wood, particularly slash, can be quite flammable. Excessive amounts of slash can also impede travel by large animals such as deer, tree planters, and hikers. The third group is particularly important, because much of the outcry against clearcutting and related activities can be traced to the public's aversion to slash. Indeed, many people find snags unaesthetic too.

THE RED-COCKADED WOODPECKER

The fires that burn across much of the southeastern United States each year are a powerful force of natural selection and have shaped many components of the ecosystem, not just the pines described in Chapter 4. Cavity-nesters that live in these forests are quite vulnerable to fires, because dead trees lack the thick, protective bark of live trees and burn readily when a ground-fire sweeps by. To avoid this problem and its lethal consequences for young in the nest, one species, the red-cockaded woodpecker, excavates its nest cavities in live pine trees, not dead ones, and this unusual ability makes it an important component of the fauna. Flying squirrels, bluebirds, various bats, chickadees, and

even other kinds of woodpeckers, all depend, to varying degrees, on the red-cockaded to provide cavities for them (Ligon 1970, Dennis 1971). Obviously, it is not in the interest of the red-cockadeds to provide extra cavities for other wildlife, so they usually live in clans of two to nine individuals and try to defend their cavities against the onslaughts of other species. It is possible that social factors, not competition for cavities, is the primary reason for living communally. These clans are not composed of a random assortment of birds; they consist of one pair that does the breeding and the rest are younger birds, usually offspring from previous years, who help raise the young (Ligon 1970, Lennartz and Harlow 1979).

Excavating cavities in living trees, especially pines, is not an easy task, because conifers protect themselves from injury by producing resin to help keep insects and disease organisms out. For a red-cockaded to successfully excavate a cavity, the tree has to be large enough for the woodpecker to drill through the resinous sapwood into the heartwood and still have enough room to construct a cavity. (Incidentally, the resin produced around a red-cockaded hole blankets the outside of the tree, making the holes extremely conspicuous, but apparently this resin blanket may deter at least one kind of predator, snakes, that cannot easily crawl across it [Dennis 1971]). In most cases, red-cockadeds select trees which have a heartwood infected with red heart fungus to make excavation easier (Conner and Locke 1982). Red-cockadeds need big and preferably infected pine trees; thus they require old trees, roughly 75 years for loblolly pines, and 95 years for longleaf pines (Jackson et al. 1979, Connor and O'Halloran 1987, Delotelle and Epting 1988, Hooper 1988). Here lies the rub. With typical rotation lengths ranging from 25–50 years on virtually all the industrial forest lands, there is very little pine forest left that will support the red-cockaded woodpecker. In a few decades, this species has gone from being one of the characteristic birds of the coastal plain pine forests, the most abundant woodpecker, to an endangered species with scattered, relict populations (U.S. Fish and Wildlife Service 1985).

Much of the responsibility for arresting this decline currently rests with the USDA Forest Service, because their lands support approximately 70% of the remaining red-cockadeds, even though they only own 5% of the southeastern pine forests (U.S. Fish and Wildlife Service 1985). Exactly how they are going to do this has not been determined. The most conservative approach, establishing large, no-cut forest reserves around existing colonies, will certainly not be undertaken. The strategy proposed by the U.S. Fish and Wildlife Service (1985) calls for lengthened rotations, leaving individual old trees distributed in younger stands, setting aside remnant patches of old trees, or a combination of these methods. Additionally, 125 acres of foraging habitat (sites with at least 40% of the area in pines older than 60 years) would be needed. Unfortunately, these approaches require a considerable economic compromise and would probably be adopted only on publicly-owned land, thus leaving the red-cockaded woodpecker an uncommon, localized species. Even this may be an optimistic scenario, because extinction is a real possibility (Ligon et al. 1986). For example, one analysis of the genetic variability and demographics of red-cockadeds predicted that at least 509 breeding pairs would be needed for a viable population that would not succumb to inbreeding and other problems that affect small populations (Reed et al. 1988). None of the existing populations are that large.

SUMMARY

Dying, dead, and down trees are important components of forest ecosystems, because during the process of death and decay they are inhabited by an extraordinarily diverse succession of organisms ranging from woodpeckers and other cavity-users, to myriad invertebrates, fungi, and microorganisms. Not only are dead trees critical microhabitats for many species, but they are also large reservoirs of organic matter and hence play a role in nutrient cycling. Timber extraction, especially intensive, short-rotation management, tends to minimize the number of snags and logs in a stand, and thus explicit policies of retaining dead wood need to be implemented for managed stands. Many specific recommendations have been formulated about what kinds of snags to leave and how many, but as a preliminary generalization, 5–10 large snags per hectare (2–4 per acre) seems adequate to maintain most wildlife populations. Maintaining a continuous supply of snags indefinitely may necessitate having small patches of old forest distributed among younger stands, perhaps a 0.1 ha patch on each 2 ha of forest. There are alternatives, such as killing trees to meet immediate needs, or providing nest boxes, but allowing a few large, old trees to die naturally is preferable under most circumstances. Some costs are associated with snag maintenance programs, notably some reduction in tmiber extraction and safety risks, but they do not outweigh the considerable value dead wood has in a forest ecosystem.

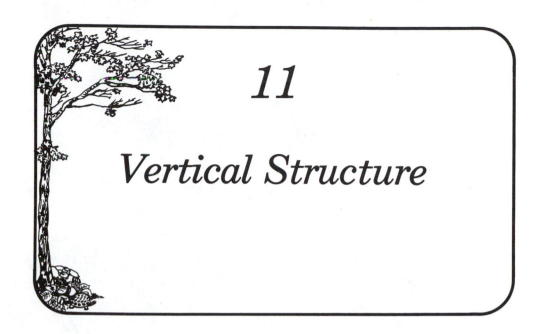

11

Vertical Structure

Forests are tall; indeed it is height, as much as anything, that defines a forest. Many forests are so tall that layers or strata of vegetation are readily apparent to even the most casual observer. From the sunny, breezy crowns of the dominant trees to the cool, damp litter that carpets the forest floor, a forest presents a continuum of different environments and thus the creatures that thrive in the treetops often bear little resemblance to those that never leave the ground.

It is easy to visualize a hypothetical well-stratified forest in which plants of many species and statures distribute their foliage evenly among several strata (Figure 11.1). However, there are probably far more stands that diverge from this idealized image than conform to it. In some dense stands of spruce or fir, the trees are so tightly packed that little light can reach the forest floor to support a rich community there (Figure 11.2). Many pine forests are swept by ground fires so frequently that a robust understory seldom develops (Figure 4.1). In contrast, stands of aspen and birch have such an open canopy that the understory is often an impenetrable thicket of small conifers (Figure 12.2).

The image of a highly stratified forest is very much linked to the image of a climax forest and is therefore subject to all the controversy that the concept of climax entails (Halle et al. 1978). Nevertheless, two generalizations can be made: First, a forest that is uneven-aged will usually have fairly well-developed stratification; and second, a forest with very little stratification will usually be quite even-aged. These two statements leave a large third group, a rather gray one: even-aged forests with moderate to pronounced stratification. Such stands are often incorrectly assumed to be uneven-aged. To avoid this error, many forest ecologists would prefer to speak of even-height and uneven-

Figure 11.1   Vertical structure of a hypothetical tropical moist forest in which the foliage is distributed fairly evenly among many strata. (The details of vertical layering in tropical forests is a very complex story; for more information see Bougeron 1983, Halle et al. 1978, or Richards 1983.)

Figure 11.2   In a dense, young stand of shade-tolerant trees such as spruce or fir, there is often very little ground vegetation.

height stands rather than even-aged and uneven-aged. Here we will use all four terms but will avoid using even-aged and uneven-aged when we are really referring to height. To understand how these three situations can arise, we have to consider the process of forest stand development.

Forest Stand Development

Various elements of stand development have been described in previous chapters where succession was discussed; here succession is reviewed from the perspective of stratification. Forests originating after a major disturbance will usually be even-aged for many years, and they may or may not have much vertical stratification, depending on patterns of dominance and competition among the major tree species (Figure 11.3A) (Oliver 1980, 1981). For example, often a disturbed site will be occupied by a set of species with such marked differences in shade-tolerance and growth rates that clear strata are soon

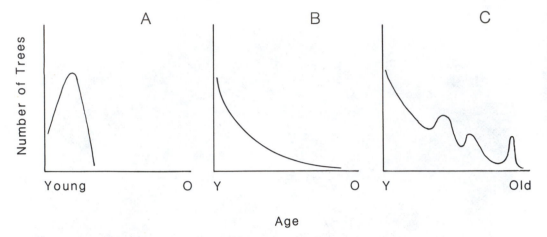

Figure 11.3 Three hypothetical age distributions showing: A, an even-aged stand; B, an uneven-aged stand; and C, a stand in which an uneven-aged distribution has some distinctive anomalies.

formed; i.e., fast-growing trees will form a high stratum over one or more strata of slow-growing trees inexorably coming up from below (Figure 12.2). Alternatively, sometimes one or two shade-tolerant species will fully occupy a site from the beginning and effectively exclude most competitors for a long time (Whipple and Dix 1979, Alaback 1982) (Figure 11.2); such stands will have little vertical stratification. In either case, if another large-scale disturbance does not set back the clock of succession too soon, individual trees or groups will eventually die, creating holes in the canopy beneath which small trees can grow. Over time, this mosaic of very small disturbances will create a forest which is uneven-aged and has well-developed stratification (Figure 11.3B) (Tesch 1981). Sooner or later, it may be decades or millennia, a severe disturbance will initiate the whole process again.

Like most descriptions of forest succession, the preceding summary portrays disturbance regimes as one of two extremes, either a catastrophic disturbance that destroys most of the trees over a large area, or the tree-by-tree mortality that creates small gaps. However, in some forest types an intermediate degree of disturbance may be common (Figure 11.3C) Craig Lorimer. (1980) analyzed the age and diameter structure of a 400-year-old forest of hemlock, beech, tulip-poplars, and birch in North Carolina and found that the forest's apparently uneven-aged structure was, on close examination, characterized by several distinct peaks in the age distribution. The peaks for various species coincided with one another, and these were also periods in which growth (as measured from tree-ring width) was relatively great. Lorimer concluded that a series of partial disturbances was responsible for these pulses of regeneration and growth. Specifically, he estimated that during the last 250 years there had been eight light disturbances, most likely wind storms, each of which had removed less than 10% of the canopy trees.

Conifer stands, which are often thought to be typically even-aged, may also have a broad age distribution because of unpredictable, external events (Franklin and Hemstrom 1981). In Arizona, Alan White (1985) studied an old ponderosa pine stand with a wide, albeit very irregular, range of ages; almost half the trees in the stand had become established in one 30-year period, while only two of the trees originated during an earlier 40-year span. This irregular distribution might be linked to the necessity of meeting four conditions before an age-class of seedlings could become established. Initially, (a) a hot ground fire to prepare a seed bed free of grass competition, and (b) a good seed crop were needed; next, a period free of (c) droughts, and (d) severe fires was required to allow the seedlings to survive their sensitive stage.

The Herb and Shrub Perspective

Vertical stratification may be primarily the result of changing patterns of dominance among trees, but this is not the whole story. Disturbances that are confined to one stratum, notably fire and grazing, also profoundly affect stratification, particularly by their effect on herbs and shrubs. For example, ground fires are almost an annual event in some forests, and they can effectively limit the ground vegetation to species of plants that are tolerant of being regularly burnt (Chandler et al. 1983). Grasses often thrive in such a regime, but few small woody plants can tolerate it. One major way plants adapt to fire is to have a large portion of their structure underground so that the above-ground structure can be expendable.

Like fire, the effects of herbivorous animals are often confined to a single stratum. Large populations of some mammals can easily inhibit the development of herb and shrub strata and tree regeneration. For example, introduced wild boars reduced herb cover to 5% of its normal amount in some parts of the southern Appalachians (Bratton 1974, 1975). Furthermore, they shifted the structure of some herb populations toward small, nonflowering individuals and changed the composition of the herb community as a whole toward species with deep or poisonous roots. Similarly, 40 years of sustained browsing by white-tailed deer had a striking effect on the structure of a Pennyslvanian forest (Whitney 1984). Plots measured in 1929, when deer browsing was very low, had average small sapling densities of 3,915 per hectare in hemlock-beech stands and 21,984 per hectare in hemlock stands; by 1978, a marked increase in deer population had reduced these densities to 28 and 110 per hectare, respectively. Substantial canopy defoliation by insects, notably caterpillars of species such as spruce budworm and gypsy moth, can certainly have a short-term effect on vertical structure. However, unless the outbreak is severe enough to kill many trees, it is unlikely to produce a very marked change in the development of lower strata or affect animal populations (DeGraaf 1987).

Despite the impacts of disturbance factors such as fire and herbivores, it is trees that largely determine how well shrubby and herbaceous plants fare. This is simply because the development of the canopy profoundly shapes the environment close to the ground. Obviously, the amount of light reaching the forest floor is largely a function of how effectively the trees arrange their leaves to intercept it (Horn 1971). As a generalization, shade-tolerant tree species are far better at this and cast a more complete shadow than shade-intolerant trees. In temperate deciduous forests, many herbs mitigate this problem by growing very quickly during the spring, before the tree leaves have emerged (Muller 1978). This is not a complete solution, because even in a leafless forest as little as 25% of the total radiation striking the forest actually reaches the ground (Hutchinson and Matt 1977). The success of shrubs and herbs may also be dependent on available soil moisture (Collins et al. 1985); indeed, moisture may be more important than light in some forests (Anderson et al. 1969). Water availability is less likely to be problematic under open canopies that allow a relatively large portion of the rain to reach the ground than under a thick canopy in which much of the water is intercepted and returned to the atmosphere.

From an ecological perspective, it may be adequate to evaluate the development of understory plants in terms of their size and abundance, but from an evolutionary standpoint it is also important to consider their success in sexual reproduction. Plants may be able to survive under a dense canopy, but this does not necessarily mean they will have adequate resources to produce flowers and seeds (Pitelka et al. 1980).

Although trees are certainly dominant over herbs, the relationship is not entirely one-sided; patches of forest herbs can inhibit the establishment of tree seedlings. In a West Virginia old-growth forest of hemlocks and hardwoods, the distribution of tree seedlings was affected more by the herbaceous cover than by soil pH, light, or soil moisture (Maguire and Forman 1983). In this case, the overall negative relationship between herbs and seedlings had some interesting subplots: Partridgeberry patches actually contained more maple and hemlock seedlings than expected, whereas seedlings were especially scarce in hay-scented fern patches. Allelopathy, which involves one plant inhibiting another by releasing toxins into the soil, might be behind such negative relationships (Whittaker and Feeny 1971, Horsley 1977).

The Epiphyte Perspective

One could argue that trees, shrubs, and herbs are life forms that create forest stratification, or, conversely, that forest stratification has led to the evolution of these different life forms. It is probably not an argument worth pursuing and is, to a certain extent, irrelevant to a large group of plants, epiphytes. Because these plants live attached to other plants, their vertical distribution in a forest is determined more by their relationships with trees than by their structures (Barkman 1958).

In boreal and temperate realms, the dominant epiphytes are lichens and mosses. Sometimes these plants are just an inconspicuous crust, a gray-green patina on the bark of a tree; sometimes they festoon the branches in pendant masses that have given rise to the name, old man's beard. Epiphytes really come to the fore in tropical rain forests where the epiphytic lifestyle is not limited to mosses and lichens. Far above the dark floor of tropical forests, ferns, orchids, bromeliads, and other plants representing 68 vascular plant families manage to find sufficient moisture and nutrients (Benzing 1983). The critical feature is that water be abundant enough to make clinging to tall trees a feasible strategy. In the Uluguru Mountains of Tanzania, an elfin cloud forest with an annual rainfall of 3 m had 13,650 kg of epiphytes per hectare (Pocs 1982).

The term epiphyte does not usually include parasitic plants that tap the vascular systems of their hosts to rob them of nutriments. On the other hand, the traditional view of epiphytes as innocuous hitchhikers may be changing (Benzing 1983). If nothing else, epiphytes occasionally become so thick that they break off branches and they may even significantly increase the chance of the whole tree being blown over (Strong 1977). Furthermore, epiphytes inevitably collect and use nutrients before they reach the ground where ground-rooted plants have access to them. This nutrient piracy is not well understood, but there is some evidence that it may be important. For example, some trees have canopy root systems to access nutrients in the mat of dead organic matter collected by epiphytes (Nadkarni 1981). Whatever their role in nutrient cycling, epiphytes add another element of diversity to the vertical stratification of a forest.

THE IMPORTANCE OF VERTICAL STRUCTURE

It does not require a great deal of ecological insight to realize that ecosystems in which the vegetation is tall and well-stratified will generally support a more diverse biota than ecosystems in which most of the vegetation is concentrated in a relatively thin plane. To take an extreme example, toads, lilies, and other low plants and ground-dwelling animals can be found in both fields and forests, but warblers and oaks are largely confined to forests. This effect is more subtle when comparing different forests, but it is reasonable to assume that the same principle applies: Forests with well-developed vertical structure probably have more species than forests in which most of the vegetation is in one stratum. (See the box for a description of how the relationship of bird diversity and foliage distribution has been analyzed quantitatively.)

FOLIAGE HEIGHT DIVERSITY

The connection between species richness and vertical stratification was first formally described by Robert and John MacArthur (1961), who noted a positive relationship between the number of bird species inhabiting various habitats and an index of the vertical distribution of foliage. This index, commonly called *foliage height diversity* (FHD), was calculated from the Shannon diversity formula using the number of strata and how evenly foliage is distributed among the strata; see Figure 11.4 and Aber (1979) for examples. The relationship between FHD and bird species diversity (often abbreviated BSD) has been explored many times since the MacArthur's original work (Recher 1969, Wilson 1974, Moss 1978*b,* Dickson and Segelquist 1979, Elliott 1987). Their results have generally been corroborated, although there have been notable exceptions as well, especially among tropical forest birds (Lovejoy 1974, Pearson 1975). Unfortunately, this work has not been widely replicated with other taxa. A survey of mammals along a transect from savannah to tropical dry forests in Venezuela found a correlation between species richness and FHD (August 1983). An English study in which the variety of beetles and bugs was compared at different stages of succession (Southwood et al. 1979) tends to support the FHD effect. Early in succession the number of insect species in a plot could be predicted quite well from the number of plant species, but in later stages the number of vertical layers in the vegetation also had to be taken into account to explain the variation in insect diversity.

It is interesting to note that the Shannon diversity formula is not affected by the *absolute* population size of various species and similarly FHD is unaffected by the absolute amount of vegetation in any stratum; only the *relative* amount of vegetation is of consequence. For example, if the profile curve for a tropical moist forest shown in Figure 11.4 was shifted to the right so that the average vegetation density was about 50%, instead of about 40%, the FHD value would remain the same. At least one study—a comparison of birds and trees in two tracts of mixed-conifer forests in Arizona—indicates that the absolute amount of vegetation may be of some importance to the bird community (Franzreb and Ohmart 1978). In this study, one tract had never been cut, and in the second study tract, about 70% of the trees (84% of the basal area) had been recently removed. Despite the cutting, some characteristics of the vegetation, notably tree species richness and FHD, were very similar in the two plots. Quite remarkably, bird species diversities in the two plots were also very similar—2.75 and 2.70 in 1973, and 3.19 and 3.14 in 1974—and thus this study indirectly corroborated the relationship between BSD and FHD. But the story does not end here. The total volume of foliage in the uncut plot was about 7.5 times greater than in the cut plot (113,984 m³/ha vs.

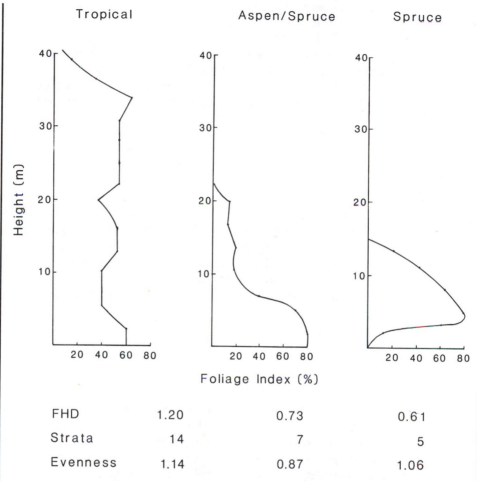

Figure 11.4 Hypothetical vegetation profiles from three forest types; see Figures 11.1, 12.2 and 11.2, respectively. The *x*-axes can be any index of the amount of foliage; two of the most commonly used indices are percentage of the sky that is obscured as one looks up through a stratum, and percentage of a white board that is obscured as one views it from a distance. Below each profile three descriptive parameters are listed: the foliage height diversity (FHD) as calculated from the Shannon diversity index (see Appendix 3), the number of strata (this is analogous to species richness), and an index (*J*) of how evenly the foliage is distributed among the strata.

15,270), and this difference probably explained why the density of breeding pairs was significantly higher in the uncut plot (633 vs. 544 in 1973; 866 vs. 712 in 1974).

The FHDs calculated in Figure 11.4 are somewhat unorthodox in that the strata are all of equal height, 3 m in this case. Often, researchers divide the forest into strata that roughly correspond with obvious classes of vegetation, such as herbs, shrubs, understory trees, and overstory trees, and thus have strata with unequal dimensions such as 0–0.6 m, 0.6–7.6 m, >7.6 m (Dickson and Noble 1978). Others have used less arbitrary units, but still scale the strata to give more emphasis to vegetation near the ground (e.g., 0–0.5, 0.5–1, 1–3, 3–5, 5–10, 10–15, 15–20, 20–25, >25 m [Elliott 1987]). Obviously, a 0 m–1 m stratum is much more likely to be critical to a certain species of bird than a 22–23 m stratum, because the former is a unique assemblage of ground plants and the latter is just one of many layers in the tree canopy. In a Louisiana bottomland hard-wood forest, 31% of the bird sightings were in the lowest stratum (0–0.6 m), even though it represented only 2% of the forest's height, and the top stratum (>7.6 m) had 71% of the forest's height but only 33% of the sightings (Dickson and Noble 1978). Thus, there is reasonable ratio-nale for having arbitrary strata that correspond to a stand's structure. On the other hand, every forest type has a different vertical structure and a different bird community, and it would be better proof that the relationship between FHD and BSD is universal if independently defined strata were used. When the strata conform closely to different vegetation life forms, foliage height diversity may actually be more an index of physiognomic diversity than of the distribution of foliage.

Niches

The relationship between biological diversity and vertical structure is based primarily on differentiation of niches—the roles or places species occupy in the function and structure of an ecosystem. Similar plants and animals use many strategies to coexist in the same area: habitat differentiation is one of the most common, and, on a microhabitat scale, vertical separation is espe-cially widespread (Schoener 1986). A classic example of this comes from Rob-ert MacArthur's (1958) work in Maine's coastal spruce forests where five spe-cies of warblers share the same tree crowns by concentrating their activities in different parts of the trees. Scores of studies, from Arizona to Zaire, have shown similar results (Balda 1969, Sutton and Hudson 1980) and the effect is by no means limited to birds. White-footed mice in Virginia live below deer mice (Harney and Dueser 1987); saddle-backed tamarins (small monkeys) in Bolivia live below red-chested mustached tamarins (Yoneda 1984); vertical sep-aration plays a major role in facilitating the coexistence of eight species of anole lizards in Puerto Rico (Rand 1964). In fact, some anoles have gone a step further and evolved vertical separation to minimize competition among males, females, and juveniles of the same species (Schoener 1967).

The insect fauna often changes dramatically from top to bottom in a forest, perhaps because insects are likely to be quite sensitive both to plant structure and to microclimate (Southwood et al. 1979, Strong et al. 1984). The shifts can be both subtle; e.g., one species of mosquito or termite replacing another member of the same genus (Abe and Matsumoto 1979), or rather coarse; e.g., flying forms of insects predominating in the upper canopy (Sutton et al. 1983*a*). Such changes in the insect fauna may be of importance beyond the interests of insect taxonomists. In a study of insect grazing in Australian rain forests, it was found that foliage consumption was significantly heavier close to the ground, perhaps because the lower, shade leaves on a tree are likely to be more palatable than the sun leaves that must have a thicker cuticle to avoid excessive water loss (Lowman 1985). As with all generalizations, there are exceptions, and not all forest animals sort themselves out vertically; in a Panamanian rain forest, 148 species of barklice showed very little vertical separation (Broadhead and Wolda 1985).

One tends to think of animals selecting a stratum because it is the best place to forage, but this is not always the case, especially among birds that can readily fly up and down. Great tits choose a song perch height at which they are well-hidden from predators but easily heard by mates and rivals (Hunter 1980), while male tropical cockroaches seeking a mate have to be high enough to smell the pheromones wafted up from the female (Schal 1982).

When forest and wildlife managers think of vertical stratification, they are likely to focus on how trees of different ages provide a multistory habitat for many animals. This is only part of the whole picture. The effects of vertical stratification on the diversity of the entire biota are very much linked to differences in the statures of various plant species. People often speak of the size range from mouse to elephant, but the span from a ground-hugging moss to a cloud-piercing redwood is far greater. This, of course, is also a form of niche differentiation—quite an important form given that being tall is a major way plants compete for light. Height competition is especially interesting among temperate and boreal herbs that die back each winter and have to replace their entire above-ground mass in the spring. In places where the density of herbs is very high, plants have to allocate a large portion of their resources to stems and other support tissue so they can be tall (Givnish 1982). Conversely, if competition is slight the plants can concentrate their efforts on growing leaves. Compare two orchids from the eastern United States: The moccasin flower has ground-hugging leaves and lives in acidic, often dry, forests, while the showy lady-slipper has to compete in the rank vegetation of swamps and approaches a meter in height.

The best examples of vertical separation among plants come from epiphytes, because many species have very specific microhabitat requirements that influence their vertical distribution on a tree (Hale 1965, Topham 1977, Smith 1982). Light is obviously an important factor, but so too are subtle features such as pH of the bark and the presence of rain-tracks where water flows down a bole. In one experiment, lichen colonies transplanted from the base of a tree to just 1.3 m above the ground could not tolerate the different conditions and died (Brodo 1961). Generally, the species richness of epiphytic lichens tends to decline going up a tree, presumably because the environment is drier and the plants are dependent on a meager pool of nutrients derived more from rain and dust than from fallen organic matter and leachates. On the other hand, vascular epiphytes often do better higher in the trees where light is less limited (Pittendrigh 1948).

Rhinos, Orchids, and Fruit

Although well-stratified forests are usually biologically diverse because many plants of various structures and an array of arboreal animals can coexist there, vertical separation of niches is only part of the story. This is illustrated by three examples of how vertical stratification affects the interactions of plants and animals.

1. Of rhinos and deer

It is somewhat counterintuitive to think that vertical stratification is of enormous importance to animals that are too large to get off the ground, but this is the case with various large forest mammals such as deer, duikers, Sumatran and Javan rhinos, serows, tapirs, and okapis. Many of these species have rather sparse populations because the availability of high-quality browse may be limited by a lack of light in the forest understory. It is really only in grasslands that one finds the dense, low vegetation that can support huge herds of ungulates such as wildebeests, zebras, and bison. This said, patterns of disturbance in many boreal and temperate forests can produce large areas of early successional ecosystems that support substantial deer populations. Distribution of food is the crux of the matter, but concealment from predators is also a significant issue that links the well-being of ground-dwelling animals and ground vegetation (Thomas et al. 1979d, Bowman and Harris 1980). Large forest animals merit special attention, because most of them are popular sources of meat and sport and many of them, especially the tropical forest species, are threatened with extinction.

2. Of lichens and orchids

Epiphytes and their animal associates provide another good example of the importance of vertical structure. First, the variety of epiphyte species themselves is far greater than most people realize. For example, over 300 species of lichens have been recorded from British oaks; one mature individual was home to 52 species (Rose 1974). Even more surprising to temperate zone inhabitants is the fact that about 10% of all vascular plant species are epiphytic, some 30,000 species in all (Benzing 1983). Well over half this number comes from a single family, the orchids.

It is intuitively obvious that a tropical forest tree supporting a huge and varied mass of epiphytes will also have a commensurately richer fauna. To take one narrow study as an example—a survey of the aquatic biota that lived in the pockets of rainwater collected by three kinds of Jamaican bromeliad—68 species were recorded, including such seemingly unlikely creatures as crabs and tadpoles (Laessle 1961). To give some idea of how bewildering the diversity can be, even in this small study the investigator had to turn to 18 other specialists to help identify his samples.

A temperate or boreal tree enshrouded in a thin layer of lichens will also be significantly enriched by the epiphytes' animal associates, because with populations reaching into the thousands per square meter (Andre 1983), these creatures are diverse, numerous, and perhaps even picturesque:

> [Trees are] clothed in a coating of crustaceous lichens, painted on in intricate patterns of softly blended colours, and forming a pasture ground on which many different sorts of small insects browse happily. (Cunningham 1907)

Actually, there may well be more kinds of animals that hide in and among lichens than kinds that eat lichens (Gerson and Seaward 1977). Some animals are small enough to swim about in the lichen's surface film of water and hardly need to hide (e.g., nematodes and protozoans). Others, such as many moths, beetles, and grasshoppers, are large enough to go a step beyond simply hiding in the lichens, and can imitate them, in some cases carefully seeking only the species of lichens that they resemble most closely. A few caterpillars and lacewing larvae take a more direct approach to being cryptic by sticking lichen fragments to their back. Even some vertebrates, such as the gray treefrog, green salamander, and Malagasy bark gecko, have taken up the lichen camouflage strategy (Richardson and Young 1977).

3. Of flowers and fruits

For the most part, it is not in the interest of plants to provide food for animals, but there are two well-known plant products specifically designed to be consumed, nectar and fruit. Everyone knows that the purpose of nectar is to reward bees, hover flies, and other insects, birds, and mammals for dispersing pollen from plant to plant. It is less widely known, but equally true, that the mass of fleshy pulp in which many plants place their seeds is designed to reward animals for dispersing seeds away from the parent plant (Howe 1986).

A large number of tropical birds and mammals are very dependent on fruit—various parrots, toucans, fruit bats, hornbills, cotingas, monkeys, and hundreds of other species (Howe 1986). In temperate forests, obligate fruit consumers are quite rare, but many, perhaps most, birds and mammals occasionally eat fruit (Willson 1986). Even such unlikely animals as coyotes and pine martens are facultative frugivores. Pollinators are much more diverse than frugivores because their ranks are dominated by insects. In fact, there may be a positive feedback loop at work here; pollinators that are relatively faithful to a single species of plant may play a critical role in plant speciation (Crepet 1983).

Pollination and frugivory (fruit consumption) are frequently cited examples of coevolution between plants and animals, but they are not particularly tight systems (Jordano 1987). Most flowers are pollinated and most fruits are consumed by quite an array of animals and, conversely, each pollinator or frugivore usually exploits many plant species (Wheelwright et al. 1984, Feinsinger et al. 1986). Nevertheless, it is fair to say that a forest in which flowers and fruits are offered by many plants in many strata is likely to support a relatively rich fauna.

MANAGING FOR VERTICAL STRUCTURE

A simple "how-to" message can be distilled from this chapter: To manage a forest stand for vertical diversity, one should implement the kind of fine-scale, uneven-aged management that produces uneven-height forests—in other words, selection harvesting. A far more complex message about how to enhance the vertical diversity of even-aged stands through thinning and other techniques could be developed here, but these methods will be covered in the next chapter.

Although the details of how to implement selection harvesting merit some attention—e.g., size of canopy gaps created (Collins and Pickett 1988); extent of soil disturbance (Reader 1987)—it is the "how-much" issue that requires especially careful consideration. In Chapter 6, this issue was discussed in the context of creating a forest landscape that is spatially heterogeneous, one in which many types of natural disturbance—from individual tree falls to large fires—are emulated. To recapitulate briefly, it was suggested that selection harvesting be widely employed to represent the small end of the spectrum, particularly in regions (e.g., tropical forests and many temperate deciduous forests) where small-scale disturbances were probably the norm before human intervention.

On the whole, forest managers find even-aged management much easier to implement and can thus justify it on an economic basis. Furthermore, the discovery that some uneven-height forests are actually even-aged has increased the number of forests for which even-aged management can be justified ecologically too (Oliver 1980, 1981). However, there is also ample justification for uneven-aged management, because there are many types of forest that would, under natural conditions, routinely remain free of major disturbance long enough to develop an uneven-aged structure. Moreover, in this chapter we have seen that uneven-height forests are not only an important point on the continuum of spatial scale but are also the type of forest with the greatest vertical diversity.

Even though even-aged management tends to concentrate all of a forest's resources into a single stratum (Peterken 1981), one could argue that a forest comprised of many even-aged stands will also provide vegetation at many different heights as long as the whole forest has a balanced age structure. The young stands will have well-developed herb and shrub layers, middle-aged stands will approximate an understory, and so on. This is true, but a quick comparison of the herbs and shrubs growing under a forest canopy with those growing in the open will plainly show that the similarities are quite superficial. You do not need to be a botanist to distinguish between the grasses, goldenrods, and buttercups of a meadow and the lilies, ferns, and mosses of a forest floor. A diverse forest landscape needs *both* the herb and shrub strata associated with mature, uneven-aged forests and the low strata found in the early successional stages of an even-aged management program.

The need for using a broad approach can be illustrated by two studies conducted in loblolly pine forests in Texas and Louisiana. The Texas study compared the breeding bird populations in even-aged and uneven-aged stands and concluded that the absence of well-developed vertical layering in some even-aged stands limited both bird density and diversity (Dickson and Segelquist 1979). The Louisiana researchers compared the amount of ground forage available for consumption by white-tailed deer in pine stands with and without a midstory of deciduous trees (Blair and Feduccia 1977). The deciduous midstory inhibited the development of forage plants for the deer and was too high to be accessible to the deer; thus anyone more interested in deer than in bird diversity would be inclined to use controlled burning to remove the mid-story and reduce vertical stratification.

To think of this as a case of having to choose between game and nongame management oversimplifies the issue. It is true that traditional wildlife management has often focused on providing robust ground vegetation because so many game animals—deer, turkeys, quail, rabbits, etc.—are ground-dwellers. However, there are also myriad nongame species that can benefit from a program of enhancing understory plants, not least of which are the plants themselves. For example, management schemes designed to improve fruit production (sometimes called soft mast in this context) will benefit fruit-producing plants, many birds and mammals, and some turtles and invertebrates (Johnson and Landers 1978, Stransky and Roese 1984).

To put it another way, any forest stand, no matter what its configuration of strata, will support a suite of organisms, but the more diversified its stratification, the more diversified its biota will probably be. This does not mean that every stand should be well-stratified. Instead, it means that a forested landscape should have stands with a variety of vertical structures: some with close-knit canopies that leave little light and water for a sparse understory; some in which open-crowned intolerant trees shelter a thick copse of tolerant saplings; some in which foliage is quite evenly layered from the forest floor to the tallest crowns. This said, forest stands with well-developed stratification are often far too limited in regions where intensive forest management is the rule, and thus efforts to increase their extent will usually be of benefit to wildlife diversity in general.

THE WILD DAFFODIL

I wandered lonely as a cloud
That floats on higher o'er vales and hills,
When all at once I saw a crowd,
A host, of golden daffodils,
Beside the lake, beneath the trees,
Fluttering and dancing in the breeze.

When William Wordsworth penned these famous lines, it is unlikely that he contemplated the relationship between the daffodils and the trees beneath which they fluttered and danced. Nevertheless, the relationship is of great importance, at least to the daffodils, and is of considerable interest, especially to John Barkham, a British ecologist who spent ten years studying this relationship and other aspects of the ecology of the wild daffodil, *Narcissus pseudonarcissus* (1980a, 1980b). It is described here because it is a very well-documented example of how managing trees can have ramifications for species in other strata.

Barkham worked just 25 kilometers southeast of Wordsworth's beloved Grasmere in Brigsteer Wood, parts of which have seen the coppicing of common hazel and common ash since at least 1239 A.D. (Coppicing is a management system that focuses on regenerating the understory from shoots that emerge from cut stumps. Above this understory a scattering of larger trees such as oak, so-called standard trees, are often grown too; this is called coppicing-with-standards. Coppicing was once widely practiced in Britain but declined dramatically in the nineteenth century because of a poor market for small stems—primarily faggots for fuel and sticks for wattle fencing, tool handles, bean poles, etc. [Peterken 1981]. If in the future some forests are managed for biomass fuel production, coppicing may be revived.)

Each year Barkham visited fifty 4 m² plots, counted all the daffodil shoots, and recorded whether or not they produced a flower. He found that the daffodils were greatly influenced by forest management practices such as

cutting, thinning, and planting. For example, on a coppice site the number of shoots per 4 m^2 went from about 300 in 1970–72 to over 500 in 1976, a change attributable to the understory being thinned in the winter of 1972–73. However, after 1976 the number of shoots began to decline as competition from other plants, particularly brambles, began to limit the daffodils. The number of plants to flower followed an even more dramatic course, rising from essentially zero before coppice thinning, to about 50 in 1974, over 100 in 1975, and falling back to near zero by 1976.

This same pattern of growth and decline was repeated on other sites where the overstory or understory was either thinned or cleared. Generally, it took up to four years for the daffodils' positive response to culminate; from 4–7 years after the disturbance was a period of decline as competing vegetation flourished; and by 8–10 years the decline in population density leveled off.

Manipulations of the trees were not solely responsible for changes in the daffodil population. Year-to-year variation in weather was also very important, although even the effects of sunlight, temperature, and rainfall were modified by the canopy. For example, daffodils growing on shaded sites were more successful during relatively sunny summers, while those in open areas did better during cloudy summers.

Forest management had another indirect effect on the daffodils through its influence on the distribution of animals, notably roe deer, which were attracted to the flush of vegetation that followed thinning or canopy removal. The deer did not actually consume daffodils, but their hoof-scraping and trampling brought the daffodil bulbs closer to the surface where they were more likely to be attacked by slugs. Moreover, bulbs near the surface did not perform as well in general, producing fewer shoots and flowers with a lower rate of successful pollination.

Admittedly, these findings are largely of academic interest. The wild daffodil is not an endangered species, there is not a great variety of animals that rely on it for food or shelter, and a latter-day Wordsworth would still find much to inspire him in the forests of England's Lake District. Nevertheless, the daffodils do nicely illustrate the complex interplay of vegetative strata, an interplay that should give a forest manager pause to think about how forestry practices influence the forest's vertical structure.

SUMMARY

All forests are tall, at least compared to other ecosystems, and although all forests have recognizable layers of vegetation—an overstory, understory, shrub and herb strata are present in most—there is enormous variation in how well the vertical structure is developed. In some forests, much of the vegetation is concentrated into a canopy composed of the dominant trees. Other forests, particularly uneven-aged forests and even-aged forests in which there are no clear dominants, often have their foliage distributed fairly evenly. The vertical structure of a forest is important, because in a well-stratified forest a diverse array of plants and animals can coexist, occupying different niches that are often defined by vertical separation. There are many obvious and well-known examples of this (e.g., some birds are always in the canopy while others seldom leave the ground), but there are also many less conspicuous examples such as the vertical distribution of epiphytes and all their associated animals.

To manage a forest for vertical diversity can be quite straightforward; it simply requires uneven-aged management, i.e., selection harvesting. Enhancing the vertical diversity of even-aged stands is also feasible, but generally more complex. Not every stand has to have a well-developed vertical structure, but across a forest landscape it is important that a significant portion of the stands be managed for an uneven-aged structure; exactly what that portion should be will vary among forest types.

12

Intensive Silviculture

Alan J. Kimball

and

Malcolm L. Hunter, Jr.

At one time, most forests were a home for primitive people whose lives were thoroughly enmeshed in the forest, making them an integral part of the ecosystem. Now such relationships can only be found in a handful of remote regions, and elsewhere the shift to agrarian and industrial lifestyles has dramatically changed the relationship people have with forests. With the march of time, forests became first, obstacles to agriculture, and later, resources to be utilized by industry. In a few areas, the evolution has taken yet another step and forests have been reduced to small, isolated oddities to be preserved. Where people have recognized the wisdom of managing trees, rather than mining them, forestry has become a respected profession and silviculture has approached the intensity of agriculture and produced remarkable results. Intensively managed conifer plantations can produce three to seven times the usable wood per hectare produced by unmanaged stands (Farnum et al. 1983). Global patterns of deforestation, ever-expanding human populations, and increasing wood consumption are forcing people everywhere toward more careful planning and more intensive silviculture. Some wildlife advocates look upon this change with a skeptical eye, believing that intensive silviculture, like intensive agriculture, invariably leads to reduced species richness and simplified stand structure (Grime 1979, Gamlin 1988). There is some truth to this generalization, but this need not be the case.

There are three basic considerations that are fundamental to the relationship of silviculture to biological diversity.

1. Practicing intensive silviculture on productive sites close to population and processing centers can relieve the pressure on the remaining forest land by producing a given amount of wood from a relatively small area of land. (This of course assumes that levels of wood consumption do not expand commensurately.) Silviculture can thus play a pivotal role in providing biological diversity, forest reserves, and dispersed recreation in a world of ever-shrinking forests (Hyde 1980, Wyatt-Smith 1987).

2. Although silviculture is usually directed toward increasing the production of marketable wood, this need not be the case. The tools and practices of silviculture can be used to produce whatever vegetation patterns or goods and services are deemed appropriate by the forest owner. The goal can be pine sawlogs, Kirtland's warblers, improving degraded stands, or overall wildlife diversity.

3. Even where intensive silviculture does produce very simple stands or monocultures, these can be integrated into whole landscapes as valid components of overall landscape diversity. For example, even-aged pine plantations may not be internally very diverse, but they do provide suitable habitat for some specialized species, such as pine warblers, and thus make a modest contribution to diversity at the landscape scale.

It is clear that all three of these points are relevant to the idea of managing forest landscapes for wildlife. However, because most silvicultural treatments are prescribed on a stand-by-stand basis (albeit with constant reference to the overall goals of ownership), this chapter focuses on the second point and explores how the various practices of intensive silviculture can work for or against biological diversity at the scale of a single stand.

Silvicultural Systems, Coexistence, and Biological Diversity

Silviculture is the theory and practice of controlling forest establishment, composition, structure, and growth (Spurr 1979). Harvesting techniques (also called regeneration methods because they are the first step in establishing the next stand) are coupled with subsequent intermediate treatments, such as thinning and weed control, to form silvicultural systems. A silvicultural system is thus a program of management for at least the entire life of an even-aged stand, or for an entire rotation for an uneven-aged stand. Silvicultural practices are typically named by what is done to the stand or site, but alternatively, one can focus on the ecological results, rather than on the treatments themselves. Harvesting techniques and the natural regeneration methods (clearcut, seed tree, shelterwood, coppice, selection) are often the beginning points of intensive silviculture, and their effects on wildlife diversity are covered in the preceding chapters by considering the patterns and changes they produce. In particular, the chapters devoted to species composition (Chapter 4), age structure (Chapter 5), spatial heterogeneity (Chapter 6), and vertical structure (Chapter 11) focus on the ecological results of regeneration treatments. This chapter focuses on the results of forest protection, artificial regeneration, and intermediate treatments.

Following any disturbance in a forest, a host of plants occupy the available space, whether by sprouting or from wind-blown, animal-dispersed, or stored seed, and thereafter succession is paced primarily by patterns of changing dominance. To a large degree, intensive silviculture focuses on shaping these patterns of dominance and coexistence in young stands. Thus, to prescribe silvicultural systems that will enhance or maintain diversity, it is vital to understand what factors will allow several plant species to coexist on the same site. Coexistence in the competitive world of plants depends on three primary factors: spatial variation inherent in a site, the ability of plants to partition available resources, and site productivity (Grime 1979).

1. *Coexistence through site variation:* Even within small stands under intensive management, micro-site differences can produce concomitant variation in the arrangement of herbs and trees alike. One example is the micro-relief of the forest floor produced when a large tree falls, leaving a pit where the root mass was and a mound where the log was assimilated into the soil. This leads to variations in drainage and nutrient concentrations and thus to differences in herb and seedling success. Species adapted to wet sites can grow side by side with more exacting species, and thus species richness is greater than it would be on more homogeneous sites. (Some forms of site manipulation used by silviculturalists can also create heterogeneity at a micro-scale.)

2. *Coexistence through resource partitioning:* Forest plants sometimes escape competitive exclusion by partitioning the available resources through phenology and/or stratification. Perhaps the best known example of this comes from many herbs of northern deciduous forests that compensate for their low stature by completing most of their growth in the spring, before the overstory leaves shut out the sunlight. (Foresters routinely manipulate competition in a stand by controlling the degree to which any one stratum dominates the others.)

3. *Coexistence through stress and disturbance:* Competition is largely a feature of productive sites, because here the species best suited to the site grow rapidly and crowd out other species. On poorer sites, all of the plants are stressed, and therefore it is much more difficult

for any one plant to exclude another. By reducing overall plant biomass production and thus preventing or slowing competitive exclusion, moderate levels of stress and disturbance can help maintain
species richness (Grime 1979). (Foresters often try to limit the effect
of stress and disturbance on crop trees.)

The challenge for intensive silviculture is how to work with these natural
patterns of spatial and vertical partitioning, plus manage levels of stress and
disturbance, in ways that foster, rather than reduce, biological diversity. To
consider these further we will review the practices common to intensive silvicultural systems and describe the manner in which they may enhance or reduce
diversity. Unfortunately, some of this discussion must be rather speculative
for lack of scientific documentation, even though foresters have been managing forests for centuries. A descriptive outline of silvicultural treatments derived from Smith (1986) is provided in the box for those unfamiliar with silvicultural terminology.

MANAGING STRESS AND DISTURBANCE

Intensive silviculture involves significant inputs of labor and capital and thus
will usually be practiced only on the most productive sites. Nevertheless, circumstances do arise—notably a scarcity of forests and a high demand for
wood—that compel people to bring marginal sites under intensive management, even though this may require reducing stress through expensive measures such as fertilization, drainage, or irrigation. On both marginal and productive sites, once an investment in intensive management has been made
there is a strong incentive to protect that investment by protecting the trees
from insect outbreaks, disease, wildfire, or overgrazing, and by protecting the
site from erosion or nutrient depletion.

Improving the Site by Reducing Stress

The productivity of forest sites can be improved by terracing, irrigation,
drainage, and fertilization. Terracing and irrigation are the most intensive of
these practices and are not widely used by foresters. The exceptions are tree
nurseries and agroforestry regimes that produce food and fodder as well as
wood; these are beyond the scope of this book. In contrast, fertilization
and drainage are much more widely used by foresters and will play an ever-
increasing role as intensive silviculture becomes more widespread.

Fertilization

Forest ecosystems are not closed systems. Nutrients enter via weathering of rocks, precipitation, and dust, and depart through leaching, erosion, and harvesting. Nutrient removal by harvesting has been of particular concern, because as the demand for wood products has increased there has been a steady increase in the degree of utilization of each tree harvested and a steady decline in the time between successive harvests. Such changes cause an increase in the removal of nutrients and a decrease in the amount of time available for nutrient replenishment, and thus lead foresters to be concerned about the degradation of site quality (Young 1974, Daniel et al. 1979). These concerns may be especially pertinent where warm, moist climates foster the rapid breakdown of leaf litter, logs, and other organic matter, thereby eliminating the forest floor as a buffer against nutrient depletion. Forest fertilization can be implemented to replace the nutrients removed in harvesting, to improve the productivity of marginal sites, and to enhance the productivity of sites near wood processing facilities. However, fertilization is an expensive practice, especially when conventional agricultural fertilizers derived from petrochemicals

---

SILVICULTURAL TREATMENTS

A Descriptive Outline

I. Regeneration Treatments: done to establish a stand

  A. Artificial Regeneration: establishing a stand by transplanting or seeding

   Site Preparation: removing debris and/or competitors before planting or seeding: can be done manually, mechanically, chemically, or by burning

   Direct Seeding: establishing a stand by planting or spreading seed: seed can be broadcast or planted in spots

   Planting: establishing a stand by setting out seedlings, transplants, cuttings, or stumps

  B. Natural Regeneration: establishing a stand from seed produced on the site, seed brought to the site by wind or animals, or from the sprouts of root stocks already present on the site

   Even-aged Stands: stands in which there are no more than two age classes

    Clearcutting: establishing a new stand by first removing all of the previous stand; can be done as strips or patches

Seed Tree:	establishing a new stand from seed derived from widely spaced residual trees
Shelterwood:	establishing a new stand by gradually removing the existing stand so that new seedlings become established under the protection of the older trees
Coppice:	establishing a new stand from sprouts derived from the previous stand
Coppice with Standards:	mixture of seedlings and sprouts
Uneven-aged Stands:	stands in which there are at least three age classes
Selection:	maintaining an uneven-aged stand by repeated cuttings in a stand at regular intervals; may be organized as single tree or group

II. Intermediate Treatments: done between regeneration periods

  A. Stands *not* past the sapling stage:

Release Treatments:	freeing young trees from competitors manually, mechanically, chemically, by burning, mulching, grazing, or cultivation
Weeding:	removing any competitor, regardless of position
Cleaning:	removing overtopping competitors of same age
Liberation:	removing overtopping competitors of older age

  B. Stands *past* the sapling stage:

Thinnings:	reduce the number of stems per hectare to reallocate growth, anticipate and capture mortality, and stimulate understory
Low thinning:	from below removing weakest first
Crown thinning:	within the main canopy (high thinning)
Dominant thinning:	above the main canopy
Pruning:	remove limbs to improve wood quality and alter microclimate
Prescribed Burning:	controlled use of fire to control understory composition and fuel accumulation
Sanitation:	remove insect and disease-bearing material
Salvage:	remove dead trees while still marketable

and manure are used. Forest fertilization in the future is more likely to rely on nutrient-rich waste products such as ash from biomass boilers, paper mill sludges, and sludge from municipal sewage treatment plants.

Fertilization and biological diversity

1. *Fertilization of stands will generally accelerate succession.* Decreasing nutrient stress will lead to more rapid plant growth and intensify competitive exclusion, and thus tends to accelerate succession (Allen 1987). The effect of this on wildlife diversity could be profound because of the great importance of a stand's successional status on the abundance and diversity of the species dwelling within it (see Chapter 5, Age Structure).

2. *Fertilization of stands with open canopies usually results in a vigorous response of the herbaceous and shrub strata.* If there is little competition from overstory plants, the understory can respond to fertilization more fully; thus the response will generally be greatest when fertilization follows recent thinning. Irrigating two New York forests with municipal wastewater produced a response in the herbaceous stratum that apparently caused an increase in the diversity and numbers of breeding birds for two breeding seasons (Lewis and Samson 1981). An application of urea to a loblolly pine plantation in Mississippi increased the amount of deer forage for two growing seasons in both thinned and unthinned stands (Hurst et al. 1982). (See Chapter 11 for a discussion of vertical structure and biological diversity.)

3. *Fertilizer and lime can alter conditions for forest invertebrates, bacteria, and fungi.* Research in northern Europe has shown that increasing the pH of the forest soil shifts the microflora toward bacteria and thus affects the many invertebrate species that feed on soil microorganisms, especially nematode worms (Heliovaara and Vaisanen 1984). Fertilization can also affect invertebrates—some positively, some negatively—by improving tree vigor (Heliovaara and Vaisanen 1984). For example, fertilization may make Scots pine stands more resistant to attacks of defoliating insects by facilitating the trees' production of defensive chemicals. On the other hand, the increased turgor pressure of fertilized trees may lead to an increase in populations of sap-feeding insects.

4. *Applications of municipal waste, industrial residues, and ash from biomass boilers may increase the exposure of wildlife to heavy metals and toxins such as dioxin.* Industrial sludges and municipal waste have sometimes been found to contain levels of dioxin and

other toxins that laboratory tests suggest are unsafe, although documentation of impacts in natural settings has been lacking. For example, the application of dioxin-contaminated paper industry sludges to red pine plantations in Wisconsin produced detectable levels of TCDD in wild animals, but there were no discernable effects on histopathology, reproduction, or populations (Martin et al. 1987). Similarly, applications of municipal wastewater and sewage sludge to forested sites have led to accumulations of lead and cadmium in the tissues of small mammals (Anthony and Kozlowski 1982), and cadmium and zinc in plants and cottontail rabbits (Dressler et al. 1986) but again, there were no detectable effects on the organisms' health. Caution may also be needed when the ash from burning wood for power generation is applied to forested sites, because some tree species, for example, aspen, accumulate heavy metals in their bark, and these can be greatly concentrated in the ash.

5. *Applications of municipal wastewater to forested sites can increase understory biomass yet cause reduced plant species richness.* A study of mixed oak and oak-pine stands in Pennsylvania irrigated with municipal wastewater found that while the overall biomass of the understory increased, there were 32% fewer plant species present (Dressler and Wood 1976). Because the understory response was concentrated in a very few species that were not palatable to deer (24% fewer palatable species), the net result was a decrease in usable forage, even though the forage produced had increased nutrient concentrations. Competitive exclusion and physical damage due to ice breakage were thought to account for the loss of species on the irrigated sites.

Drainage

Millions of hectares of forests are growing on sites that are too wet for maximum growth. By reducing the stress produced by waterlogging, drainage should result in deeper root systems and better wind-firmness of the crop trees. While drainage technology is well known to agriculture, it has not been widely used in silviculture because of the lower value of forest land and forest products, not because of any overwhelming technical difficulties. Where land and forests are in high demand, drainage is an accepted part of forest management. In Finland, large areas of peatlands have been drained by shaping the peat into alternate mounds and trenches to allow planting trees (Vaisanen and Rauhala 1983). In recently glaciated terrain such as the northeastern United States, the drainage of many sites could be improved by removing glacial deposits that impede water flow.

Drainage and biological diversity

1. *Draining a forest site can change it into a different type of ecosystem.* A wooded swamp is not the same as an upland forest, and a different set of species will live there. Thus, extensive draining can deprive the landscape of a unique, and often a particularly rich, ecosystem. A study of breeding birds in the Connecticut River valley of Massachusetts found that the most poorly-drained sites had the most abundant and diverse breeding bird populations (Swift et al. 1984). In southern Georgia, the mixed hardwood stands typical of wet sites supported 17 bird species that were absent or rare in the adjacent slash pine flatwoods (Johnson and Landers 1982). The waterfowl capacity of some watersheds in Finland has been reduced by 45% by a combination of agricultural and forestry drainage over a 100-year period (Haapanen and Waaramaki 1977). Even modest degrees of drainage that do not change a stand enough to consider it a new ecosystem can still eliminate some species, particularly plants, that have very specific requirements.

2. *Drainage projects can have significant off-site effects.* The water from drainage projects must go somewhere—either onto another piece of land or into an existing water body, where it may have a negative effect. For example, the water from forest drainage can acidify existing impoundments (Haapanen and Waaramaki 1977). Drainage projects must be planned within a watershed or landscape framework to assure that sedimentation, chemical, and thermal effects are carefully considered.

Improving the Site by Protection from Disturbance

Forest protection is usually designed to prevent or lessen the interruption of tree growth brought about by disturbances such as wind, wildfire, insects, or disease. Protection affects forest diversity at both the landscape and stand scale, because many forests arise from large-scale disturbances (Oliver 1981) and are continuously shaped by small-scale disturbance events (Pickett and White 1985). Forest protection, by preventing something from happening, is every bit as much an ecological force as forest harvesting.

Wildfire

As society has grown more urban, people's understanding of wildfire has waned, but the image of a dreaded threat persists. In the U.S., the image is very clear; it is Smokey the Bear standing in a charred landscape bereft of trees and animals. The truth is, when fuels, weather, and terrain are just right, wildfires *can* wreak havoc of proportions exceeded only by a nuclear holocaust. However, the vast majority of wildfires are actually low-intensity burns producing modest levels of disturbance in the forest, and it is quite reasonable to also perceive wildfires as a force of renewal and rebirth (Komarek 1966).

While forest management will be hampered unless the threat of wildfires has been reduced, some forest ecosystems, such as the hard pine forests of the southeastern United States, or the eucalyptus forests of southwestern Australia, are composed of tree species that depend on recurring fires to maintain their dominance of the landscape (Chandler et al. 1983). Policies of fire exclusion in such fire-dependent ecosystems can lead to a reduction of shade-intolerant tree species, a loss of herbaceous understories, and a gradual build-up of dangerous fuels, either in the form of highly flammable litter or as an unnatural midstory that can carry a surface fire into the forest canopy. Wide-scale and effective wildfire suppression programs can thus change the role of fire from stand maintenance to stand replacement, and thus increase the very hazard they are designed to prevent (Clark 1988). Land managers attempting to reestablish historical patterns of repeated, light fires through the careful use of prescribed burns have been thwarted in some ecosystems by the buildup of fuels and increased tree density resulting from years of suppression (White 1986) and by regulations restricting the use of fire as a silvicultural and wildlife management tool. Somewhere between complacency and fear is an attitude of reasonable and judicious suppression of wildfire. While it is unrealistic and foolhardy to let all wildfires burn freely, suppression of all wildfires at any cost is likewise economically and environmentally unjustifiable.

Wildfire protection and biological diversity

1. *Wildfire exclusion can change the vertical structure of forest stands.* Fire exclusion can lead to the development of a mid-story of trees and shrubs in stands that would normally consist only of an over-story of large trees with thick, fire-resistant bark, and an understory of grasses and other smaller plants that can regrow quickly after a fire. For example, in the western U.S., fire exclusion can allow white fir and incense-cedar to come in under ponderosa pine (Biswell, 1972), and in the southeastern U.S. shade-tolerant hardwoods will occupy the understory and choke out the herbaceous stratum unless there is regular burning (Komarek 1966). Thus, while fire exclusion may initially foster increased vertical diversity by allowing a new stratum to develop, over time there will be a decrease in fire-tolerant, early successional trees and their herb and shrub associates.

2. *Wildfire exclusion can reduce the availability of early successional habitat.* Studies of fire histories for the United States and Canada have shown that wildfires are common to virtually all temperate for-

ests and recur at frequencies ranging from 2 years for ponderosa pine with grassy understories in the Sierra Nevada (Wright and Bailey 1982) to over 1,000 years for high elevation conifers in New Brunswick (Wein and MacLean 1983). Prior to the land-clearing activities of European colonists, many North American species (notably many kinds of game such as deer, ruffed grouse, and turkey) were quite dependent on forest openings resulting from lightning fires, windstorms, and the fires set by native Americans. Kirtland's warbler is still dependent on fire because of its tie to dense young stands of jack pine (Radtke and Byelich 1963). Wildfire is also a vital part of maintaining the diversity of the woodlands and brushlands of eastern and southern Africa, and some eucalyptus forest landscapes in Australia (Chandler et al. 1983).

Insects and disease

Forest trees provide such a rich and varied substrate that thousands of organisms have evolved to exploit them. Indeed, some insects and infectious diseases have been so successful that they have come to play an important role as agents of stress and disturbance, and thus they help keep a forest diverse in time and space. Some, such as the mountain pine beetles and shoe-string root rot, produce small patches of disturbance, while others, such as spruce budworm, can defoliate whole stands in a matter of weeks. Furthermore, when infectious diseases and insects are introduced into forests removed from their source of origin, they can have quite devastating consequences. Chestnut blight and Dutch elm disease have effectively eliminated their new hosts from once major roles in North America. Forests are also subject to an increasing array of environmental diseases related to air pollution (Bormann 1985). In 1984, nearly one-half of the forest land in West Germany was affected by dieback, and in 1982 one-fifth of the total maple forest in Quebec was hard hit by sugar maple decline (Desgranges et al. 1987). The current decline of high elevation red spruce in New York and New England may also be linked to atmospheric pollution (Johnson and Siccama 1983).

Despite their important ecological roles, insects and diseases are not usually welcome in a managed forest and considerable effort goes into controlling them. Nonetheless, no program of forest protection can do more than lessen the impact of a few major pest species. Chestnut blight, beech bark disease, Dutch elm disease, fusiform rust, and shoe-string root rot are just a few examples of major diseases that have plagued forest managers. Gypsy moth, southern pine beetle, mountain pine beetle, and white pine weevil in North America; pine moths and bark beetles in China (McFadden et al. 1981); and siricid wood wasps in New Zealand (Spradberry and Kirk 1981) account for millions of dead and devalued trees each year. Even where forest managers have resorted to massive aerial spraying of insecticides, insect defoliation sometimes is the major factor influencing forest harvesting over a vast region.

In contrast to fire protection efforts (which have often been so effective that forest managers have had to reintroduce fire) it is the techniques used to control insects and disease, rather than their success, that is often questioned in relation to wildlife health and diversity.

Insect and disease control and biological diversity

1. *Insecticides can have toxic effects on nontarget vertebrates.* In the 1940s and 1950s, DDT and several other new insecticides were developed, primarily to control insects that posed agricultural and human health problems. Some of these insecticides were so inexpensive and effective that forest landowners also began to find it economical to protect trees from insects. Unfortunately, some of the insecticides made from chlorinated hydrocarbons did not readily break down in the environment; they persisted in the tissues of animals and were passed through the food chain in ever increasing concentrations, a process known as biological magnification. Although a wide variety of animals became victims of these insecticides, the plight of eagles, peregrine falcons, brown pelicans, and ospreys was perhaps the best known because: (a) They are at the top of the food chain; (b) they are vulnerable to a sublethal effect, eggshell thinning; (c) their low populations and limited reproductive capacity make them especially susceptible to disturbance; and (d) they are large, popular birds. Eventually, this destruction sparked Rachel Carson's (1962) *Silent Spring* and a succession of other pleas to ban the use of persistent, broad-spectrum insecticides. In many developed countries, these bans have been implemented, and most of the insecticides now in use are not particularly toxic to vertebrates and usually degrade within a few weeks. There is still considerable call for concern, because many of the most destructive insecticides are still being used in developing countries where environmental regulations are less stringent and funds for more costly alternatives are scarce. (In most developed countries, only the *use* of these chemicals was banned, not the *manufacture,* and thus exports continue largely unabated.) Finally, there is always the possibility that there could be long-term, chronic effects associated with insecticides thought to be relatively benign.

2. *Most insecticides kill a broad range of nontarget insects.* Even some of the biological pesticides, such as those based on the bacterium *Bacillus thuringiensis,* often kill a whole order of insects, not just the target species. Most people do not think of insects as having

great value in their own right, but this is a very narrow viewpoint. Certainly anyone concerned with maintaining biological diversity has to consider insects, because this is where most of the diversity lies; just how much is difficult to say, because most species of insects have not been described. The ecological roles of insects are also of enormous importance; two that affect other groups of organisms are described next.

3. *Insecticides can reduce the food supplies available for insectivores.* An insecticide that killed nothing but insects could still affect other animals that depend on insects for food. For example, spraying carbaryl to control spruce budworm has been shown to alter the foraging behavior of warblers (Hunter and Witham 1985) and the behavior and growth of ducklings (Hunter et al. 1984).

4. *Insecticide applications can lead to reduced pollination and fruit set.* Many groups of insects pollinate flowers—butterflies, bees, and hover flies are most important—and if populations of these insects are decimated, the sexual reproduction of many plants can be inhibited (Johansen 1977). This problem is recognized by many farmers who import hives of bees to pollinate their crops, but there is not much concern for the flowering plants of forests and the animals that rely on their fruit crops (Hansen and Osgood 1984).

5. *By limiting the abundance of suitable habitat for pest species, integrated pest management can affect wildlife habitat diversity.* Carefully planned programs coupling silvicultural treatments with insecticide applications (IPM, or integrated pest management) show promise for reducing the dependence on chemicals for insect control in several ways; one of them is avoiding conditions that foster a pest outbreak such as large areas of uniform habitat (Huffaker 1980, Knight and Heikken 1980). Thus, a landscape in which the species composition and ages of forest stands is very diverse is a landscape that should be safer from insect pests and provide diverse wildlife habitat. However, an IPM program that eliminated, not just minimized, old stands would work against the interests of all the species associated with old forests. Although the landscape patterns produced by IPM are quite different from those produced by an insect outbreak, they too can be conspicuous and significant. During a budworm suppression program, rules against spraying near water leave unsprayed stands of spruce/fir along riparian zones that show as gray strips of standing dead trees among the green where pesticide application was successful. Hastily built road networks made to salvage high-risk stands fan out through the forest connecting clearcuts, and the intensive culture of young stands designed to diversify the age structure of the forest appear as carefully spaced, tiny dots of green.

6. *Exotic insects and infectious diseases can stress native trees and compete with native biota.* Exotic pests have far more potential to reduce forest diversity than have native pests that have coevolved with their hosts. Since 1969, Great Britain has suffered a tremendous loss of hedgerow elms from Dutch elm disease (Osborne 1982). While dead elms initially benefit woodpeckers and nuthatches by increasing foraging sites, as they fall the loss of elms may eventually reduce the nest sites required by secondary cavity nesting species such as kestrels, stock doves, barn owls, and tawny owls. Hopefully, aggressive protection measures can prevent the introduction and spread of devastating exotics such as Dutch elm disease and chestnut blight in the future.

7. *Air pollution is threatening biological diversity over vast areas.* Oxides of nitrogen and sulfur, heavy meals, and ozone from smoke stacks and exhaust pipes are reaching forests and lakes far downwind from their place of origin, causing stress to species with varying abilities to respond. One manifestation of stress in forest stands is attacks by secondary pathogens such as bark beetles, aphids, and root rots (Klein and Perkins 1988). As the trees weaken and die, the thinned canopy allows a stronger herbaceous and shrub stratum to devlop and that leads to changes in the avifauna as well. Sugar maple stands in Quebec affected by dieback had 16% fewer breeding pairs in the canopy and 47% more breeding pairs in the shrub stratum than healthy stands (Desgranges et al. 1987). While there was no difference in the number of species foraging in the overstory, affected stands had 20% more species foraging on tree trunks and 25% more species foraging in the shrub stratum. Whether this trend represents a stress-induced, temporary change in structural diversity or is the harbinger of the loss of canopy trees and species remains unknown, but this should not deter the formulation of effective pollution abatement policies. On a much broader scale, excessive carbon dioxide and the greenhouse effect may be causing global warming and other climate changes that will stress forest ecosystems in fundamental ways.

STAND ESTABLISHMENT

One of the most fundamental tasks in forest management is establishing a new stand either naturally, by careful handling of the previous stand, or artificially, by means of direct seeding or planting. In either case, the harvesting of a mature tree or stand is the first step in establishing the next crop. Chapter 6 discussed the various regeneration cutting methods; here we will focus on other treatments that usually follow cutting if a desirable stand is to be established.

Site Preparation

To establish a new forest stand, it is often necessary to:

1. weaken competitors and expose a mineral soil seedbed for natural regeneration;
2. remove competing vegetation and logging residues for tree planting; or
3. remove residues and prepare a mineral soil seedbed for artificial, direct seeding.

Because the objective in each case is to reduce the amount of living and dead plant material on the site, site preparation is a form of disturbance (Grime 1979). This may seem somewhat paradoxical, because one of the main reasons foresters undertake site preparation is to accelerate succession and the rate at which a crop of trees is produced. The answer is simple; although site preparation acts as a disturbance agent to set back succession a few years, its net effect is to hasten the time at which trees dominate the site again.

The impact of site preparation on the structure and composition of the resulting stand depends on the same list of factors that determines the response of the forest to any disturbance, be it fire, windstorm, clearcutting, or site preparation. A list gleaned from three sources (Grime 1979, Oliver 1981, Pyne 1984) of factors controlling post-disturbance response includes:

1. the character and intensity of the disturbance;
2. the size of the area disturbed;
3. the growth rates of the colonizing and residual plants;
4. the regeneration mechanisms of invading species;
5. the coincidence of weather, seed crop, and disturbance;
6. the density and multiplication rates of seed predators and competing plants;
7. the spatial pattern of disturbance and surrounding stands;
8. time since the last disturbance;
9. character of the last disturbance;
10. the productivity of the site; and
11. the season of the year in which the disturbance occurs.

A successful site preparation program depends on manipulating one or more of these factors at a cost that is in line with the management priorities of the forest owner. There are three basic types of site preparation—mechanical, chemical, or prescribed burning—and some of their characteristics are described here.

Prescribed burning for site preparation

Fire is one of our oldest land management tools; for millennia people have used it to favor the types of plants and animals they prefer. Broadcast burning (as opposed to burning laboriously-piled debris) is an effective, fast, and inexpensive site preparation tool for disposing of logging residues or undesirable vegetation. Fire lines and a holding crew are necessary to use fire safely, but because heavy equipment, if needed at all, is only used at the periphery of the treated area, the per-hectare cost of burning is much less than that of mechanical site preparation techniques, and there is far less chance of serious soil disturbance. Fire has an advantage over herbicides in that it can dispose of logging slash and unwanted vegetation that could hinder planting crews. By timing a prescribed burn to coincide with the proper weather conditions, fuel moisture, and soil moisture, a predetermined percentage of mineral soil can be exposed for natural reseeding (Sandberg 1980). Finally, fires can return nutrients to the site and raise soil pH (Wright and Bailey 1982). The most significant limitation on the use of fire for site preparation is not safety, cost, or predictability, but concern for air quality downwind. Another factor limiting the efficacy of fire for site preparation is that, unlike mechanical or chemical means, fire is a natural occurrence in the forest, and many weed species are adapted to it. Most burns will have only a transitory effect on competing plants and may actually stimulate the germination of some additional species from reserves of seed buried in the forest floor.

Mechanical site preparation

Logging residues and competing vegetation can be made less of an obstacle to both natural regeneration and tree planting crews by some expensive but effective methods, such as crushing, windrowing, or plowing. Additionally, these methods often expose mineral soil seedbeds suitable for direct seeding. Unfortunately, equipment heavy enough to topple and crush trees, or to plow through brush and logging slash, is heavy enough to cause soil compaction and erosion; pushing debris into long piles or windrows is especially likely to strip off the topsoil and dump it in debris piles along with the slash. Windrows often contain far more soil than woody debris and can incorporate 10% of the site's nutrient reserves, as much as would be removed in six bolewood harvests (Morris et al. 1983). Moreover, because many of the plants that compete with newly established seedlings reproduce from buried seeds, sprouts, or rhizomes, the benefits of mechanical site preparation are often short-lived.

Chemical site preparation

When undesirable plants occupy a site slated for regeneration, they can be killed with herbicides. Herbicides can have some significant advantages over other methods of site preparation. Most importantly, they are much less likely to cause soil damage than are mechanical methods, and they involve far less risk of offsite damage than fire. Furthermore, they can be much more selective than either a prescribed fire or a bulldozer. Indeed, by choosing the right herbicide, dosage, timing, and application technique, herbicides can be as selective as any manual means of vegetation control. Also, some herbicides have a residual effect, killing roots and stems, not just the current foliage (Walstad and Kuch 1987). On the other hand, herbicides have no effect on slash and may actually increase the obstructions and fire hazard by increasing the amount of dead, dry fuel. Finally, there is widespread concern about the wisdom of releasing any type of toxic chemical into the environment.

Site preparation and biological diversity

1. *Site preparation will generally accelerate succession.* By decreasing competition for tree seedlings, site preparation will lead to more rapid growth of the seedlings and hasten the point at which they dominate the site. The effect of this on wildlife diversity could be very profound because of the great importance of a stand's successional status on the abundance and diversity of the species dwelling within it (see Chapter 5, "Age Structure") and its effect on nutrient cycles.
2. *Light site preparation can increase wild fruit and forage production.* Light site preparation that only slightly disturbs the soil and thus does not kill root stocks may increase diversity (Swindel et al. 1983), production (Moore and Swindel 1981), and nutritional value (Stransky and Halls 1976) of the forage plants consumed by large herbivores such as deer. Clearcuts on which slash was first burned and then chopped with heavy, bladed drums pulled by a bulldozer prior to planting longleaf pine had more herbaceous food plants three years after treatment than unprepared sites (Buckner et al. 1979). Shrubs take longer than grasses to reach full development; fruit production by shrubs on burned and planted sites in Georgia did not peak until four years after burning and planting, and the number of species fruiting was greatest 6–10 years after treatment (Johnson and Landers 1978). Mechanical treatments that result in the least soil disturbance favor fruit production because they did not seriously retard shrub development (Stransky and Halls 1980).

3. *Intensive site preparation reduces fruit production and over time reduces the availability of understory plants.* Site preparation initially stimulates herb and grass production, but the more intensive practices foster rapid tree canopy development and thus shorten the period that the herb and shrub strata are well-developed (White et al. 1975). Intensive mechanical site preparation such as chopping, or slicing with a bulldozer blade, and then windrowing and burning followed by disking, can set back the development of the fruiting shrubs enough to cause reduced fruit production for at least three years after treatment (Stransky and Halls 1980). While heavy disturbance may reduce fruit production and understory biomass, it does not necessarily reduce plant species diversity.

4. *Site preparation that eliminates cull trees, snags, and logs can reduce biological diversity.* Removing unsaleable or rotten cull trees and snags from clearcut sites can have lasting effects on wildlife diversity (see Chapter 10, "Dying, Dead, and Down Trees"). The loss of trees and snags resulting from dragging heavy chains between two bulldozers through Colorado pinyon-juniper woodlands reduced breeding bird densities by one-half and bird species richness by two-thirds, and the effects were discernible for 8–15 years after treatment (O'Meara et al. 1981). On intensively prepared sites in the flatwoods of Florida, the lack of arboreal refuges for tree-dwelling snakes and lizards reduced reptile species richness for at least four years (Enge and Marion 1986). Leaving some untreated areas, individual snags, and cull trees within site-prepared areas may provide refugia for arboreal species. Undisturbed cypress domes within clearcut and site-prepared pinelands in Florida increased the densities of winter birds in nearby clearcuts (Swindel et al. 1983).

5. *Site preparation burns may increase plant species diversity on the disturbed site.* There is often an ample and diverse pool of buried seeds on forest sites, some of which have persisted for decades until a releasing disturbance occurs (Radosevich and Holt 1984). Site preparation burns typically remove the dry litter and varying amounts of the organic horizons along with the logging slash, and thus stimulate the germination of stored seed species such as blueberry, huckleberry, raspberry, and pin cherry. The post-fire community may thus be more complex than the pre-fire community, due to a combination of the germination of stored seed species, the resprouting of top-killed plants, and the proliferation of wind- and animal-disseminated species.

6. *Mechanical site preparation may foster greater plant species richness than chemical techniques.* Research in eastern Canada (Marceau 1981) and in the southeastern United States (Martin 1981) indi-

cates that mechanical site preparation can increase competition with the planted trees by stimulating sprouting and spreading of rhizomatous species. In addition, stored seed species may be encouraged as well. Mechanical techniques may thus foster greater species richness than chemical techniques that produce far less physical disturbance and more root system mortality.

Artificial Regeneration

Probably no other silvicultural practice has the potential to influence landscape diversity as much as artificial regeneration, because it gives a forester the greatest scope for creating entirely new types of stands—almost anything that a given site can support, even stands comprised of exotic species or trees that have been improved by genetic engineering. There is considerable incentive to be creative too; artificial regeneration and tree improvement programs may yield production levels 200–600 percent higher than can be achieved by natural stands (Farnum et al. 1983). Despite the potential for improved profits, the initial costs of artificial regeneration are high, and thus it operates under rather strict economic constraints, constraints that favor pure stands of one species and one age class. In other words, artificial regeneration often produces stands with impoverished species diversity and simplified vertical structure. However, this need not be the case; plantations can be established and maintained in ways that overcome some of the internal simplicity that sometimes earns them the label "biological deserts" among people concerned with forest diversity.

Planting allows foresters to control the species, age, spacing, and even the seed source of a stand. Some species can be reproduced asexually by means of sprouts, cuttings, or stump culture, and thus whole stands can be a single, carefully chosen clone. Perhaps the most intensive and controversial use of planting is to convert one kind of stand to another. Stand conversion is an expensive and heavy-handed practice, but it is often prescribed when a good site is dominated by an unproductive or undesirable stand. Industrial foresters who want a highly productive crop of a valuable species on land that is reasonably accessible to a mill are the most common practitioners, and the results are usually even-aged stands of conifers. Sometimes the main impetus to stand conversion is a desire to correct the effects of past cutting practices. Many stands have been so degraded by high-grading (cutting the best and leaving the rest) or coppicing for fuelwood that the best remedy is simply to start over.

Direct seeding has often been proposed as an alternative to planting, because it would seem less costly to establish new stands by just sprinkling the desired mix of seed from the air. Unfortunately, success is often limited by adverse weather, seed predation by birds and small mammals, or competition from pioneer herbs and shrubs. Several studies have documented the effects of small mammals on both seeds and seedlings with the common conclusion that very intensive mechanical site preparation, very heavy seeding rates, and perhaps even poison bait feeding stations are needed to overcome the problem (Radvanyi 1970, 1973, 1974).

Artificial regeneration and biological diversity

1. *Closed-canopy plantations often lack vertical structure and therefore have relatively low species diversity.* The plant species diversity of plantations tends to peak as the young trees develop enough height to become another habitat layer distinct from the herbs and shrubs; then tapers off as the tree canopy closes. The abundance and diversity of small mammals (Atkeson and Johnson 1979) and birds (Childers et al. 1986) generally follow this same pattern, except that some older stands with greater foliage height diversity and more open canopies can provide more niches than young plantations (Jessop 1982). A study of pine plantations in New Guinea found that whole groups of bird species found in surrounding natural stands were missing in the plantations (LaMonthe 1980). Bird and mammal species diversity was likewise found to be much lower in plantations in Indonesia than in either undisturbed stands or in partially cut stands (Wilson and Johns 1982).

2. *The arrangement of plantation trees can be manipulated to enhance vertical structure and thereby support more species.* One key to maintaining wildlife diversity in stands past the establishment phase is maintaining an open canopy. Plantations typically include many more stems per hectare than will ultimately be harvested as final crop trees. The intervening trees are used as trainers for the crop trees (better form, smaller limbs), to help shade out competing plants, and as a source of early financial return through thinnings. Planting at wider spacings can be done in ways that do not reduce the final product volume but still allow for greater plant species richness and greater vertical diversity early in the rotation (Lewis et al. 1985). Wild turkey and deer habitat in pine-hardwood forests of the southern United States can be enhanced by spacing trees at 3 × 3 meters or wider (Wenger 1984).

   Whether wide spacings are used or not, it is still possible to build vertical structure into a planted stand by establishing the planted trees at time intervals designed to provide vertical diversity (see Figure 12.1). Age classes could be established as individuals, groups, or alternate rows. When starting with a recently cleared area, the major difficulty involved would be keeping the unplanted

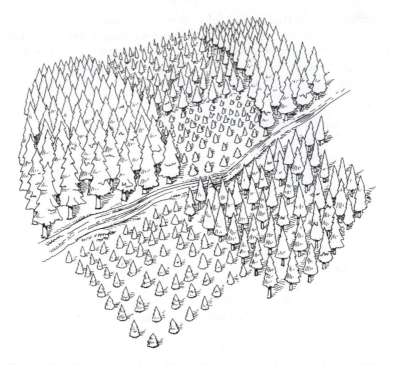

Figure 12.1 Vertical structure can be built into a planted stand by establishing the planted trees at time intervals either as individuals, groups, or alternate rows.

portion of the site free of established competition until each age class was established. This could be done chemically or mechanically or by establishing a cover crop. Interplanting legumes in sycamore plantations in northern Alabama improved the growth of the crop trees by controlling weeds and providing available nitrogen (Haines et al. 1979).

3. *Young plantations may support more plant species and produce more food for herbivores than closed-canopy natural stands, but the reverse may be true of older, unthinned plantations.* Plant diversity and production of forage (fruit and vegetation) for quail, turkey, and deer were studied following the conversion of Virginian second-growth hardwood-pine forests to loblolly pine plantations (Felix et al. 1983, 1986). Very young plantations had more species and greater production of shrubs and herbs than older natural or planted, closed-canopy stands. By the time the plantations reached 15–20 years of age, however, both the diversity and production of herbs and shrubs had fallen behind the natural stands and remained behind until the first thinnings were made. In the southeastern

United States, the conversion of hardwood stands to conifer plantations has been found to negatively affect songbird populations, because the subsequent reduction in the diversity of mast producing
species results in temporal gaps in the production of mast (Noble
and Hamilton 1975).

4. *Planting old fields to trees can reduce available openland habitat.*
Old grass fields planted to one or two coniferous species have become synonymous with "plantation" in many people's minds. Old
field plantations tend to produce less shrubby understory growth
than previously forested sites and thus have led to the perception
that plantations inherently lack vertical structure (Marion and
Harris 1982). In some landscapes, old field habitats are becoming
rare as housing development and forest succession squeeze them out
or fill them in. The consequences for openland wildlife are obvious;
for example, widescale afforestation in southern Scotland is thought
to have reduced the productive foraging habitat for golden eagles
(Marquiss et al. 1985). Where wildlife density, diversity, and visibil-
ity are desired, the decision whether to plant and how to plant a field
should be made with care.

5. *Establishing a mixture of species can improve both the species richness and vertical structure of plantations.* Spreading seed mixtures
from the air has been advocated as a way to foster diversity because
of the inherently patchy distribution in the resulting stands in both
composition and structure (Noble and Hamilton 1975). Likewise,
there is no reason that several species could not be planted together.
Where the final crop is designed to be a mixture of species that have
similar growth rates and shade tolerance, it is advisable to plant
each species in blocks or double rows to reduce the tendency for
one species of the mixture to squeeze out the others. Alternate row
plantings and simple tree-by-tree or "bucket" mixtures often wind
up as nearly pure stands due to competitive exclusion. In Central
Europe, this tendency is overcome by carefully blending the mixture
to include species that differ enough in their inherent growth rates
to give stratified mixtures (Smith 1986) (see Figure 12.2). A random
mixture of European larch, red pine, eastern white pine, and Norway
spruce planted in Connecticut sorted into four strata within 33 years
of planting (Smith 1986).

INTERMEDIATE TREATMENTS

As global reserves of old-growth forests are depleted and competition for wood
increases, forestry is entering an era of trying to match growth to harvest
without the subsidy that virgin stands have provided in the past. Attention is

Figure 12.2 Establishing a mixture of species can improve both the vertical structure and species richness of plantations. This is especially true if the mixtures are carefully constructed to include species that differ enough in their inherent growth rates to give stratified mixtures such as the aspen/fir stand depicted.

shifting from trying to locate and access the next stand of saleable wood to intermediate practices that accelerate the rate at which the desired trees can dominate and grow on the site.

Release Treatments

Release treatments, i.e., freeing seedlings and saplings from noncrop competitors, is one of the earliest treatments undertaken in the life of a stand. Young trees are very vulnerable to competition, especially from mature trees that shade the small trees from above and remove water and nutrients from below. After a substantial harvest, most of the competition will come from the same stratum, from other seedlings, and from herbs and shrubs. Foresters often seek to minimize this competition because they wish to increase the chances of desirable trees surviving and to hasten the time when the crop trees will dominate and fully use the site. Occasionally the most serious competition is among the young trees themselves, and a forester will decide to remove a portion of the trees to favor rapid growth of the remaining individuals. By controlling competition during the period when a stand is becoming established, release treatments have an enormous effect on the stand's species composition, the relative dominance of different plants, and the density of the trees.

Release treatments must be carried out in ways that kill the competing plants, or reduce the density of the crop trees, without destroying or unduly stressing the crop. Thus, they tend to be quite selective and their ecological impacts are relatively subtle; nevertheless, the effects can be important and long-lasting.

Release treatments and biological diversity

1. *Release treatments accelerate succession by helping the crop trees dominate the site sooner.* Because each stage of succession has a special set of dominant wildlife species, collapsing the time scale shortens the time that a site is usable habitat for species of the young forest. A study of clearcuts treated with herbicides in Norway related a 30% reduction in bird densities to a reduction in invertebrate food (Slagsvold 1977). Similarly, clearcuts planted to spruce and released with aerial applications of herbicides in Maine had a reduced density and diversity of breeding birds for one to three years following treatment (Santillo 1987). The earlier in the life of a stand that a release treatment is carried out, the greater the potential to negatively affect species diversity. Conversely, delaying re-

lease treatments prolongs the time that a site may be suitable for openland wildlife.

2. *Herbicides can have a longer residual effect on target vegetation and plant species richness than cutting or burning.* Cutting and burning usually provide only a temporary release, because they kill only the above-ground portions of competitors and have little effect on the root system. Herbicides that either are active via the soil (picloram, simazine, and hexazinone) or are translocated down into the root systems (glyphosate and 2,4-D) can kill the root stock as well, and thus have a longer residual effect (Walstad and Kuch 1987).

    Many, perhaps most, of the plants that will eventually dominate a community are already present at the time of disturbance either as plants or propagules, or become established very early in the life of a stand (Radosevich and Holt 1984). Because secondary succession is largely a shifting pattern of dominance among persistent species (see Figure 3.3b), release treatments that kill (as opposed to those that merely cut or top-kill) competitors at an early stage could therefore reduce the pool of species on site for the entire rotation. If the treatment occurred so early that species that persist from one disturbance to the next as stored seeds did not have an opportunity to produce seeds, then the impact on plant species richness could persist into a second rotation.

3. *Herbicide applications can be modified to maintain or build diversity.* Picloram-based herbicides have been used successfully in Kentucky to produce edge habitat and clearings in closed-canopy stands. While small mammals were found to be twice as abundant in the resultant openings as in the forest (McComb and Rumsey 1982), the response of songbirds varied by species and was significantly related to dosage rate (McComb and Rumsey 1983b). A study of recent clearcuts in Maine that were treated with aerially applied herbicides found that the swaths inadvertently missed by the helicopter pilot provided important refugia for brushland mammals and birds (Santillo 1987). Such swaths do not have to be accidental. By alternately turning the spray boom on and off, a helicopter pilot can consciously create virtually any pattern of treated and untreated vegetation desired. A series of research plots in Maine has demonstrated the ability of helicopter pilots to create 1 ha to 1.5 ha patches of varying habitat by using different herbicides at different rates (McCormack et al. 1978), or to put down alternate treated and untreated strips as narrow as 2.4 m wide (McCormack and Banks 1983).

4. *Where the crop species is overstocked, early spacing treatments can help maintain diversity.* Sometimes a young forest is so dense that it forms a thicket that rapidly dominates the site, excluding most

other plants (Figure 11.2). Early attention to proper spacing can accelerate the growth of the individual crop trees and, at the same time, open the closed canopy enough to allow a richer flora to persist longer. Furthermore, if the treatment involves cutting individual stems or injecting individual trees with herbicide, it is possible to leave some noncrop trees, perhaps fruit-bearing species important to many animals.

5. *Liberation treatments can foster habitat for cavity-nesting wildlife.* Sometimes after a stand has been high-graded, leaving old, undesirable trees that are overtopping the young crop, the crop can be liberated by killing the old trees. If, instead of being cut, these old trees are killed and left in place using the various methods described in Chapter 10, the supply of snags is increased. Because snags have a finite life, the supply of cavities could be prolonged if some cull trees were left alive to die and become snags later.

6. *Early release treatments can have lasting effects on primary cavity nesters that rely on shade-intolerant tree species that form soft snags.* Throughout the boreal forest, chickadees depend on soft snags of aspen and birch for nesting and winter shelter. If release treatments are conducted early in the life of a spruce-fir stand, either the supply of suitable snags will be much reduced, or the snags available will be too small to afford usable cavities. Because the soft-rotting trees are early successional species, the supply will remain limited until another disturbance occurs.

7. *Concentrating release work around individual crop trees makes the work more efficient and helps build and maintain stand diversity.* When saplings are released from competing trees by cutting, girdling, or herbicide-injection it often makes little sense to remove all the undesirable trees from the stand. Needlessly treating portions of the site where there are few crop trees wastes effort and diminishes the stand's tree species diversity. In addition, crop tree release treatments can be the first step toward initiating a coppice-with-standards approach to building stand structure and maintaining tree species diversity where there is a market for pulpwood or fuelwood. Staggering the time of harvest of the low-value products will further increase the vertical structure in the stand (see Figure 12.3).

Thinning

Once a stand of forest trees is established on a productive site, the crowns of the trees expand until they touch. As the trees occupy all the available growing space, competition causes the death of some trees, and then the surviving trees compete for the growing space made available by the demise of their

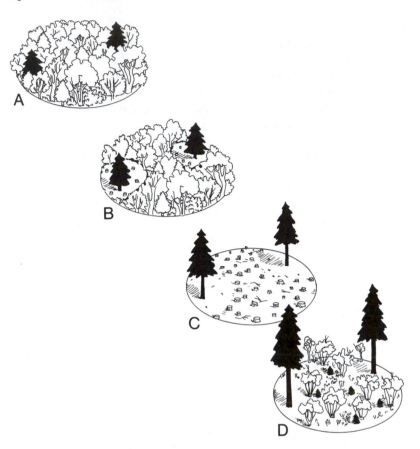

Figure 12.3 Concentrating release work around individual crop trees in very young stands (A) makes the work more efficient and helps maintain species richness (B), and can be the first step toward a coppice-with-standards approach to building vertical structure while maintaining species richness and raising forest products (C and D). Using this approach, it is possible to use a single site to raise both a small crop of high-quality sawtimber trees (standards) grown on a long rotation, interspersed among trees of lower value and sprouting species (coppice) that are harvested on a far shorter rotation for fiber or fuelwood.

neighbors. Foresters often implement thinnings to encourage the natural tendency for a few large trees to ultimately occupy the space that once supported many small trees.

The process of competition-induced mortality occurs incrementally as the crowns of some trees are overtaken and passed by the crowns of more competitive individuals. The gaps resulting from this kind of mortality thus tend to be small and beneath the main crown canopy. Although mortality is recurrent, the standing crop biomass increases over the life of the stand, because disturbance of the active foliage is minimal. Thinnings that mimic this process by removing trees from beneath the main crown canopy in anticipation of mortality are termed low thinnings (or thinning from below) (Smith 1986). Crown thinning and dominant thinning refer to removing trees from within or above the main crown canopy respectively (thinning from above). By reducing photosynthetic surfaces, crown and dominant thinnings result in a temporary reduction in the energy fixed and biomass accumulated by a forest stand. If these thinnings reduce the biomass accumulated by a forest, why are they among the most prevalent practices of intensive silviculture? Smith (1986) suggests several reasons:

1. Thinnings can favor the growth of selected crop trees whose greater value will compensate for the reduced overall biomass produced.
2. Thinnings made in even-aged stands can result in a financial return early in the rotation from the sale of pulpwood or fuelwood.
3. Thinnings made in uneven-aged stands, in concert with the regular regeneration cutting cycle, can increase the amount of material sold or used at each stand entry.
4. Thinnings can salvage the biomass of trees that die, or are about to die, from competition.
5. Thinnings can foster grasses, forbs, and shrubs beneath the forest canopy for wild and domestic animals.

Thinning too early and/or too heavily can reduce stand growth, leave stands vulnerable to windthrow, and foster an understory robust enough to compete for moisture and nutrients. Thinning too late or too lightly will lead to overstocking, reduced growth of the crop trees, and increased losses to mortality. Between these two extremes, a forester has to choose among several levels of stand density that all produce about the same amount of growth (Langsaeter, 1941 cited in Daniels et al., 1979). The same stocking guides that help foresters prescribe thinnings to maintain timber production can be used to prescribe thinnings designed to maintain or enhance within stand diversity (see box).

Thinning and biological diversity

1. *Thinnings can change vertical diversity and thus alter species richness.* Because many forest birds occupy very specific microhabitats among the strata of a forest (Dickson and Noble 1978, Connor et al. 1983c), the simply structured, often shallow canopy of an even-aged stand will have a simple avian community (Chapter 11) compared to a mature forest in which natural thinning reduces competition and allows sunlight to reach the forest floor (Harlow et al. 1980, Lanyon 1981). In contrast to low thinning and dominant thinning (which by definition remove elements of vertical structure), crown thinning

Figure 12.4 is an example of the stocking charts prepared for foresters working in the northeastern United States. Foresters are able to evaluate stand density and to prescribe thinnings by plotting stands on the chart. The B line is usually the desired goal of thinnings made to foster individual crop trees without sacrificing total timber production. Maintaining stands at densities below the B line would mean that the intervening spaces among crop trees are *not* stocked with trees. Because the C line represents a level of stocking that will result in the stand reaching the B line within ten years; thinning to levels below the C line would assure at least ten years of canopy opening and would thus foster the development of more vertical diversity within the stand. A single, heavy thinning early in the life of the stand might allow more shade-intolerant tree species and those other species that depend on them to persist within the stand, thereby increasing species richness. More frequent, somewhat less severe, treatments might still build vertical structure.

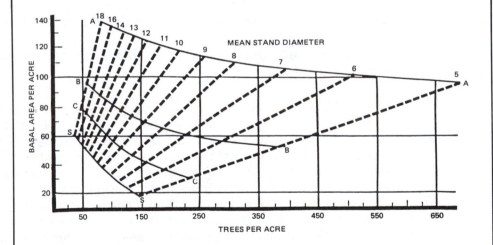

Figure 12.4   Stocking chart for even-aged northern hardwoods, based on number of trees in the main canopy, average diameter, and basal area per acre. For timber objectives, stands above the A line are over-stocked. Stands between lines A and B are adequately stocked. Stands between lines B and C should be adequately stocked within 10 years. Stands below the C line are definitely understocked (from Carl et al. 1982).

can create an open canopy that enhances the development of herb and shrub strata and promotes the development of deeper crowns on the crop trees (see Figure 12.5). In plantations and natural stands comprised of one species, early and repeated crown thinnings are often the most practical way to maintain plant species richness and vertical diversity, especially where short rotations and even-aged management systems preclude stands from developing the fragmented canopies of old age.

2. *Crown thinnings can influence the quantity and quality of understory forage available for herbivorous animals.* Thinning at intervals of five years or less has been suggested as a way to increase forage production; a longer interval might allow the tree crowns to expand and fully shade the site again between treatments (Conroy et al. 1982). Although the protein, phosphorous, and calcium content of forage grown under heavy shade can be greater than for forage grown under full sunlight, shade can also reduce the digestibility of forage by causing a high cellulose content, and the reduced digestibility can more than negate the advantage (Blair et al. 1983). Increasing the light from 8% to 45% of full sunlight will result in more leaf biomass produced, and the forage will tend to have increased levels of soluble carbohydrates, digestible energy, and digestible dry matter.

3. *Thinning can contribute to reduced forage production over time.* Thinnings can lead to the development of a mid-story of shade-tolerant species that may actually intercept even more light than the original canopy, and a reduction in forage production may result (Blair and Feduccia 1977). Thus, while thinnings may build vertical

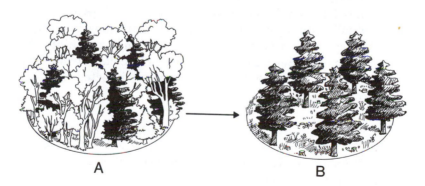

A   B

Figure 12.5 Crown thinning can create an open canopy that enhances the development of herb and shrub strata and promotes the development of deeper crowns on the crop trees. Crown thinnings can also impact species composition.

structure that benefits songbirds, they can simultaneously work to the detriment of terrestrial herbivores (Dickson 1981).

4. *Heavy crown thinnings can significantly alter the microclimate within a forest stand.* The forest canopy intercepts solar radiation, precipitation, and wind in ways that markedly reduce the exposure of forest-dwelling plants and animals to climatic extremes. The same microclimatic changes that allow a lush herbaceous response to canopy openings can expose sensitive species or individuals to stress or death. To avoid freezing temperatures, overwintering monarch butterflies in Mexico avoid openings and areas that have been radically thinned (Calvert et al. 1982). Black-tailed deer in British Columbia are dependent on dense stands of evergreens to escape the deep snows and cold winds of winter. Moderate thinnings can encourage ample summer food for the deer and help build deep crowns that improve the interception of snow. Heavy thinning can result in a broken canopy that affords little thermal protection and snowpacks deep enough to severely stress the deer (Nyberg et al. 1986). (Also see "The Wild Daffodil" section, Chapter 11.)

5. *Low thinnings and dominant thinnings can reduce vertical structure and species richness.* If either low thinning or dominant thinning is done thoroughly, it is possible to remove significant habitat components for those species that live in the lower canopy or that prefer perching and nesting sites above or below the main canopy. This is especially true in even-aged, stratified, mixed species stands where the different strata of the forest are dominated by different tree species (see Figure 12.6).

6. *Low-quality trees can be removed from competition in ways that augment habitat diversity.* When possible, removing low-quality stems in groups will foster a greater understory response than removing individual trees (McComb 1982). Killing individual low-quality stems by either girdling or herbicide application will provide forage and nesting sites for cavity-using wildlife (McComb and Rumsey 1983*a*); girdled southern red oaks rot more slowly and remain standing longer than herbicide injected oaks (Conner et al. 1983*b*).

Pruning

Pruning involves removing the lower limbs from the boles of the crop trees and is done to:

1. promote the growth of knot-free lumber;
2. improve air movement through a stand, thereby lessening the potential of fungal diseases;

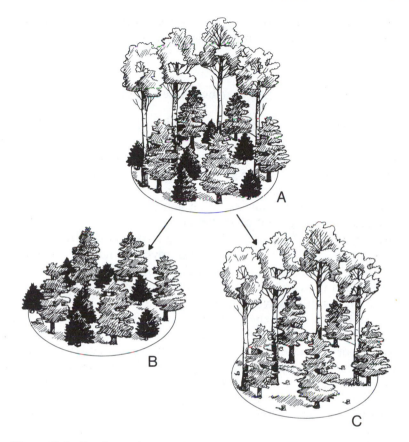

Figure 12.6 Dominant thinning (B) and low thinning (C) reduce the vertical structure in a stand. In stratified, mixed stands the species composition can be negatively affected as well.

3. improve access;
4. reduce the chance of a surface wildfire climbing into the tree crowns; and
5. improve the appearance of a stand.

When the limbs removed are live limbs close to the ground, there is a reduction in the forage available for browsing wildlife, e.g., snowshoe hares, and a simplification of the vertical foliage diversity that is important to songbirds. This effect can be pronounced where low limbs are routinely gathered for feeding domestic animals. When pruning is delayed until the lower limbs have died from shading, there is less impact on the structure of the stand, although there is a reduction in low-level perches and feeding sites for birds. Pruning impacts on songbirds can be minimized (and the work made more efficient) by restricting pruning to selected crop trees within a stand where doing so would not unduly compromise other objectives.

Prescribed Fire in Forest Stands

Prescribed fire is used in timber management to eliminate heavy accumulations of litter that could fuel a destructive wildfire, to control the composition of the forest understory, to recycle nutrients, to control certain insects and diseases, to maintain grasslands and grassy understories, and to decrease the water used by plants on the watersheds (Komarek 1966, Pyne 1984). There can be a narrow window between achieving desired results with a prescribed fire and losing control of it. Escaped fires can kill more than just brush. In Michigan, the Mack Lake fire was set to benefit Kirtland's warblers but escaped and killed one firefighter, destroyed 44 homes, and burned 24,000 acres (Simard et al. 1983). Even when the fire itself is contained, the smoke from a prescribed fire can exacerbate respiratory ailments or even cause multivehicle accidents on highways downwind.

By controlling the frequency, intensity, and pattern of prescribed fires, land managers can direct succession over large areas at low cost and with limited labor. The effects of fire can be approximated but not duplicated by other means, because some species depend on fire to open pitch-sealed cones or to break the dormancy of buried seeds. Fires also are unique because ash tends to increase soil pH and provide a pulse of readily available mineral nutrients for plant growth (Pyne 1984). Prescribed fire is a traditional tool in the management of some forest types, but it has far broader potential for managing wildlife habitat diversity than is currently being exploited.

Prescribed fire in forest stands and biological diversity

1. *Prescribed fire can be used to control the degree of vertical diversity produced by thinning.* Thinning can stimulate the growth of understory plants (see "Thinning" section), but if many of these are shade-tolerant tree seedlings or tall shrubs, they will eventually dominate the understory and shade out many herbaceous species (Blair and Feduccia 1977). Prescribed burning can be used to top-kill some or most of this woody mid-story and thus maintain an understory with greater species and structural diversity (Komarek 1963). This principle is regularly used in the management of he hard pines in the southeastern U.S. There, one- to two-year burning schedules are used to benefit quail and turkeys and three- to five-year burning schedules allow more woody sprouts which benefit deer (Marion and Harris 1982). Unduly extending the interval be-

tween burns would let the sprouts grow out of reach of the deer, could lead to increased fuel accumulations, and might let some of the understory stems develop bark thick enough to resist surface fires. An attempt to restore oak savannah conditions in Minnesota was only partially successful, because oaks greater than 25 cm dbh already had bark thick enough to survive surface fires (White 1986).

2. *Special habitat elements such as den trees and fruit trees can be destroyed during prescribed fires.* A fire will burn whatever is burnable; there is no discrimination in favor of uniquely valuable stems. Prescribed fire in the ponderosa pine forests of Arizona can cause a 45% net decrease in snags greater than 15 cm in diameter in the first year after burning. Snags greater than 50 cm in diameter or that are in the advanced stages of decay are especially vulnerable because of the debris that accumulates at their base (Horton and Mannan 1988). The endangered red-cockaded woodpecker nests almost exclusively in live pine trees that would normally be almost invulnerable to light, surface fires (see Chapter 10). However, the resin that flows from the cavity entrances forms a column of pitch that makes the cavities more vulnerable to low-intensity surface fires. Attempts to protect nest trees by raking away litter accumulations prior to burning have been only partially successful, and not all nests are discovered in time. Where prescribed fire is to be used in proximity to special habitat elements, it may be necessary to construct a wider than normal fireline and to burn more frequently to prevent heavy fuel accumulations (Connor and Locke 1979).

3. *Burning at different times of the year promotes habitat diversity.* Because the moisture content of forest soils and forest litter varies with the season, and because various plant groups differ in phenology and bear their growth points at differing heights above the soil surface, the effects of prescribed fires on different plants will vary with the season. Burning between growing seasons maintains the diversity of flowering forbs and shrubs in longleaf pine forests, but the diversity decreases progressively as burns occur later in the growing season (Platt et al. in press). In the southeastern U.S., repeated summer burns will eliminate most sprouting hardwoods and favor grasses, while winter burns in that same region can be used to foster sprouts for deer forage (Stransky and Harlow 1981). A study of winter burning in the flatwoods of north Florida found that the frequency and biomass of the herbaceous species increased, and the number of woody species in the understory was not reduced; thus there was an overall increase in understory species diversity (Moore et al. 1982).

4. *Prescribed burning can influence the quality and quantity of forage for herbivorous animals.* Because grazing animals seek out forage

growing on recent burns, hunting cultures have often used burns to concentrate game. Fires make new growth more nutritious, more digestable, and, by consuming dead materials, more accessible. Prescribed burns have produced increases in forage production of up to five times preburn levels (Hurst and Warren 1980) and increases of twice preburn levels have been found to persist for four growing seasons (Hallisey and Wood 1976). The new growth following a burn can contain more nutrients, less lignin and crude fiber, and as much as three times the crude protein compared to forage from unburned stands (Stransky and Harlow 1981). Unfortunately, unlike the improvement in forage production, improvements in forage quality seldom last more than one growing season.

5. *The frequency and intensity of prescribed burns can influence the fruit production of the forest understory.* Fruit production is often temporarily reduced when shrubs are set back by fire. A study in slash pine plantations in southern Georgia compared different prescribed burning regimes and concluded that while the maximum number of species fruiting occurs 6–10 years after a fire and some species take 4 years to reach peak production, a 3-year interval between burns yielded the best overall fruit production (Johnson and Landers 1978).

SUMMARY

Intensive silviculture is becoming increasingly important to a world of shrinking resources and expanding human population. Whether there will eventually be enough forests to support all of the wildlife species that depend on them and the ever-increasing human demand for forest resources remains an open question. The outcome will ultimately depend both on curbing that demand and on the ability of foresters to increase the productivity of those forests closest to population and processing centers. There is nothing inherent in silvicultural techniques or systems that will either automatically augment or deplete the biological diversity of a stand. As this chapter has shown, silviculture can do either; it is the forest owners and their agents that determine to what end these silvicultural techniques are used. Unfortunately, there is no way that the careful management of a single stand or small forest property can individually overcome landscape scale patterns imposed by the collective result of hundreds of individual decisions. While silviculture focuses on individual stands, the landscape remains the scale at which the fate of wildlife species is ultimately determined.

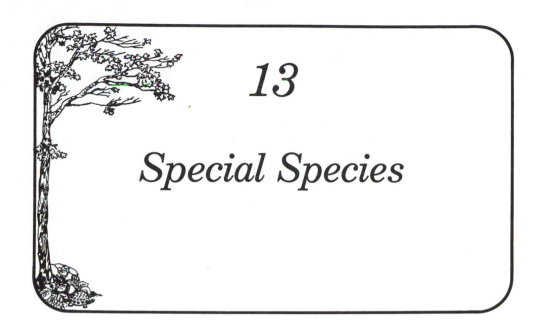

13

Special Species

Biological diversity, age structure, spatial heterogeneity—these are conceptual entities, intangible ideas that have some significance for the professionals that use them, but mean little to the general public. Even ecosystems are rather vague and ill-defined for most people. It is easy to associate a visual image with "oak-pine forest," but to define a forest precisely, to decide where one type of forest ecosystem ends and another begins, is difficult even for an expert. In sharp contrast, an individual plant or animal is very real and definitive. People can relate to it; they can admire its beauty; they can desire to possess and use it; they can even form emotional relationships ranging from antipathy to love and sympathy. From perspectives based on individual organisms, it is only a small step to being interested in whole species *and* forming value judgments about their relative worth. For example, a brief encounter with a bald eagle, even a few dramatic photographs in a magazine, could easily convince someone that bald eagles merit special attention as a species. Through such impressions we come to hold a few species in high esteem, while a few others are singled out for disfavor; most species are either ignored, or lumped into large categories—songbirds, flowers, weeds, and vermin—that are based as much on human values as ecological or taxonomic considerations.

The activities of wildlife managers generally reflect societal values and thus are often oriented toward the management of individual species, rather than conceptual goals such as biological diversity. Specifically, wildlife managers direct most of their attention toward highly valued species. Pest species also receive considerable attention, although most of it comes from people who work with insecticides, herbicides, rodenticides, etc. and not from conventional wildlife managers working on vertebrate animals in natural ecosystems. Wildlife management that falls under the rubric, "saving wildlife habitat," is an activity of consequence for most species, although it too is often catalzyed by concern for a handful of popular species.

There is also a logistical reason for emphasizing the management of species; species or populations are logical units for management; all their members are relatively similar and, broadly speaking, will respond to management in the same way. This makes it much more straightforward to define objectives and strategies for a species-oriented management plan than for one with broader goals like maintaining diversity. In this chapter we consider the relationship between species-oriented management and management aimed at maintaining biological diversity in some depth, and then review some of the basic ideas behind managing habitat for special species—sometimes called "featured species management" (Holbrook 1974).

Why Are Some Species Special?

The veracity of George Orwell's (1946) well-known quote: "All animals are equal, but some animals are more equal than others" is not debatable, although some people would rather recognize it only as a description of the way things are, not as a prescription for the way things should be. There are three basic reasons that people consider some wildlife species "more equal than others": (a) The most fundamental reason is a species' utilitarian value: does it provide food, clothing, shelter, medicine, etc.? Ecological value—i.e., is the species known to be a critical component of an ecosystem that provides clean water, food, clothing, etc.?—can be considered a subset of utilitarian value. Unfortunately, the ecological value of only a few species is really understood, and thus it is relatively unusual for this factor to be considered. (b) Another reason is based on values that are more subjective or esoteric such as those associated with aesthetics, religion, and sport-hunting. A bird-watcher, a shaman, and a duck hunter may seem to share few values, but they are all attracted to some intangible essence that certain kinds of wildlife possess and others lack. (c) The third reason is somewhat more complex, because it is based on the potential, but undescribed, values of all the species about which virtually nothing is known. This, of course, is the vast majority of species. If, in the absence of good information, one assumes that all species could have undiscovered utilitarian value, the prudent strategy is to set priorities by considering those in greatest danger of becoming extinct to be most special. In short, the third set of values is based on the rarity or vulnerability of species and the assumption that they all have potential value.

Obviously, there is much overlap among these categories: Many species sought for their sport or aesthetic value have economic utility; many endangered species have great aesthetic value. However, the order in which they are presented here—utilitarian value, subjective value, and rarity value—reflects the priorities that society uses when managing these resources. First come the utilized species, such as the trees we need for lumber, fuel, and paper, and the fish we commercially harvest to eat; then come the species we enjoy for more subjective reasons, the beautiful, exciting, or inspiring species; and finally, there are the rarities that we may lose if we are not careful.

SPECIAL SPECIES MANAGEMENT AND DIVERSITY

If some species of wildlife are special and thus subject to more attention and management, how does this affect the broader goal of maintaining biological diversity? When you think of trees as special species of wildlife—special because of their enormous utilitarian value—it is obvious that special species management undertaken on a commercial scale can have profound effects on biological diversity, some positive, some negative. This whole book is a testament to the important relationship between timber management and biological diversity, and we need not dwell on the idea further here.

What about special species management in the context of traditional wildlife management with its focus on vertebrate animals, especially game animals? Habitat management for game animals is usually far less intensive and extensive than management driven by commerce. To take an extreme case, compare the impact of implementing a cutting pattern designed to favor ruffed grouse on 40 acres (see Figure 5.7), versus introducing commercial logging to southeast Alaska or the Amazon basin. Because game management will tend to create small islands of special habitat in a sea of forest managed primarily for timber, it will usually diversify the landscape and thus enhance wildlife diversity. For example, when managers of turkey habitat in the southeastern United States call for fewer hardwood stands to be converted to pine plantations (Shaffer and Gwynn 1967), biological diversity in general will benefit. On the other hand, most of the southern forests managed for deer, quail, and turkeys have no red-cockaded woodpeckers, and a management regime exclusively focused on game species can overlook many nongame species. Moreover, it is unlikely, but certainly conceivable, that a large-scale program of managing habitat for a game species could actually work to the detriment of maintaining wildlife diversity.

Management for the third type of special species, those threatened with extinction, clearly has a pivotal role in managing for diversity. If a species is lost, the biological diversity of the whole planet is diminished. Even if a species is only extirpated locally, this has a broader effect because the loss of even a single population reduces the planet's genetic diversity to some degree.

Coarse Filters and Fine Filters

The interaction between managing for endangered species and managing for biological diversity is well-illustrated with a metaphor used by The Nature Conservancy (1982): coarse filters and fine filters (Noss 1987). The coarse-filter approach to saving biological diversity involves maintaining a variety of ecosystems; it assumes that a representative array of ecosystems will contain the vast majority of species in a region; TNC has estimated 85–90%. (This is effectively the approach advocated in Part II of this book, particularly in the chapters on species composition and age structure.) The problem with the coarse-filter approach is that some species are almost certain to slip through the pores, because no ecosystem classification scheme could be comprehensive enough to capture every species. Consider the Virginia round-leaf birch, a tree that occurs in one small valley and has a total population of only a few dozen individuals (Preston 1976). This species is so rare that if there were a hypothetical Appalachian Deciduous Forest Ecosystem Reserve System with representative stands of yellow birch-beech-sugar maple, red oak-basswood-white ash, white oak-red oak-hickory, etc., the odds of any one of these reserves containing a population of Virginia round-leaf birch are infinitesimally small. In theory, one could avoid this shortcoming by defining a Virginia round-leaf birch ecosystem type and including an example in the reserve system, but, of course, this would be circumventing the whole idea of a coarse-filter system.

Here is where the fine-filter approach comes to play. The fine-filter approach is directed toward individual species known to be endangered and catches them even though they passed through the coarse filter. A Virginia round-leaf birch reserve would be a perfect example of the fine-filter approach at work. The limitation of the fine-filter approach is expense and lack of information. In most parts of the world the status (population size, distribution, and security) of most species is unknown, and even if a comprehensive list of endangered species could be assembled, there would be insufficient money to deal with them all. In the current state of affairs, fine filters work reasonably well for large vertebrates; that is to say there are management programs in place for most endangered species; their effectiveness is another question (Figure 13.1). Unfortunately, smaller vertebrates and plants are bypassed in most parts of the world, and invertebrates and microorganisms are largely ignored almost everywhere.

To summarize, if properly implemented, the coarse filter–fine filter combination should work well: (a) to save the vast majority of species in representative ecosystems; and (b) to save many of the more conspicuous endangered species.

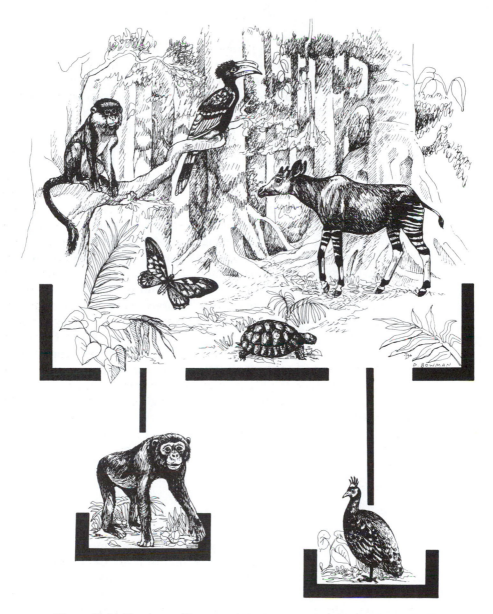

Figure 13.1 The coarse-filter approach of maintaining biological diversity is particularly appropriate in the tropical forests of developing nations—a Zaire rainforest is portrayed here—where information about each species and conservation funds are very limited. The fine-filter approach of saving special species such as the bonobo chimpanzee and Congo peacock is usually implemented only for large birds and mammals.

Tana River Monkeys

A somewhat different perspective on the coarse filter–fine filter approach is provided by a story from East Africa. The Tana River in coastal Kenya is lined by remnant stretches of riverine forest; patches of this forest totaling just 25 km² are the sole home of two endangered monkeys: the Tana red colobus and the Tana crested mangabey (Marsh et al. 1987). These same forests are a source of very large logs from which dugout canoes are carved; a survey of canoes along 25 km of river found 75 canoes made from six species of tree. There was considerable variation in how long it took to carve out different species and how long they lasted on the river before rotting; the most commonly used tree was a type of fig that could be carved in just two weeks but usually rotted after little more than a year. Unfortunately, this same fig tree was the major source of food for both species of monkey, and the removal of several large trees per year could significantly diminish the monkeys' tenuous prospects for survival. Fortunately, a reasonable management alternative was discovered. One of the most common trees in the forest, a species of ebony, is not especially important for the monkeys and is very suitable for making canoes; it requires slightly longer to cut, three weeks, but lasts for three years.

The Tana River story is an important lesson in how even subtle changes in the structure of a forest can have harmful consequences. On the other hand, if the Tana River monkeys were not confined to a relict of their former habitat, they would probably not be very vulnerable to some selective cutting for canoe logs. Thus, even here the bigger picture—deforestation and fragmentation—is ultimately more important. As with the Virginia round-leaf birch, the message is that a two-tiered approach is often needed. The ecosystem, coarse-filter, or macro approach should be foremost, but at the same time one cannot ignore individual species—the fine-filter or micro approach—especially when species known to be endangered are involved.

Keystone Species

The Tana River monkey story also contains a message about how important it is to consider the interactions among individual species, and how this affects their ecological importance. Some species have roles, as prey, predator, symbiont, or competitor, that accentuate their ecological importance beyond what one might predict from their abundance or biomass. Anyone entering a redwood stand would quickly recognize the critical role redwood trees play in this ecosystem; the importance of such a dominant species is patently obvious. More subtle and interesting is the crucial role that small, inconspicuous species can play, such as the mycorrhiza-forming fungi discussed in Chapter 10. Species that are not dominants, but still hold critical roles in ecosystems, are called keystone species because, like a single small stone supporting a whole arch, they have a central role on which the integrity of the whole ecosystem relies. The best-documented examples of keystone species come from some marine ecosystems where a single predator species can affect the whole community structure, often increasing diversity by preventing any one species from becoming dominant (Paine 1969). In terrestrial systems, ecologists have speculated that trees that produce large amounts of fruit during the dry season, when food for fruit consumers is generally scarce, may be keystone species (Terborgh 1983, 1986). Conversely, the fact that some seeds cannot germinate without passing through the gut of an animal may make some animals, elephants for example, into keystone species (Lieberman et al. 1987).

Needless to say, anyone concerned about biological diversity should pay some attention to keystone species, because they are crucial to ecosystem integrity, and if they are lost, there is likely to be a snowballing effect that reduces diversity substantially. However, our understanding of most ecosystems is so limited that we often do not know which species may hold keystone roles; generally this can be determined only by experimentally removing a species from an ecosystem. Fortunately, it is likely that most keystone species are common enough to be maintained by the coarse-filter approach and do not require special attention. It may, however, be dangerous to assume this, especially on islands where there are many species with limited ranges and low populations. Recalling the story of the Mauritian tree that could only germinate from seeds consumed and defecated by a dodo (Chapter 2), imagine the consequences if most of the island's trees had required seed passage through a dodo's gut. In this case, the failure to use a fine-filter approach to save the dodo could have had catastrophic effects on the whole island's biota.

Indicator Species

The term "indicator species" refers to species that have such narrow ecological tolerance that their presence or absence is a good indication of environmental conditions. Traditionally, the term has been used most often by foresters who note the distribution of certain understory plants as an index of site conditions such as drainage and fertility (Cajander 1926). More recently, environmentalists have used the term to describe species that are very sensitive to environmental degradation and thus serve as a "miner's canary" to warn of a problem. The well-known demise of brown pelicans and peregrines in response to use of persistent pesticides, and the consequent efforts to restrict these chemicals, is a good example. In the context of managing forests for wildlife diversity, the term could be used to refer to species that require certain kinds of forests—based on species composition, age, or area—or that need specific features within a forest. For example, spotted owls are a good indicator of large, old conifer stands, and woodpeckers are often a good indicator of snag abundance. Unfortunately, while an appealing concept, practical use of indicator species has many limitations, primarily because it often relies on erroneous assumptions about organisms and their relationships with their environment and other species (Landres et al. 1988). Therefore, it is preferable not to think in terms of managing indicator species, but rather to focus on monitoring indicator species to provide a rough index to the abundance and condition of various types of ecosystems. From this perspective, indicator species are a tool to facilitate the coarse-filter approach.

Flagship Species

Ideas like keystones and indicators may carry a bit more weight with the general public than biological diversity and ecological integrity, because they can be tied to individual species. However, to really form an effective link between species-oriented management and management for diversity, it is best to find a flagship species, a species that can galvanize substantial public support and indirectly facilitate the wise conservation of a whole galaxy of species. The tiger provides an excellent example of this. In the 1970s, world-wide concern about the fate of the tiger—arguably the most spectacular animal on earth—precipitated a program to establish tiger reserves throughout the Indian subcontinent (Sankhala 1978). Dry, deciduous woodlands on the Deccan plateau; moist, forested slopes in the Himalayan foothills; mangrove swamps in the delta of the Ganges; and tropical rainforest-riverine grassland interfaces in the Assam Valley are just a sampling of the subcontinent's ecosystems that are being managed to save the tiger from extinction. With a range that reaches from Siberia south to Indonesia and, until recently, west to Turkey and Iran, the tiger shares habitat with a large portion of the wildlife of Asia. It is likely that, at some time in the future, it will be possible to look back and identify dozens of species, perhaps hundreds, that would have probably gone extinct if they had not been able to "ride the coattails" of the tiger.

Like the tiger, many flagship species are large, conspicuous animals with extensive home ranges and fairly broad habitat needs, and thus their popularity becomes the basis for wildlife management on large tracts in a variety of ecosystem types. However, under some circumstances, the flagship species process can work very differently, almost in reverse. The spotted owl is a classic example of this (Chapter 5), for it is concern about an ecosystem type, the old-growth forests of the Pacific Northwest, that makes the owl a flagship species around which people can rally.

Special Populations

Most individual organisms are a product of sexual reproduction and thus are a unique mixture of genes taken from their mother and father. No one would seriously propose that maintaining all these unique mixtures—all these individuals—should be a goal of conservation, but maintaining most of the building blocks—the genes—is often discussed. As mentioned in Chapter 2, efforts to maintain genetic diversity often involve intensive breeding programs and other methods more appropriate to the managers of zoos and botanical gardens than the managers of forests, and thus lie beyond the purview of this book. However, it is important to note that for special species every population—every group of regularly interbreeding individuals—is a special gene pool that merits attention. To take two extreme examples, the wolves that live in the deserts of southern Iran are very different animals from the wolves of the Canadian arctic, and the Scots pines of fog-shrouded Scotland are very different from the one growing on sand dunes in Mongolia. Thus, saving Scottish pines and Canadian wolves is not adequate to protect these species. The issue is not always so clear; are the wolves in Iran's Elburz Mountains sufficiently different from the ones in the Zagros Mountains to deserve special attention? In the absence of detailed genetic information it is difficult to answer such questions, but the conservative approach would be to assume that they are quite different and maintain both. Obviously, a more fundamental approach, maintaining the entire native biota of all regions (all parts of Iran in this case), would accomplish the same thing. It is when this is impossible and decisions about the fate of certain species and their constituent populations must be made that these become difficult issues.

Minimum Viable Populations

Two conservation issues—endangered species and the isolation of populations due to habitat fragmentation—have led to the question: What is a minimum viable population (MVP), or more fully, what is the smallest population that has a reasonable chance of long-term persistence? A literal interpretation of the Biblical story of Noah gives a trite answer—two, one male and one female—but many wildlife managers, population biologists, and geneticists consider the question more interesting, and they have produced a growing body of literature describing the use of population viability analyses (PVA) to determine MVP for different populations. (We can only skim the subject here; see Soulé 1987 for a comprehensive account.)

Population viability is a matter of probabilities—the chances of a population going extinct because of factors that can be conceptually lumped under terms such as randomness, stochasticity, uncertainty, or chance variability. Mark Shaffer (1981) has categorized these factors into four types:

1. *Demographic randomness:* Random variation in individuals' chances of dying or reproducing can lead to extinction. For example, a small population could have such an unbalanced age and sex structure that most of the individuals are too young or too old to reproduce, and the few sexually active individuals are all of one sex.

2. *Environmental randomness:* Changes in climate, and the distribution of prey, predators, parasites, or pathogens, affect the chances of extinction. Human-induced changes such as the introduction of a toxic chemical would also constitute environmental stochasticity.
3. *Catastrophes:* Floods, droughts, major storms, etc. can occur at unpredictable intervals and precipitate a species' extinction.
4. *Genetic randomness:* Loss of genetic variability through inbreeding and genetic drift can increase the likelihood of extinction.

Obviously, these factors all interact (e.g., a catastrophic event could easily leave a population with an unbalanced age structure); this is especially important because the smaller the population the more vulnerable it is to each of the factors.

Using information about these factors and a species' basic biology and ecology, it is possible to construct models that make predictions such as: For population X to have a 95% probability of persisting for 200 years, it must have an effective population size of at least 675 individuals. (The *total* population size is usually considerably larger than the effective population size; how much so depends on the structure of the population, age at sexual maturity, and other factors.) One early analysis concluded that, as a generalization, an effective population size of at least 50 breeding individuals would be needed to avoid extinction due to inbreeding depression during a short period (Frankel and Soulé 1981). Survival for centuries and millenia would require an effective population of at least 500 breeding individuals because of the likelihood of catastrophic events and the need for having genetic variability with which to evolve adaptations to a changing environment.

These generalizations have become known as the 50:500 rule and are often cited, but more recent work has emphasized the need for specific analyses, rather than reliance on a coarse rule of thumb, and some newer estimates of MVPs range from hundreds to millions of individuals (Soulé 1987). It is also wise not just to think about minimum numbers, but rather the range of demographic, genetic, and environmental conditions through which a population is likely to persist. There remains much work to do on the population viability concept (particularly with regard to the way small populations are distributed across the landscape to form a larger entity, sometimes called a metapopulation [Gilpin 1987]), and this will continue to be an important interface between biology and conservation (Salwasser et al. 1984).

MANAGING SPECIAL SPECIES

Having decided that a species merits special management, that it should be featured in a program of habitat management, how does one go about doing it? This is classical wildlife management that began millennia ago when people used fires to maintain grasslands as habitat for the kinds of animals they liked to hunt. Over time, these practices were refined especially by European game-keepers, and with the advent of ecology in the beginning of this century, game-keeping has evolved into professional wildlife management. This long history of managing bird and mammal populations and their habitat has given wildlife managers a large body of experience, and when this knowledge is applied to endangered species, wildlife managers assume a pivotal role in the efforts to maintain biological diversity.

A number of syntheses of wildlife management practices have been written—Aldo Leopold's 1933 *Game Management* was the first of note—and thus we will only briefly outline them here. There is no set procedure to follow when embarking on a habitat management program for a featured species; here is one possible sequence of steps:

1. Decide on the desired population size. A vaguer objective, e.g., "increase population size," is often used, but, in light of problems associated with habitat fragmentation and population viability (Soulé 1987), it is desirable to formulate this objective carefully.

2. (a) Determine the species' typical habitat and average density in that habitat type; and (b) provide a sufficient area of that habitat for the desired population. Sometimes people try to distinguish between the habitat a species prefers and that which it really must have to survive, but as explained by Ruggiero et al. (1988), this is a simplistic dichotomy that ignores the complex and dynamic nature of a species' habitat relationships.

3. If suboptimal habitat must be used, assess and manipulate habitat quality. For animals, these usually fall in three categories: food, water, and cover. (Food and water are self-explanatory; cover refers to vegetation, rocks, logs, and other structures in which an animal can find shelter from severe weather or can hide from predators or prey.) For plants, physical features of the site such as moisture, light, and nutrients are likely to be critical. In theory, one of these factors will usually be limiting the size of the population. For example, a deer population may have adequate food and water but be limited by the amount of shelter from winter storms. In practice, the situation is often more complex because of interactions among factors; e.g., perhaps if the abundance of high-quality deer food were increased, the need for shelter would be diminished. One often hears this problem stated in another manner; instead of addressing the habitat's limiting factors, managers speak of increasing the habitat's carrying capacity, i.e., its ability to sustain a given population density.

4. If populations are still below the objective, consider additional factors that might be limiting populations—such as disease, parasites, predation by human hunters and other predators, environmental

contaminants, etc.—and deal with the ones that appear most critical
and feasible to control. Although a certain population density may
be the objective, it is often advisable to assess management success
in terms of evolutionary fitness, as estimated from survival rates and
reproductive success, because fitness is a much better index of a
population's well-being (Van Horne 1983).

The simplicity of this overview of special species management is not
meant to imply that it is a very easy enterprise. Only the concepts are easy;
the implementation can be very difficult, especially with limited funds. Very
few management agencies are entirely satisfied with their ability to manage
even a handful of the most important species, let alone all the creatures that
are not important enough to receive major attention but too important to ne-
glect. Songbirds are a good example of this problem; they are so well-known
and popular that it is difficult to ignore them, but they usually do not have
enough utilitarian value to justify expensive management (despite many attempts to document and promote them as consumers of insect pests). Applying a special species approach to songbirds is not very practical because there
are so many kinds, scores or even hundreds of species in most regions, each
with its own set of requirements. It was with this problem in mind that wildlife
biologists proposed the guild concept of management.

The Guild Approach

A guild is a group of species that exploit the same class of environmental
resources in a similar way (Root 1967, Terborgh and Robinson 1986). For example, insectivorous birds that live in deciduous forests and forage in the canopy might constitute a guild. In theory, all the members of a guild should
respond similarly to a change in the habitat, and this would allow a manager
to focus attention on just one species, a representative of the whole guild, and
be confident that the other species would respond in the same manner (Severinghaus 1981). In practice, the method has some limitations because various
members of the guild are each unique and use different components of the
ecosystem (Verner 1984). For example, researchers in Oregon assigned 19 bird
species to five guilds and compared their population densities in old-growth
and managed conifer forest (Mannan et al. 1984). Only in one small guild of two
species did all (both) guild members show a similar response to the difference in
habitat.

Inevitably, the guild concept would work at a crude scale; if all the trees
in a forest were cut, it would no doubt have a dramatic effect on all members
of a guild of canopy-foraging birds. However, it is the impacts of subtle
changes that are most difficult to predict, and the guild concept will have less
to offer here. Guilds and a similar concept, life forms (Thomas et al. 1979*e*),

may be most useful as a way to organize management of a large number of species, for example, on U.S. National Forests where there is a mandate to maintain viable populations of all native vertebrates.

Information for Special Species Management

Anyone who decides to embark on a special species management program and goes looking for a how-to-do-it book is likely to be disappointed. There are remarkably few works that provide managers with easy to follow recipes for achieving their objectives. There are probably two reasons for this; one is a scarcity of wildlife habitat management literature in general; the second is a matter of applicability.

The literature of wildlife management is dominated by articles about the ecology, behavior, and physiology of those animals that commonly become featured species—mainly game fish, birds, and mammals, and endangered species. Most of these articles discuss the management implications of their results, but studies that comprehensively document the effects of habitat management are quite uncommon because of the difficulties of habitat management research. Such research usually requires: (a) expensive experimental manipulations of habitat (this often forces biologists to piggy-back their work onto a regular operation, like a commercial forest harvesting program, and to bend their experimental design to fit it); (b) manipulating sizable areas of land, (this makes it hard to find similar areas for controls and replicates); and (c) a very long time—time to document baseline conditions and year-to-year variations, time to measure a manipulation's effect, and time to monitor how the effect changes as succession proceeds. Not all habitat management research need be as elaborate as implied here, but as a generalization, it is very difficult to thoroughly study the effect of habitat management, and thus there is a dearth of good information about it, especially in the primary (i.e., research journal) literature. Typically, one is forced to extrapolate from a handful of one- or two-year studies from widely scattered locations, each of which compared one or two sites that were manipulated to one site that was not.

The second reason for a scarcity of practical, how-to-do-it guides is the difficulty of developing such guides that are broadly applicable. Many featured species occur over such a wide geographic range, and in so many types of habitat, that it would be very hard to make detailed recommendations for management that would be valid for a species throughout all the places it occurs. Take North America's most popular big-game animal, the white-tailed deer; managing spruce stands for winter forage and cover in southern Canada—using prescribed burns to create an understory of deer browse in the southeastern United States—and maintaining riparian strips in the semi-arid west to maintain white-tailed deer in a landscape dominated by open land and mule deer— are very different activities (Crawford 1984). Not only are the biological and

ecological issues complicated, but so are the political, social, and economic issues that determine the style of natural resource management in different
areas. Such complexity seems to daunt most people, and few have taken up
the challenge of synthesizing specific habitat management guidelines for forest
species. Following are a few examples that have been produced.

Ruffed grouse: Gullion, G. W. 1984. *Managing Northern Forests for Wildlife.* The Ruffed Grouse Society, Coraoplis, Pennsylvania. 72 pp.

American woodcock: Sepik, G. F., R. B. Owen, Jr., and M. W. Coulter.
1981. *A Landowner's Guide to Woodcock Management in the Northeast.*
Maine Agricultural Experiment Station Miscellaneous Report 253. 23 pp.

Snowshoe hare: Brocke, R. H. 1975. *Preliminary Guidelines for Managing
Snowshoe Hare Habitat in the Adirondacks.* Trans. Northeast Section
Wildlife Society Fish Wildlife Conference 32:46–66.

Black-tailed deer: Nyberg, J. B., F. L. Bunnell, D. W. Janz and R. M.
Ellis. 1986. *Managing Young Forests as Black-tailed Deer Winter Range.*
British Columbia Ministry of Forests, Land Management Report 37. 49
pp.

Ring-necked pheasant: The Game Conservancy. 1981. *Woodlands for
Pheasants.* The Game Conservancy, Fordingbridge, Great Britain. 72 pp.

Also see chapters on mule deer and elk in Thomas (1979) and on deer and
elk, salmon and trout, spotted owls, and bald eagles in Brown (1985).

The U.S. Fish and Wildlife Service has produced recovery plans for over
200 endangered plants and animals that live in the United States; many of
these are forest species and some of the recovery plans deal, at least briefly,
with issues of forest management. Examples may be found among the following:

Bald Eagle	Hawaiian Vetch
Chapman's Rhododendron	Kauai Forest Birds
Columbian White-tailed Deer	Kirtland's Warbler
Delmarva Peninsula Fox Squirrel	Mississippi Sandhill Crane
Eastern Timber Wolf	Puerto Rican Parrot
Florida Panther	Red-cockaded Woodpecker
Green Pitcher Plant	Small Whorled Pogonia
Grizzly Bear	Virginia Round-leaf Birch
Hawaiian Forest Birds	

(These are available from The Fish and Wildlife Reference Service, 6011 Execu-
tive Blvd., Rockville, Maryland 20852.)

Finally, there are both biological monographs and broad syntheses of information for many game species; typically these include a general account of habitat management.

Bump, G., R. W. Darrow, F. C. Edminster, and W. F. Crissey. 1947. *The Ruffed Grouse*. New York State Conservation Dept., Albany. 915 pp.

Halls, L. K. (ed.) 1984. *White-tailed Deer*. Stackpole, Harrisburg, Pennsylvania. 870 pp.

Hewitt, O. H. (ed.). 1967. *The Wild Turkey and its Management*. The Wildlife Society, Washington, D.C. 589 pp.

Prior, R. 1968. *The Roe Deer of Cranborne Chase*. Oxford University Press, London. 222 pp.

Rosene, W. 1969. *The Bobwhite Quail*. Rutgers University Press, New Brunswick, New Jersey. 418 pp.

Wallmo, O. C. (ed.). 1981. *Mule and Black-tailed Deer of North America*. University of Nebraska Press, Lincoln. 605 pp.

Thomas, J. W., and D. E. Toweill (eds.). 1982. *Elk of North America*. Stackpole, Harrisburg, Pennsylvania. 698 pp.

SUMMARY

People find it much easier to identify with individual species than with concepts like biological diversity, and thus wildlife management has traditionally focused on species, especially those that society considers special. Special species may have utilitarian value because of their economic or ecological importance, or subjective value for less tangible reasons like aesthetics. Among species with unknown or unexploited value, those that are most threatened with extinction will generally be considered of highest value. There are several ways in which managing for special species interacts with managing for biological diversity. First, one can consider managing for diversity at an ecosystem level to be a coarse-filter approach; some species will inevitably slip through a coarse filter and need to be saved from extinction by a fine-filter—species-oriented—approach. Second, certain species play subtle but critical, keystone, roles in ecosystems and must be maintained for the sake of ecosystem integrity. Third, some species merit being monitored simply because they are good indicators of the condition of different ecosystems. Finally, species that are exceptionally popular can serve as flagships and catalyze conservation efforts for whatever ecosystems they inhabit.

The chapter concluded with three short sections: (a) an outline of how species-oriented management can be structured (provide enough habitat to support the desired population; improve the habitat's carrying capacity by increasing food, water, or cover as necessary; control other limiting factors such as hunting); (b) an explanation and critique of the guild system for managing many species at once; and (c) an introduction to the literature of special species habitat management.

Part IV

Synthesis and Implementation

It is one thing to envision idealized forest landscapes on which a balanced array of forest stands provides habitat for a wealth of species. It is another thing to synthesize disparate elements of this vision into a cohesive plan that can be implemented on a real landscape — a landscape with ecological, social, and economic constraints.

In Chapter 14 we will examine some of the principles that form a foundation for forest management planning and describe two tools that are essential to forest planning: ecosystem classifications and forest models. Finally, in Chapter 15 we will consider the costs and benefits of managing forests for biological diversity and explore the question of who should pay the costs: landowners, wildlife users, consumers of forest products, or taxpayers.

251

14

Management Plans

To a casual observer, much of the world must seem to be dominated by great sweeps of monotonous terrain—deserts, prairies, tundra, and ocean that stretch to the horizon in a uniform blanket. In these places, it takes a trained eye to see much more, to recognize the subtle shifts in soil moisture, topography, salinity, and so on, that produce a mosaic of different environments inhabited by a varying array of organisms. In contrast, the patchy nature of most forested landscapes is much more obvious, because forests in different successional stages are so prevalent, and because so many forests are part of a quiltwork of pastures, croplands, and wetlands. Frequently, one can also readily detect differences in the forests' species composition that reflect site conditions; e.g., waterlogged organic soils of swamps; dry, sandy outwash plains; steep, rocky slopes.

Because of the patchy nature of forested landscapes, each one is unique, and thus an effective forest management program has to be individually tailored to each situation. To reiterate a quote noted in the Preface to this book:

> I have read many definitions of what is a conservationist, and written not a few myself, but I suspect the best one is written not with a pen, but with an axe. . . . A conservationist is one who is humbly aware that with each stroke he is writing his signature on the face of his land. Signatures of course differ, whether written with axe or pen, and this is as it should be. (Aldo Leopold, 1949)

Professional forest managers put their signature on the land by writing management plans and overseeing their implementation. In this chapter we will explore some principles for writing management plans that integrate ideas for maintaining biological diversity.

BALANCE, CONSTRAINTS, AND OPPORTUNITIES

For a forest landscape to be diverse, it is not sufficient that it have a balanced age structure, be spatially heterogeneous, and be composed of a variety of forest types. These three factors and others must be integrated into a balanced design. To take a very simple example, imagine a hypothetical landscape equally divided between coniferous and deciduous stands, and between stands younger and older than 40 years old. If all the young stands were deciduous and all the old stands were coniferous, then species requiring old deciduous forests and young coniferous forests would be without suitable habitat. Balancing an assemblage of stands in terms of their three basic chararacteristics—species composition, age, and area—is conceptually easy (Table 14.1).

Even making a design flexible enough to accommodate natural variation would not make it unduly complex. If oak stands reach maturity at an older age than pine stands, then their rotations can be lengthened, or if large-scale disturbance is more common in pine stands, then they can be managed at a coarser scale (Table 14.2).

The diversity of forest landscapes can be estimated using the same indices that ecologists use to measure species diversity by simply treating each kind of stand (e.g., young, small-scale, pine stands) as if it were a different species and using its total area to represent its abundance (Romme and Knight 1982, Hunter 1986, O'Connor and Shrubb 1986). For example, using the Shan-

TABLE 14.1   A Balanced Distribution of Age Class, Dominant Species, and Management Scale for a 9,000 ha Hypothetical Forest Composed of Oak and Pine Stands

	Young	*Intermediate*	*Old*
Oak			
Small-scale	500 ha	500	500
Medium-scale	500	500	500
Large-scale	500	500	500
Pine			
Small-scale	500	500	500
Medium-scale	500	500	500
Large-scale	500	500	500

TABLE 14.2　A Modification of Table 14.1 to Account for Differences in the Rotation
Ages and Natural Disturbance Regimes of Oak and Pine Stands

	Young (< 50 yrs.)	Intermediate (50–100 yrs.)	Old (> 100 yrs.)
Oak			
Small-scale (Individual selection)		1,500 ha of uneven-aged forest	
Medium-scale (< 1 ha)	500	500	500
Large-scale (1–10 ha)	500	500	500
	< 40 yrs.	40–80 yrs.	> 80 yrs.
Pine			
Small-scale (< 1 ha)	500	500	500
Medium-scale (1–10 ha)	500	500	500
Large-scale (> 10 ha)	500	500	500

non diversity index (Appendix 3) the landscape represented in Table 14.2
would have a value of 2.71. If this same landscape had fewer kinds of stands
(lower richness) or many differences in the areas of various kinds of stand
(lower evenness), it would have a lower diversity index; for example, as restruc-
tured in Table 14.3, the landscape has a diversity index of 2.45.

Note that the landscapes of Tables 14.2 and 14.3 would have a higher
diversity index if the small-scale oak stands were divided into young, intermediate, and old even-aged stands, rather than existing as 1,500 ha of uneven-
aged stands. However, this would undermine the whole rationale for having
large, uneven-aged stands managed by selection cutting. This points out the
danger of using arbitrary indices; they should not be allowed to blindly drive
management, especially if they contradict a management plan based on com-

TABLE 14.3　A Modification of Table 14.2 Showing a Less Diverse Array of Stands*

	Young (< 50 yrs.)	Intermediate (50–100 yrs.)	Old (> 100 yrs.)
Oak			
Small-scale (Individual selection)		1,500 ha of uneven-aged forest	
Medium-scale (< 1 ha)	800	600	—
Large-scale (1–10 ha)	150	850	820
	< 40 yrs.	40–80 yrs.	> 80 yrs.
Pine			
Small-scale (< 1 ha)	560	670	40
Medium-scale (1–10 ha)	940	80	800
Large-scale (> 10 ha)	950	560	—

*There are only 14 kinds of stand, compared to 16 in Table 14.2, and total stand areas are less even (range
40–1,500 ha).

mon sense and sound planning (Salwasser et al. 1984). Diversity indices are just a convenient tool for comparing alternatives.

There is a second danger with diversity indices because they give people the ability to "create" diversity on paper. In other words, by simply subdividing classes of ecosystems more finely (40–80 years becomes 40–60 years and 60–80 years), it is very easy to diversify a forest landscape on paper. There is nothing wrong with fine subdivisions as long as there is some ecological basis for them, but they must be carefully considered when it comes to comparing apples and oranges. For example, imagine you wished to compare the age class diversity of a group of spruce stands to a group of aspen stands (all the stands are even-aged, the spruce stands ranging from 1 to 120 years, and the aspen stands from 1 to 40 years). Would it be best to define age class intervals of 40 years and thus recognize three age classes of spruce stands (<40, 40–80, 80–120 years) and only one age class of aspen (< 40 years)? Obviously, this would make the group of spruce stands seem much more diverse. Alternatively, it might be better to have flexible age class definitions and recognize both four spruce age classes (<30, 30–60, 60–90, and 90–120 years) and four aspen age classes (<10, 10–20, 20–30, 30–40 years). There is no absolute answer; again, the important thing is that diversity indices be used to clarify and organize thinking, not to make decisions.

After constructing an ideal array of stands, the next step is to arrange it on the landscape. Again, it is fairly straightforward to arrange stands in hypothetical spatial patterns that will accommodate both wildlife species that need edges and those that favor forest interiors. Figure 14.1 shows one such arrangement; species that need edge habitats will be found in clusters of small- and medium-sized stands located within a matrix of larger stands that will harbor forest interior species (Franklin and Forman 1987). However, as explained in Chapter 8 on "Islands and Fragments," in many regions, perhaps most, it is more realistic to think of the whole forest as isolated in a landscape dominated by human activities. Such settings call for a very different approach to arranging stands, one in which forest interior species are managed in a core buffered from external effects, and the edge species live on the periphery of the forest.

There is also an issue as to the scale of landscapes to consider. At a minimum, a landscape unit should be large enough to include a complete array of stands and many replicates of each type. Ideally, a balanced array of stands would be maintained across a very extensive landscape, but as the scope approaches thousands or millions of square kilometers, some difficulties arise. First, there are limitations imposed by ownership patterns; these are discussed below. Second, at a regional scale, problems of definition will occur; e.g., is a young pine stand at the north end of the unit the same as a young pine stand at the south end? In some circumstances, for example maintaining a viable population of spotted owls in the Pacific Northwest, it is necessary to surmount these problems and think in terms of regional landscapes.

Young Intermediate Old

Figure 14.1   A hypothetical landscape in which some of the stands from Table 14.2 are arranged; species that need edge habitats will be found in the central clusters of small and medium-sized stands, and forest interior species will be well-distributed in a matrix of larger stands.

Hypothetical distributions are easy; the challenges arise when one comes to placing such concoctions on a real landscape with all of its constraints—ravines, rivers, roads, property lines, etc. (Figure 14.2; Salwasser and Tappeiner 1981). Of course, these same features, especially the natural ones, contribute to the diversity of the forest landscape and thus should be considered as much an opportunity as a challenge. Next we will explore two types of constraints briefly, transportation systems and sensitive areas, and a third, ownership patterns, in some detail.

Figure 14.2 A hypothetical landscape in which differences in the sensitivity of different sites (steep slopes, riparian zones, etc.) are accomodated by differences in the style of management.

Transportation Systems

With the demise of river drives, roads have become an essential element of managed forests, and because their construction and maintenance is so expensive, they are a major constraint in forest management planning. There are at least three ways that transportation systems affect efforts to create a diverse forest landscape for wildlife, or conversely, ways that efforts to create a diverse landscape affect transportation systems:

1. The small-scale management needed to create small stands and uneven-aged stands is more expensive than large-scale, even-aged management because it requires (a) having harvest operations spread over a relatively large area, and (b) making regular returns to a stand. If carefully planned, this may not increase total vehicle miles, but it does necessitate having a large network of maintained roads.

2. Transportation costs will encourage timber managers to concentrate their efforts to maximize timber production near mills. This means that they will be less likely to make major concessions for wildlife management in these areas. Conversely, they should be much more willing to incorporate wildlife management objectives in areas remote from mills.

3. Dissecting a forest with roads may diminish its value as wildlife habitat. Firstly, there is evidence that roads exacerbate the problem of forest fragmentation (Oxley et al. 1974, Mader 1984, Small and Hunter 1988). Also, roads open to the public may facilitate over-hunting and disturbance in general (Thomas et al. 1976); indeed, in many regions of the world, forest roads open the way to settlement and conversion of forests to agricultural land.

Sensitive Areas

Some sites are so sensitive to disturbance that one needs to carefully weigh the idea of removing timber from them. In areas characterized by steep slopes, wet soils, shallow soils, riparian zones, and other fragile features, conventional harvesting can cause severe damage to a site. Sometimes the damage is virtually irreparable, degrading an ecosystem's productivity for decades, even centuries. There are logging techniques that are relatively gentle, ranging from "high tech" approaches such as lifting logs with balloons or helicopters, to "soft tech" methods like employing horses to twitch logs over a blanket of snow. By using such methods, and a plan that emphasizes small-scale harvesting and long rotations, many sensitive areas could be harvested without unduly damaging the environment. On the other hand, because these techniques tend to be relatively expensive, it is often advisable to withdraw sensitive areas from harvest altogether. Every landscape should have a reasonable portion of old forests, and sensitive sites are an obvious location for these.

Locating old forests in riparian zones is especially sensible. These may be the single most important type of wildlife habitat, and by withdrawing them from harvest their intrinsic value can be readily maintained. Moreover, they can provide a system of corridors linking all the old forests of a region into a network that will mitigate the effects of forest fragmentation (Noss and Harris 1986).

Despite the attraction of locating old forests in sensitive areas, there are two reasons that this should not be the whole story. First, most sensitive areas are not extensive enough to provide adequate habitat for old forest species with large home ranges. The thousand or more hectares needed by a pair of spotted owls would stretch for over 100 kilometers (Forsman et al. 1984), far too long for a nightly patrol, if they were confined to the thin ribbon of a riparian forest. Second, by their very nature sensitive areas cannot represent the whole range of environments that should be included in a system of old forest reserves. Some areas that are prime sites for growing timber—fertile, well-drained, flat, etc.—need to be set aside too because their associated communities are not identical to those growing on poorer sites. To express it another way, in each region some tracts of land should be set aside as parks or nature reserves because it is usually not possible to readily meet all the needs of wildlife in a forest managed for timber production. Some of these reserves can be in sensitive areas that are inappropriate for other uses, but some of these should be very large—hundreds, perhaps thousands, of square kilometers—and all the region's different ecosystems should be represented among the reserves. (The selection, size, location, and management of wildland reserves is an important issue, but is largely beyond the scope of forestry and hence of this book. For a review of this topic, especially the evaluation of potential reserves, see Usher 1986.)

Obviously, there is considerable logic in concentrating intensive forest management on the most productive sites where the economic return is likely to be greatest, and there may also be some indirect benefits to wildlife. For example, it has been estimated that focusing on productive sites and other efficiencies would allow nearly 2 million hectares of forests in western Oregon and Washington to be removed from timber harvest and added to parks without any loss in overall timber production (Hyde 1980); this could triple the area of forestland in the region's state and national parks and USDA Forest Service wilderness areas. However, the negative side of implementing this scenario on a broad scale is that virtually all of a region's most productive sites would have to be committed to intensive forestry—short rotations, monocultures, etc.—and these are the same sites that probably have the greatest potential for supporting a high species diversity. Thus, while both economic and ecological considerations suggest that intensive forest management should be concentrated on productive sites, and low-intensity or no-harvest units located in sensitive areas, there should be significant exceptions to this generalization.

Ownership Patterns and the Scale of Management

Arguably, the single most important feature of managed landscapes is a human construct, the boundaries between adjacent properties. Sometimes property lines are invisible, existing only on a deed; sometimes they are bold edges that record two entirely different histories of land use. More often than not they mark the end of one forest management plan and the beginning of another.

From the standpoint of managing forest landscapes for biological diversity, boundary lines impose a significant constraint for two reasons. First, most forested properties are too small to encompass a forest landscape; they are measured in tens or hundreds of hectares, not tens or hundreds of square kilometers. Second, they do not correspond well with natural boundaries; they occasionally follow shorelines, rarely follow watershed lines, and almost never follow forest type lines.

To illustrate the problems associated with ownership size, imagine a person who owned a 50 ha woodlot dominated by 75-year-old spruce trees and surrounded by other properties with similar forests (Figure 14.3). If she wished to diversify such a forest, she might devise a long-term cutting plan that would gradually transform a uniform 50 ha stand into several smaller stands of different ages, some managed with selection cuts, some with small clearcuts. Diversifying the age structure might be particularly beneficial if early successional stands were characterized by deciduous trees. Now imagine that same 50 ha stand surrounded by a young deciduous forest (Figure 14.4). If the landowner's interest in promoting biological diversity extended only to the edge of her property, her management plan would remain much the same. However, if she was concerned about the diversity of the whole landscape, her approach would be very different. She would probably decide not to cut her forest at all or only to a very limited degree.

Although it is much more difficult to implement a large-scale management plan for a forest landscape with fragmented ownership, such landscapes are by no means a lost cause for wildlife diversity. Indeed, in the absence of large-scale planning, the fragmented ownership may produce de facto diversity because of the diversity of landowner goals. Some landowners will want to have sanctuaries in which no harvesting occurs; some will want to have plantations of fast-growing profitable trees; some will never think about the management of their woodlots—they will simply sell cutting rights whenever they need the money and there is a willing buyer. This situation is far from ideal; it is a fragile chaos that is sensitive to extrinsic factors such as wood prices, property taxes, government incentive programs, and the general state of the economy. Nevertheless, it may well be more conducive to biological diversity than having a landscape completely dominated by short-rotation, single-species plantations in which wildlife management is ignored.

Notwithstanding the possibility of de facto diversity, it is desirable for

Figure 14.3 A 50 ha woodlot dominated by 75-year-old spruce trees and surrounded by other properties with similar types of forests is not a unique component of the landscape.

landowners to look beyond the border of their property to consider the context in which their forest exists. The smaller the property, the more important the external context becomes, but even very large landowners cannot entirely ignore the context issue. For example, consider the 304,000 hectare White Mountain National Forest in northern New England. The USDA Forest Service issued a management plan in 1986 that called for routine harvesting on 46% of the forest; 51% of the forested lands would not be harvested and thus would eventually become old forests; apline areas cover the remaining 3% (USDA Forest Service 1986c. Furthermore, the Forest Service stated that most of the managed areas would also be habitat for wildlife associated with mature forests; (i.e., managed stands at a mature stage and uneven-aged stands managed by single-tree selection). Thus, they claimed:

Figure 14.4 A 50 ha stand surrounded by properties on which young deciduous forests are dominant may play a key role in landscape diversity.

... 81% of the total habitat base on the WMNF is available to those wild-life species (10% of the wildlife community) that require mature, overmature, and old growth habitats.

As generous as this might seem, it did not prevent the Forest Service from being sued by environmental groups claiming, among other things:

These timbering plans will have devastating effects on the wildlife of the forest. As their natural habitat is bulldozed out from under them, threatened species will lose one more round in their fight against extinction. (Conservation Law Foundation of New England [n.d.])

One could argue about the Forest Service's definition of what constitutes old habitats, or chastise the environmentalists for exaggerating the plight of wildlife to fight for their wilderness values, but the real wildlife issue comes down to context. If you think of the White Mountain National Forest as an independent entity, the management plan is certainly reasonable. In fact, unless a semblance of the natural disturbance regime can be maintained on the uncut parts of the forest, it is possible that the plan will result in more old stands than would have existed before Europeans colonized the area. On the other hand, if you think of the White Mountain National Forest in the context of the 13 million hectares of forest in New England, less than 3% of which have been permanently withdrawn from timber harvest, then it is much easier to expect that the National Forest should right the region's imbalance.

The issue of whether or not the White Mountain National Forest should be managed as an independent entity, or considered in the context of New England's forests, would be less important if the USDA Forest Service had some jurisdiction over private forest land; specifically, if there were a federal law mandating that private forest owners maintain a balanced age class distribution. (Although it is hard to imagine a U.S. federal forest practices act, several countries, states, and provinces—e.g., Sweden, Massachusetts, and Nova Scotia—have them.)

Government oversight would enable policy makers to superimpose a balanced forst on a landscape with fragmented ownership, but there are some pitfalls that would need to be avoided. Imagine a region in which virtually all of the ownerships were less than 50 ha and where all landowners were required to have four age classes on their properties. This would tend to produce a small-scale patchwork of forests that would support a diversity of wildlife, but species needing 20 or more hectares of uniform forest—whether it be old-growth or a recent clearcut—would be without habitat. A simple ban on clearcuts above 10 hectares could have a similar effect on any landscape, regardless of ownership size. A real example of a pitfall in government forest planning comes from Sweden, where the policy of balancing forest age distribution has made it almost impossible for landowners to have old forests, even though some might prefer woodpeckers to sawlogs. Species that do not readily fit into the government's vision of a managed forest are largely relegated to nature reserves.

The issue of ownership patterns raises a related question: When very large tracts must be subdivided for administrative efficiency, what is the optimal size of management units? Given the aforementioned arguments and those presented in Chapter 8, "Islands and Fragments," "as large as feasible" is the optimal answer. However, as long as there is adequate oversight to assure that balance is achieved for both the overall tract and each subunit, it is not particularly important that they be as large as feasible. At a minimum, they should be large enough to contain a complete array of forest stands (Table 14.2) with replicates of all the types. Ideally, they would not be delineated by arbitrary lines, but rather by natural boundaries such as watershed lines or barriers to dispersal such as expanses of water or open land.

FOREST PLANNING

Essentially, forest planning is like any other form of planning; one needs to state goals and objectives, identify and assign tasks, conduct the tasks, evaluate the results, and modify the plan in response to new knowledge and changing circumstances. Conceptually, it is quite straightforward to: (a) determine the current status of a forest; (b) envision what the forest should look like if various benefits are to be derived from it; and (c) implement an action program that will gradually, through succession and human manipulations, shift the forest toward the desired state. However, there are several reasons why this is easier said than done—several features of forest planning that distinguish it from most planning efforts. First is the diversity of goals—often conflicting goals—that can characterize forest planning. (Here we assume that timber production and maintaining biological diversity are the fundamental goals, but many forest owners give equal or greater weight to recreation, aesthetics, livestock grazing, and water resources.)

Second, the complexity of forest systems makes them far more difficult to plan than most systems. Accurately predicting and controlling the interactions of thousands of species on a dynamic landscape subject to external events such as droughts and fires, is far more difficult than most undertakings—controlling the flight of space vehicles, for example. Naturally, we do not expect very tight control of forests. If a stand of pine trees develops somewhat differently than planned, some money or a unit of habitat for some species may be lost, but this is trivial compared to the calamity of a space vehicle going awry and losing the lives of national heroes. Thus, forest planners can generally afford to think of the forest as a much simpler assemblage of a few dominant tree species, although with this simplification comes uncertainty and inaccuracy.

Finally, the time frame for forest planning is far longer than that of most human endeavors because succession is measured in tens and hundreds of years, usually far exceeding the professional life of a forest planner. This further complicates the system and makes it far more likely that some unpredictable external event will confound the program. Some events are relatively small (e.g., a tornado); some are much more far-reaching; consider, for example, the changes in technology and markets that have moved wood from sawlogs to pulpwood to biomass fuel and toward a future as chemical feedstocks. One such exigency, a change in the goals of the landowner, or society in general, would strike at the very root of the planning process.

Despite these complexities, forest planning proceeds apace; the difficulties do not diminish its necessity because, crude as it is, it is far better than no planning at all. Here we will not attempt to review the whole discipline of forest planning, but rather we will focus on two aspects that bear directly on managing forests for biological diversity: ecosystem description and forest growth and succession models.

Ecosystem Description

To undertake the first step in forest planning—describing the current state of the forest—it is necessary to have a forest classifications scheme (S.A.F. 1967, Marmelstein 1977, Bailey et al. 1978). Hundreds of classification have been devised ranging from very simple to quite complex. Among the simplest systems are those that distinguish just three forest types—softwood, hardwood, and mixed-wood—and account for forest succession by recognizing only three height classes—e.g., less than 15 feet, 15–40, and over 40 feet. More sophisticated systems recognize forest stands by their dominant species or group of species (e.g., longleaf pine or sugar maple-beech-yellow birch) and have finer age classes. To go a step further and make a forest *stand* classification into a forest *ecosystem* classification, it is necessary also to recognize the differences in physical sites that support certain types of forest. (Remember, a forest ecosystem is all the organisms in a forest—the community—*plus* the physical environment; Chapter 3.) For example, in northeastern North America one can find red spruce stands growing on coastal islands, mountain tops, and lowland flats, and a refined ecosystem classification would treat each of these separately. A more striking example comes from the lodgepole pine, a species that forms nearly pure stands of tall, straight stems (hence the common name) throughout large areas of western North America. In sharp contrast to this picture, lodgepole pine forms stands of low, scrubby trees along the coast of Washington and Oregon, where, ironically, the species was first described and given its scientific name, *Pinus contorta*. The radical difference in their environments clearly means that interior and coastal lodgepole pine stands should not be treated as the same type of ecosystem.

One way to change a forest stand classification into an ecosystem classification is by placing it within the context of a broader system of recognizing important environmental features (Walter and Breckle 1985). The broadest system commonly used is based on major climactic zones, called biomes, but biomes are generally too large and coarse (often millions of square kilometers) to help distinguish ecosystems. On a smaller scale, measured in thousands of square kilometers, are physiographic (physical geography) regions that are recognized by integrating information on climate, soils, and topography (primarily elevation). These do have considerable potential for helping to recognize ecosystems. For example, lodgepole pine ecosystems growing in sand dunes in coastal Oregon, volcanic ash in the hills of eastern Oregon, and wet flats in the Cascade Mountains (Fowells 1965), could be distinguished by their location in three different physiographic regions. Unfortunately, even physiographic regions are sometimes too large to account for all the physical conditions that might serve to distinguish two forest ecosystems. For example, a red spruce stand growing in the cool climate of a high-elevation site might be less than a kilometer from a stand growing in the valley bottom where cold air moving down the slopes accumulates. While the flora and fauna of these two stands would probably be similar, there would be some differences that a sophisticated classification system would recognize. For instance, Swainson's thrushes, bay-breasted warblers, red maples, and northern white cedars would probably be common in the low-elevation forest, whereas gray-cheeked thrushes, blackpoll warblers, mountain ashes, and mountain maples might be found only in the high forest (Noon 1981).

Succession tends to confound classification systems because it adds the dimension of time and because it is not perfectly predictable. Ecosystem classifications often avoid the problem by focusing only on the climax vegetation likely to be associated with a particular kind of site. Forest type classifications tend to treat each type as an independent unit and overlook its place in a successional sequence. Ideally, a classification would recognize the age and development (most importantly height) of each stand and where it falls within a successional sequence of forest communities. This would allow one to take a current ecosystem map and predict what the forest would be like in the future, with and without various forms of human and natural disturbance.

Classifying forest ecosystems is also confounded by the fact that many of them have had their structure altered by human activities. High-grade logging, specifically, removing all the commercially valuable trees with no provision for regenerating them, has changed the structure of some forest types to a point where they are hard to recognize. For example, the Society of American Foresters' (1967) classification of forest cover types identifies a yellow birch-red spruce type in northern New England and Maritime Canada; because of selective removal of these species and birch diebacks this type is now dominated by red maple and balsam fir in many areas. It could be argued that a new type, red maple-balsam fir, should be recognized, but if one started recognizing every different assemblage of trees as a unique type, the list would be unreasonably long, especially in areas where human manipulations have been common.

This returns us to an important point made in Chapter 3; when viewed in an evolutionary time frame, many forest communities are not long-term, stable features of the landscape; rather they are a complex assemblage of species that have come together to temporarily occupy a particular site (see Figure

3.4) (Hunter et al. 1988). Species are brought together more by a shared need for a similar environment than by interdependence that has evolved over millions of years. This said, the makeup of communities is often reasonably predictable, and thus it is reasonable to develop classification systems for them. The major point is this: It is the physical environment that provides the foundation for communities, and classification systems should reflect this fact as fully as possible.

The rationale for using a detailed ecosystem classification to facilitate the integration of wildlife management into timber management has been presented implicitly and explicitly throughout this book. To restate it here: There are millions of species of wildlife, and the most efficient way to maintain their diversity is to organize management efforts by reasonable subsets; in other words, by ecosystems. (This is the coarse-filter approach described in Chapter 13.) It is barely practicable to develop individual management plans (the fine-filter approach) for each vertebrate in an area; to do the same for all the invertebrates and plants too is completely infeasible. Managing ecosystems is an efficient way to manage wildlife, but unless the classification system used is detailed enough to at least recognize the differences between various kinds of softwood forest, it is likely that many species will be left out. Ideally, the system would be sophisticated enough to distinguish among different kinds of red spruce forest, because then one could be reasonably confident that maintaining a complete array of ecosystems would provide habitat for the vast majority of species.

Weighed against the obvious benefits are the costs of employing a detailed classification system. With the advent of powerful and inexpensive computers, it is not difficult to manipulate the complex data set that would be required. The problem is in obtaining adequate data. Here the limitations are those of remote sensing technology (Botkin et al. 1984, Dyer and Crossley 1986). Although satellite imagery has improved remarkably in recent years, it is still not usually possible to distinguish trees of the same genus; longleaf pine versus loblolly pine, for example. Furthermore, to recognize stands distinguished by environmental differences would require extensive cross-referencing with maps of soil, topography, and climate. This integration and synthesis can be effected using geographic information systems (GIS) that organize geographic data on a coordinate system with very small cells. For example, every unit of land 30 × 30 meters may be identified by a unique pair of numbers that records its location on a grid system, and this will allow various sets of descriptive data to be assigned to the appropriate cell. (GIS also offers the possibility of describing a landscape's spatial heterogeneity by aggregating adjacent cells of the same ecosystem type into units, then describing their size, shape, and juxtaposition [Krummel et al. 1987].) In short, although remote sensing technology cannot easily recognize all ecosystem types, a reasonably detailed classification can be developed by using a GIS to incorporate other information. Plans based on such a system would still have to be carefully scrutinized by wildlife managers to be certain that some critical assets had not been overlooked. These might include ecosystems too subtly different to be recognized using remote sensing, features too small or inconspicuous to be seen by remote sensors (e.g., springs or caves), and features associated with special species that merit specific attention (e.g., rare plant stations and eagle nests). With or without Geographic Information Systems, simpler classifications will be used for other purposes, and it's important that natural resource managers communicate about the relationships—cross-connections—among systems (Thomas and Verner 1986).

Forest Stand Growth and Development Models

The growing sophistication of forest planning can be attributed in large part to the development and refinement of forest models. Since the 1960s, foresters have devised hundreds of models of how individual trees grow and how stands develop in order to schedule harvests, guide silvicultural treatments, and predict timber yields (Shugart 1984). At their simplest, tree growth models are formalizations of the empirically obtained tables of stand yield that foresters have used for decades, but more recent models tend to be much more elaborate, estimating parameters such as the sizes and shapes of crowns, mortality rates, and competition between adjacent trees. Many of these models predict the long-term dynamics of tree populations and thus provide rudimentary models of forest ecosystem succession.

Recently, wildlife managers have viewed these forest models as an opportunity to develop their static models of wildlife habitat relationships into dynamic models. Very simply, their thinking has gone something like this: (a) Forest models provide reasonable simulations of how tree populations change over time through successional processes and in response to human management; (b) the habitat needs of animals can be linked to some key vegetation parameters; and (c) this linkage can produce models simulating how animal populations would change through time in response to succession and forest management. (See the box on pages 270 and 271 for an example of one model in use.)

There is much to recommend this line of thinking; it was a cornerstone of an important symposium, "Wildlife 2000, Modeling Habitat Relations of Terrestrial Vertebrates" (Verner et al. 1986), and will undoubtedly occupy many biologists and managers for years to come. The proceedings of the symposium provide an excellent review of the state of the art that need not be reiterated in detail here. (For a concise overview see Pearsall et al. 1986.) To encapsulate it briefly, dynamic wildlife habitat modeling has developed enormously, but is still in its infancy because of various limitations such as: (a) the tree parameters that dominate forest models are not necessarily key features of animal habitat (e.g., many species are influenced more by understory vegetation than by trees); (b) the home range and thus habitat requirements of many animals are of such a large scale that they do not match well the scale of individual stands and thus the scale of most forest models; and (c) it is difficult for the models to predict the quality of habitat, not just its quantity, in sufficient detail to predict population densities. Sophisticated models of wildlife habitat would cope with these shortcomings plus many others that would be even more challenging to solve. For example, how do changes in animal populations affect the rate of succession; how do gradients within an ecosystem, e.g., altitude or soil moisture, affect wildlife; or how does habitat fragmentation affect the probability of small populations going extinct?

There is one feature of these habitat models that makes them somewhat ancillary to the main theme of this book; most of them focus on managing one species or a small group of species, rather than wildlife diversity. A few deal with all species of amphibians, reptiles, birds, and mammals (e.g., Toth et al. 1986). This limitation is understandable; if it is difficult to develop a reasonable habitat model for one species, dealing with a multitude of vertebrates, invertebrates, and plants is nearly impossible. From the standpoint of maintaining biological diversity, one could recommend some directions for future modeling efforts. It would be particularly useful to refine existing forest succession models by increasing their resolution at three scales:

1. At the coarsest scale, this would mean improving the models' ability to deal with landscape phenomena such as size, shape, and arrangement of ecosystem units.
2. On an intermediate scale, models should use classification systems that are as detailed as is reasonable (essentially, using ecosystem— as opposed to forest stand—classifications, as described in the previous section).
3. On the finest scale, models should predict the distribution and abundance of what are sometimes called habitat elements (Salwasser 1986). Snags are the best example of these, and some models of snag availability have been constructed (Neitro et al. 1985). Other examples might include soil litter, mast-producing trees, and shrub or herb strata.

Modeling efforts directed at individual species can also contribute to the larger goal of maintaining biological diversity, especially if they focus on species in danger of extinction or local extirpation. Models for animals that typically range over more than one ecosystem (either because individuals need habitat elements from two ecosystems or because their home ranges are so large) would clarify issues concerning ecosystem juxtaposition. Finally, developing models for species that require a very specific environment would increase the ecosystem resolution of modeling efforts. For this purpose, plants will usually be better subjects than animals.

It is necessary to close with a caveat about using models in natural resource management. It must always be remembered that the models are only

MODELING DYNASTIES

One of many forest models developed by the staff of the USDA Forest Service has been DYNAST (The *DYN*amically *A*nalytic *S*ilviculture Technique), a system designed to compare the multiple use benefits derived from alternative forest management strategies (Boyce 1977). The model combines: (a) models of resource distribution and availability (wildlife, water, timber, etc.) with (b) models of succession and (c) various timber prescriptions (e.g., rotation age, harvest unit size, species composition changes) to predict the consequences of different management options. Gary Benson and William Laudenslayer (1986) used DYNAST to examine the effects of three alternative management strategies on mule deer, band-tailed pigeons, and pileated woodpeckers in the Tahoe National Forest of California. The first strategy called for long rotations in which harvesting did not take place until stands were 200 years old. The second strategy was designed for optimal timber production; stands would be harvested as soon as they reached the culmination of mean annual increment at 80, 110, or 120 years depending on the forest type. A mixture of the first and second strategy—some stands cut at 120 years, others at 200 years—comprised the third strategy. In Figure 14.5 the results of the three strategies are predicted.

It is notable that after 200 years, when a steady-state has been achieved, the values for three of the parameters of interest, timber harvest and habitat for mule deer and band-tailed pigeons, are about the same under all three scenarios. The exception is the pileated woodpecker; as might be expected of a bird that needs large snags, it fares much better under long rotation management. The differences among the simulations are more apparent during the period before a steady state is achieved. For example, habitat quality for band-tailed pigeons and white-tailed deer takes longer to reach maximum values under the long rotation regime than under the mixed or optimal timber production alternatives, and total values (integrated over 200 years) are higher for pigeons and deer under these two scenarios. Naturally, timber production is highest under the optimal timber scenario.

To use a model like DYNAST to facilitate management for diversity, two approaches could be employed. First, changes in the areas of different types of forest (based on age, species composition, and size) could be simulated to ensure that none of the types ever became rare. In other words, the goal would be to guarantee that the sort of balanced array presented in Table 14.2 would persist through time. Second, species-oriented simulations could be constructed for rare species to make certain that their habitat quality values never became low enough to threaten them with extirpation. These approaches correspond, respectively, with the coarse-filter and fine-filter approaches described in Chapter 12.

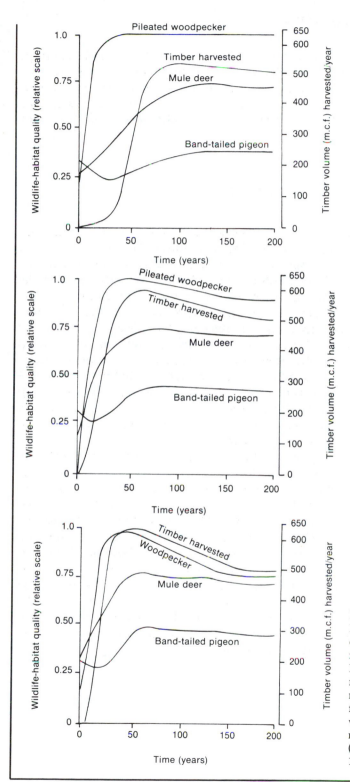

Figure 14.5 Predictions from a DYNAST model of the consequences of three different management strategies: A. Long rotations; B. Optimal timber production; C. A mixed strategy. Timber harvest is represented in thousands of cubic feet of wood harvested per year; wildlife habitat value is shown on a relative scale (0-1) for three species. (From Benson and Laudenslayer 1986)

tools, just one component in a larger planning process. The world is too large and complicated to reduce it into an all-encompassing model—even the most complex models make many simplifying assumptions—and thus one cannot rely on models to give correct decisions (Salwasser 1986). *People* must make the decisions after reviewing the syntheses of information and predictions that models provide.

SUMMARY

The diverse patchwork of most forested landscapes creates challenges and opportunities for managers attempting to maintain biological diversity. It is simple to imagine a landscape in which there is a balanced array of stands of varying species, age, and area, and in which this array is arranged on the land to maintain species of wildlife associated with both edge and forest interiors. In practice, the constraints of real landscapes make the task far more complicated. The use of various forest management techniques is profoundly affected by the distribution of sensitive areas such as riparian zones, steep slopes, and areas of shallow soils. Similarly, the expense of building and maintaining roads, especially in areas remote from a mill, often dictate the economics of forest practices. Perhaps the most important constraints are ownership lines, and in regions where ownerships are small, it is important for individuals to consider the landscape beyond their property lines if they are to play an effective role in biological diversity management.

Forest planning—managing forests on a large scale and far into the future—is somewhat more complicated than other forms of planning, but it is conceptually straightforward to: (a) determine the current status of a forest; (b) envision what the forest should look like if various benefits are to be derived from it; and (c) implement an action program that will gradually, through succession and human manipulations, shift the forest toward the desired state. Forest planning for biological diversity requires two tools, both of which need considerable refinement. First, detailed ecosystem classifications are needed to assure that a broad array of communities, and thus most species, will be maintained; coarse classifications (e.g., hardwoods, softwoods, mixed-woods) are not adequate for this purpose. Second, the models used to predict stand growth and development should be refined to make them better models of succession at both an ecosystem and landscape scale.

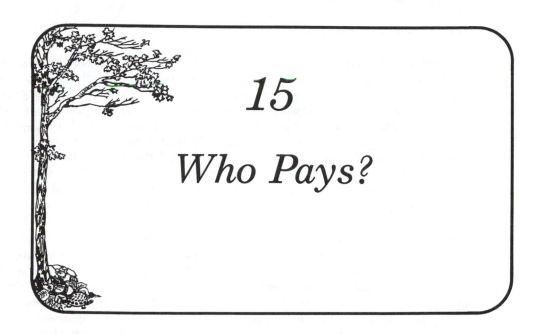

15

Who Pays?

Anyone who has ever doubted that money is the bottom line should consider how proverbs about money permeate our culture: "Time is money," "Money talks," and "Money doesn't grow on trees." And such perspectives are not limited to western life: In central Africa one hears "Money tells the truth," while a Russian adage has it that "When money talks, truth keeps silent." Another phrase, "There is no such thing as a free lunch," is particularly appropriate to the theme of this book, because unfortunately, the truth is that managing forests in a manner that maintains or restores biological diversity usually increases the cost of forest management without increasing income. The costs need not be great. As described throughout the preceding chapters, there is plenty of scope for balance, for having areas devoted to intensive production of wood products sharing the same landscape with areas where maximum timber production is modified in the interests of wildlife. However, any forest owner concerned primarily about profits will be quick to see the costs associated with compromises. In this chapter we will consider what these costs are and who might pay them.

THE COSTS

Planning

For some of the most important aspects of managing forest landscapes for biological diversity, the major cost is easy to identify, albeit often hard to quantify. It is the cost of planning, of salaries for people who understand the principles of biological diversity and can develop appropriate management plans. If a planning process already exists, as it will for virtually all large landowners, the cost of planning for landscape diversity can be largely piggy-backed onto existing plans with little additional cost. This is particularly true for one fundamental objective, balancing age structure, in which wildlife just provides an added incentive to have a steady flow of forest products. Further-more, timber managers already need to arrange different stands on the land-scape with an eye to soil fertility, sensitive sites, transportation routes, and stand juxtaposition, and considering biological diversity too will not enor-mously complicate the task, especially with computers as planning and inven-tory tools. This said, although they may be small, there will inevitably be some additional costs whenever a new element is added to a planning matrix.

To a certain extent, the extra expenses are mainly start-up costs; once a new plan or planning process is in place, it will require much less attention. However, unforeseen situations will always arise—a sudden plummet in the price of paper, a new market for a formerly underused species, massive blow-downs from a hurricane—and require adjustments to the management plan. A major threshold in the ease of management will be crossed when the manage-ment plan has been successful: i.e., a diverse forest landscape has been achieved and it is only necessary to maintain that diversity. In some cases, this objective is attained quickly and easily; too often a major imbalance of age classes will mean a wait of decades while succession travels its course.

Loss of Opportunity

Some decisions a forest manager might make—not to cut in riparian zones, to allow some old-growth stands to remain undisturbed, or not to plant an exotic species even though it will grow faster than a native one—could mean an opportunity to increase income from a given piece of land has been lost. The opportunity is not lost forever, but it is easy to think of the forest as a capital investment and of tree growth as interest; thus whenever the trees are not growing as fast as possible and being cut at the optimal time, money is being lost. (See the box for an example of how these costs can be estimated.)

The preceding paragraph may sound like an argument for always grow-ing trees as quickly as possible—high intensity silviculture and the proverbial row crop of trees—but it is not. The other side of the coin is that the cost of intensive silviculture—site preparation, planting, herbicide treatment, and thinning—can be very expensive. The cost/benefit ratios do not always favor it, and even when they do the money is not always available for investment. (This is especially true if the site is not particularly productive.) Thus, it is difficult to say how much opportunity is really being lost when a forest owner decides not to maximize wood production on a site. Even when a stand is al-lowed to age beyond the economic climax, there is effectively no loss of oppor-

ESTIMATING THE COSTS OF GROWING OLD FORESTS

Inflation, interest rates, and other financial uncertainties make it very difficult to accurately predict the income and expenditures associated with a forest management plan very far into the future. However, predicting wood production is much easier, and for many types of management decision, it is quite straightforward to use stand yield models to predict how wood production will be influenced by various alternatives. Here we will examine an example developed by Harold Wick and Rod Canutt (1979) of the USDA Forest Service, in which they estimate the consequence of managing ponderosa pine in the Blue Mountains of eastern Oregon and Washington on a long rotation. Specifically, they asked, how much would wood production be reduced if it were decided that 5% of the acreage should be providing old forest habitat (i.e., it would be over 160 years old)?

First, they assumed that the requisite area of old forest would be provided by having long rotations, rather than setting an area aside. This would mean that 15% of the area would have to be in a long rotation system, and at any given time 5% would be in the 0–80 year age class, 5% would be 81–160 years old, and 5% would be 161–240 years old. The normal regime of shelterwood cutting and commercial thinning has a rotation of 140 years (Table 15.1); under a long rotation system commercial thinnings would still occur at 40, 60, 80, and 100 years, but there would be no further treatment until 240 years.

TABLE 15.1 Standard Yield Tables for a Ponderosa Pine Type that Compare a Standard 140-year Rotation with an Extended 240-year Rotation[1]

Stand Age (Years)	Silvicultural Treatment	Wood Volume Harvested in Cubic Meters/Hectare	
		Standard Rotation 140-year	Extended Rotation 240-year
0 or 150	Overstory removal	98.0	98.0
40	Commercial thinning	34.4	34.4
60	Commercial thinning	33.9	33.9
80	Commercial thinning	40.0	40.0
100	Commercial thinning	42.5	42.5
140	Shelterwood cut	211.3	
240	Shelterwood cut		320.5
	Total	460.1	569.3

[1]From Wick and Canutt 1979)

The total wood production from the 140- and 240-year cycles is 460.1 and 569.3 cubic meters per hectare, respectively. Dividing these

figures by the number of years produces an estimate of mean annual increment for each regime:

$$\frac{460.1}{140} = 3.29 \text{ m}^3/\text{yr} \qquad \frac{569.3}{240} = 2.37 \text{ m}^3/\text{yr}$$

Obviously, the stands allowed to reach 240 years are not growing as rapidly, when averaged over the entire rotation, as the stands cut at 140
years. The relative reduction can be calculated easily:

$$\frac{3.29 - 2.37}{3.29} \times 100 = 28\%$$

This reduction in productivity takes place over only 15% of the whole
forest; 15% of 28 is 4.2%, so the overall reduction in wood production
on the forest is 4.2%. There are more sophisticated ways to make these
estimates using elaborate models of forest growth and production (Wick
et al. 1985), but the basic principle—comparing a forest in which timber
production is maximized versus one in which trade-offs for other values
are made—remains the same.

tunity if the stand is too far from other harvest operations to be cut efficiently
at that time, or in a location that is not accessible with available harvest equipment. These situations often arise in regions remote from markets, and in
mountainous areas with much steep, rocky terrain.

Despite all these qualifications, one thing is quite simple; whenever a
harvestable stand is permanently withdrawn from harvesting—because it borders a stream, because it is a critical source of snags, or because it is the last
virgin stand in the county—there is a recurring cost in lost opportunity. If the
stand is relatively unproductive and inaccessible, the loss may be trivial, but
it is still a cost. Furthermore, there are always some expenses associated with
just owning a stand, such as property taxes.

Inefficiency

On the face of it, it seems much more efficient to remove all the trees
from a valley and not to return again for 50 years, than to keep pecking away,
tree by tree. To some extent this is true. Large-scale harvesting is usually more
efficient than small-scale harvesting, but large-scale *management* is not necessarily more efficient than small-scale *management.* Consider the alternatives

of (a) entering a stand at 10-year intervals and removing 20% of the wood each time in a group selection harvest, versus (b) a 100% clearcut, followed by a controlled burn to remove slash and prepare a seed bed, hand planting seedlings, a herbicide treatment in the second year, a second treatment in year 10, and a thinning in year 20. It is quite possible that the first alternative will provide a better return on investment. Perhaps this scenario exaggerates how much less management is required following a selection cut, but the point is valid; sometimes people focus on the immediate cost of harvesting and overlook the long-term costs of regeneration. Obviously, all this assumes that the site is regenerated; if you compare small-scale management to clearcut plus laissez-faire regeneration, the latter will seem more cost-effective, at least in the short run.

THE BENEFITS

Decision makers are accustomed to systematically evaluating benefits and costs before acting, but the benefits associated with wildlife management do not readily lend themselves to such evaluations (Norton 1987, Ehrenfeld 1988). It is not easy to establish the value even of a game animal (a deer is worth far more to a sport hunter than the market value of its meat) let alone a warbler or a snag. Various techniques for placing a monetary value on wildlife have been devised—e.g., contingent valuation, hedonic pricing, safe minimum standard, cost of least-cost alternative (Freeman 1979, Alward et al. 1984, Sorg and Loomis 1985, Randall 1988)—but here we will make two simplifying assumptions. First, biological diversity is an inextricable part of environmental health, wise natural resource use, the quality of human life, and similar conceptual entities, and thus it has a broad, social value that exceeds the cost of maintaining it (Decker and Goff 1987). In other words, we *should* manage for biological diversity. Second, for forest landowners the cost of maintaining biological diversity will often exceed the financial benefits.

WHO PAYS?

Given that there are some unavoidable costs associated with creating and maintaining a diverse forest for wildlife, someone has to bear these expenses. The most obvious group is the landowners themselves, at least if you accept the idea that land stewardship is primarily their responsibility. But they are not necessarily the best choice; it may be fairer and more logical to direct some, if not all, the costs to other groups—for example, the people who use wildlife for recreational purposes, the consumers of forest products, or the general public, i.e., taxpayers. In this section we will explore the implications of having the costs borne by each of these groups.

Landowners

There are three major types of forest owners—private individuals, corporations, and governments—and each of these is likely to have a very different perspective about wildlife management on their land and its costs. Starting with individual owners, these are people who do not usually analyze the financial aspects of owning and managing a forest. Costs and benefits are hard to quantify because they often own a forest for subjective, intangible reasons that might be best summarized, "it makes them feel good." If these reasons are clarified at all, wildlife, aesthetics, and recreation are likely to be named. Their woodlot may represent a form of financial security, and they may be happy to cut some firewood or even to sell some stumpage to a logging contractor, but profits are far from their mind. Because of a lack of well-defined objectives, especially financial ones, these people seldom invest much money or personal labor in their forest to enhance its values. If they were routinely manipulating their forest, many would probably be willing to take steps that would benefit wildlife diversity. However, in the absence of any management activity, management directed at wildlife is hard to initiate. Obviously, there are many exceptions to these generalizations. They are based on research conducted in New England (Alexander and Kellert 1984) and their accuracy is unlikely to be universal, particularly in parts of the world where rural people have a subsistence lifestyle. They also may not be applicable to large private owners, who are likely to act more like a corporate owner. It would be interesting to survey landowners, giving them two alternative cutting plans—one that just removes the timber as expeditiously as possible, and one that creates a diverse forest—and ask them how much less they would be willing to accept for the cutting rights sold under the second plan.

When forest lands are owned by corporations or governments, the issue of landowners paying for wildlife management becomes less direct, because government lands are effectively owned by the taxpayers, and corporations will typically pass any increased management costs on to the consumers of their products. Hypothetically, an exception could arise in corporations whose profits are regulated by governments; utility companies, for instance. In these cases, it could be decided that additional expenses for wildlife management should be passed on to the shareholders, rather than consumers. Incidentally, managing for wildlife diversity can have a positive economic spin-off for corporations in the form of good public relations. Although hard to quantify, the consequences in terms of product sales, employee morale, and community acceptance can be significant.

Users

One can readily argue that the costs of wildlife management should be borne by the people who use wildlife, and in a narrow sense this does happen. In many countries, hunters and anglers pay license fees that support government efforts to manage populations of game fish, birds, and mammals (in the U.S. there are also taxes on sporting equipment [Kallman 1987]). Most of this activity is directed toward regulating harvests by setting and enforcing open seasons and bag limits, but it also includes a moderate amount of habitat management (Teer et al. 1983). Where wildlife habitat management really comes to the fore is in places where hunters and anglers pay landowners—sometimes governments, usually private owners—for the rights to practice their sports on a specific piece of land. In the southern United States, hunters pay hundreds of millions of dollars to lease land where they can hunt white-tailed deer, bobwhite quail, and turkeys; one corporation, International Paper Company, makes more money from selling hunting leases than from selling wood in some districts (Eubanks, pers. comm.). And hunting leases in the U.S. are relatively cheap; the right to shoot red grouse on the moors of Scotland can cost hundreds of dollars per day. There may be negative social consequences when hunting becomes a privilege of the rich (Geist 1988), but obviously such fees create an enormous incentive to manage forests for popular game species.

What about the users of nongame? How do birdwatchers, other naturalists, campers, canoeists, etc. pay for the enjoyment they derive from wildlife? (Incidentally, this is not a small group. In the U.S., 78% of adults participated in wildlife-related recreation in 1985; 74% in nongame activities [USFWS 1987].) The curt answer is that they do not pay. There are entrance fees for some wildlife refuges, some people pay membership fees in conservation groups that manage wildlife habitat, and some government wildlife agencies have established voluntary contribution programs such as income tax checkoffs, but the money raised from all these sources is a tiny fraction of the fees associated with hunting and fishing.

New mechanisms for raising revenues for all wildlife have been proposed, mechanisms that would spread the cost of wildlife management over all users. One idea is taxing items used by outdoor recreationists such as binoculars, backpacks, and tents. Just considering items directly connected to enjoying wildlife, the amounts of money involved are quite significant. A survey in the United States determined that in 1980, $635,888,000 was spent on identification books, birdseed, bird feeders, birdhouses, and birdbaths (Shaw and Mangun 1984). A 5% tax on these items would have raised over $30 million dollars. Although this is a huge amount of money compared to what is currently generated for nongame, it is still tiny compared to the nearly $200 million raised by taxes on hunting and fishing equipment. Moreover, in this same survey people who participated in wildlife-oriented recreation were asked how they felt about a special tax on equipment used for wildlife observation: 41.7% were in favor; 42.0% were opposed. It is unlikely that these people are simply stingier about wildlife than hunters and anglers. The difference is probably the distinction between consumptive and nonconsumptive use. Birdwatchers who take home only memories do not feel as compelled to pay for the experience as someone who takes home a brace of ducks to eat. They have not purchased a commodity, only enjoyed an amenity. This attitude may slowly change as the world moves from a manufacturing economy to a service economy, but in the meantime it appears unlikely that substantial sums of money can be generated for enhancing wildlife diversity from direct payments by wildlife users.

Consumers

When you buy an apple, you know that a small portion of the apple's cost went toward a retirement scheme for the farmer who grew it; another tiny portion paid for hard-hats for people in a shipping warehouse. In other words, the social costs of doing business have been internalized and are passed on to the consumer. Similarly, in recent years environmental regulations have led to the internalization of many pollution control costs. If a pad of paper bears costs associated with worker's safety, health, and retirement, and air and water pollution control, why shouldn't it also include the costs of maintaining a diverse forest? In many ways, using government regulations to mandate stewardship of our wildlife resources—and thereby passing the cost of forest wildlife management onto consumers of forest products—is simple, fair, and logical. It is also legal; governments almost always have the right to protect the public health, safety, morality, and general welfare even if this impinges on the rights of individual landowners to use their land however they choose (Cubbage and Siegel 1985). In countries such as the United States, in which the government holds property rights to all wild animals in trust for the public, it is especially easy to legally demonstrate that landowners cannot use their land in a manner that degrades a public interest. Finally, having consumers pay is also efficient, because regulations inspire great thrift in an effort to minimize the costs that must be internalized (Freeman et al. 1973).

Unfortunately, there are at least two significant flaws with this approach. The first problem is intrinsic to government regulation. By its very nature it is adversarial. It pitches industry, conservation groups, and government into an arena where debates become polarized and are resolved as much by the political adeptness of lawyers as by the knowledge and reason of resource managers. It is an arena where people win, lose, and compromise, rather than work as a team to identify, and to progress toward, common goals. Monies that could have been directed toward wildlife management are instead spent to posture, lobby, and litigate. It is one thing to have regulations that restrict behavior, "Thou shall not cut within 10 meters of a stream"; it is another thing to have regulations that require landowners to take a positive, creative approach to incorporating wildlife issues into their regular planning program. The creative approach offers the most promise and cannot be easily stimulated by regulations.

A second problem stems from the marketing of wood. If the cost of wildlife management causes a rise in product prices, this will mean a decrease in competitiveness and the risk of losing business. Competitiveness is not an issue as long as every company is faced with the same constraints, but the markets for forest products are international. If regulations in Indonesia increase the cost of Indonesian wood, then they may lose some of their markets to Malaysia and the Philippines. Hypothetically, international agreements could solve this problem; in practice, such levels of cooperation would be very difficult to achieve. This is particularly poignant when you consider the export of wood from developing countries to Japan, the United States, and Europe. In many developing countries, the need for foreign revenues is so great that officials use a lack of regulations, along with low wages and tax relief, to encourage harvesting by foreign timber corporations. Consequently, these companies ignore environmental protection techniques that are standard operating procedure in their home country and often find it cheaper to import wood than to grow it at home (Myers 1979). For example, Japan imports more tropical hardwoods than the total for all of Europe and North America, even though the country is two-thirds forested and it has been estimated that these forests could meet all Japan's projected wood demand for the next 150 years (Westoby 1982). It is sad to think of the Third World's tropical forests being degraded to keep profit margins up and prices down in wealthy countries, but it would not be easy to change this. Some attempt has been made to organize the major wood-exporting countries into a cartel similar to the oil cartel, OPEC (Myers 1979); however, in the current state of affairs, lack of international cooperation continues to flaw the logic of a regulatory (i.e., consumer-pays) approach.

Taxpayers

Biological diversity is a cornerstone of ecological integrity and a benchmark for wise natural resource management, and thus it can be readily advocated as an important social value. Stated less formally, everyone gets something out of wildlife. Most people derive some direct benefits—ranging from the sublime of aesthetic and spiritual pleasures to the mundane of wild plants and animals to eat—while only a few of the most urban people are largely isolated from wildlife. And even they are ultimately dependent on the earth's ecosystems along with everyone else. The logical conclusion is that if everybody benefits from biological diversity, then all taxpayers should pay the costs of having diverse forests. Furthermore, we all consume many forest products—tax forms, for example. It would be harder to argue that everyone should pay to support multiple-use forestry if maple syrup, witch hazel oil, and mistletoe dominated the array of forest products rather than such essential commodities as lumber, fuelwood, and paper.

There are two ways that taxpayers can pay for forest wildlife diversity: One is through the management of publicly owned forests; the other is by subsidizing certain costs on private lands.

Public land

Forests owned by the public fall under many names and have different, although overlapping, purposes. Some are intended to protect wildlife and have names like "refuges," "sanctuaries," or "reserves"; areas called "parks" are managed mainly for recreation; the forests that are principally managed for wood production are often called "forests" (here we will call these "government forests"). Whatever their name and primary goal, publicly owned forest land can be managed in a way that will further the goal of maintaining biological diversity. In many cases, this will impose no additional cost—a nature reserve consisting of an old-growth forest usually needs no special management to enhance its contribution to a landscape's biological diversity. On the other hand, a government forest being managed primarily for timber production may see some reduction in profits if a scheme for maintaining wildlife diversity is implemented. (Actually, many government forests are operated at a net loss, and "increase in losses" would be more accurate than "reduction in profits".)

This brings up the issue of public lands balancing what is happening on private land, a topic touched on in the preceding chapter's discussion of the White Mountain National Forest. One could argue that if the private portions of the landscape are dominated by short rotations and large-scale management, then the government should assume responsibility for providing those types of forests that are least profitable, notably old forests and forests managed through selection harvests. This is easily done with reserves and parks managed for wildlife and recreation, but should government forests also compensate for activities on private forests? Some would argue that government forests should be an independent entity, a model of multiple-use, and not have their ability to produce timber encumbered by the necessity of making up for the practices on neighboring land. Others would strongly disagree.

In fact, government forests do often end up compensating for a lack of certain types of forests on private land. The spotted owl, which is largely dependent on the old stands found on U.S. National Forests and Bureau of Land Management holdings, is an example of this (see Chapter 5). However, balancing is usually possible only at very large scales, because most government forests are large blocks, not parcels scattered among the private ownerships. Indeed, even the distribution of the blocks may be skewed; U.S. National Forests comprise a majority of forest land in some western states but are totally absent from six eastern states (Barton and Fosburgh 1986). In theory, the public could buy scattered tracts of forest land in regions dominated by private interests and manage them to balance the overall picture; a more realistic alternative is for the government to subsidize wildlife management on private land with voluntary incentive programs.

Subsidies

Government subsidies take many forms: price supports for commodities, cost-sharing, tax abatement, direct grants, etc. Given this variety—to say nothing of the variety of national, regional, and local governments—it is difficult to make some generalizations about how a subsidy program for biological diversity can best be devised. Each situation is unique economically, ecologically, socially, and politically, and would require a specially tailored program. This said, there are three generalizations that can be made. First, subsidy programs for wildlife management should be developed by a coalition of people, including government officials, industrialists, environmentalists, and scientists. Second, subsidies should be generous enough to cover the real, additional costs of wildlife management and thus keep prices competitive, without being so generous that the recipients can also use the subsidies to increase profit margins for shareholders. Third, broad, comprehensive programs are likely to be more efficient than narrowly focused ones. This statement requires an explanation. One could develop a program that paid a dollar for every snag left after a harvest, 10 dollars per hectare per year for stands over 100 years old, etc.—thus subsidizing each of the costly elements of a diverse forest one-by-one. However, it would usually be preferable to encourage owners to incorporate all the key aspects of biological diversity management into their regular forest management plans and then to give them a single reward based on the amount of land being managed under an integrated plan. There would be less flexibility under the latter system, but it would be much simpler and cheaper to administer for both the industry and the government.

Many forest industries already receive government subsidies that are designed to meet various objectives that have little to do with wildlife directly—e.g., encourage hiring racial minorities, promote worker safety, and, most commonly, to simply bolster the general economy. Some of these subsidies have spin-offs of consequence to biological diversity, and they should be scrutinized to determine if their impact is positive or negative. The U.S. Internal Revenue Service (IRS) provides one good example; beginning in 1921, the IRS taxed income from capital gains (e.g., sale of a house or, of more relevance to our story, sale of trees). Recognizing that a one-time gain could leave an individual with an excessive tax burden, the Congress exempted 60% of an individual's gains from taxation. In other words, if you sold some trees for $3,000 and it cost you $2,000 to grow them, the IRS would exempt 60% of your net receipts ($600 of $1,000) and tax you only on $400. Later, tax abatement for capital gains was extended (by a more complicated formula) to corporations as a subsidy to encourage corporate investment. When the government repealed capital gains tax abatement in 1987, it had a significant effect on the complex process by which interest rates, tree growth rates, and expenses (including capital gains and property taxes) are used to calculate the economically optimum age at which to cut a tree. The upshot was that removal of capital gains subsidies encouraged timber owners to grow pulpwood-quality trees on short rotations, and dissuaded them from growing large, high-quality trees (Canham and Gray 1986). It is unlikely that the IRS knew that their decision would reduce the availability of habitat for plants and animals that prefer the larger trees of a long rotation, but a lack of intent does not diminish the impact.

Another example of a government subsidy with consequences for wildlife comes from Alaska, where the USDA Forest Service sells timber from old-growth stands at prices that are arguably less than fair market values (Dixon and Juelson 1987), and inarguably so low that the Forest Service loses money on the sales. Money is lost and a subsidy occurs, because the Forest Service pays the costs of building and maintaining access roads, an expensive proposition in such rugged and remote country. Actually, selling timber at below cost is not unique to Alaska; about half the National Forests lose money on timber sales—the total was $740 million in 1982—and many are in the east and south where the forests are not remote (Stout 1985). Below-cost timber sales have become very controversial, particularly among wilderness advocates who dislike seeing large, roadless tracts being invaded, and among wildlife advocates concerned about the loss of old forest habitat. The arguments about below-cost timber sales are too lengthy to present here (Clawson 1976, Rasmussen 1985, Schuster and Jones 1985, Stout 1985, Wilkinson and Anderson 1988). Suffice it to say that however one scrutinizes these timber sales programs—short-term or long-term, timber use or multiple-use—it is difficult to make much sense of them unless you accept them as broad-based subsidies to regional economies. Many people believe that there is nothing inherently wrong with such subsidies, but this still leaves the question, are there alternative government subsidies that could accomplish the same goal without having a detrimental effect on wildlife diversity? For example, could the government subsidize the modernization of sawmills to make them still profitable with less wood? Or, to give priority to the workers' needs, why couldn't the government foster the establishment of new industries and provide workers with retraining programs and support while they are learning new skills (Freeman et al. 1973, Thurow 1980)? One also has to ask, what effects do below-cost sales of government timber have on companies trying to profitably grow timber on private lands? Could this competition limit the private companies' ability to bear the costs of incorporating wildlife goals into their management program?

There is one form of government subsidy that is quite different from the others: technical assistance. In other words, governments can help the private sector incorporate wildlife objectives into their plans simply by providing information and guidance on how to do it. Many governments have long employed foresters to assist private landowners; these foresters could be taught the principles of managing forests for biological diversity and asked to develop an outreach program using all their regular tools—workshops, informational bulletins, consultation services, etc. Among some landowners, technical assistance programs may be more cost-effective than direct financial incentives. When New England small woodlot owners were asked what types of government assistance for wildlife management they desired, equal numbers chose either property tax reduction (41%), or free technical advice (41%) (Alexander and Kellert 1984).

Synthesis

It would be easy to conclude from the preceding discussion that taxpayers—or more specifically, publicly owned land and publicly sponsored incentive programs—are the best choice for bearing the cost of managing forests for biological diversity. This may be a reasonable generalization, but there will certainly be many exceptions. In a region dominated by small private owners, a comprehensive incentive program could be very difficult to administer given the number of owners and the diversity of their approaches to management. A simple property tax abatement program designed just to keep land forested (i.e., not sold for house lots) plus some regulations and a technical advice program might be most cost-effective under these circumstances. To take a second example, it would be patently unreasonable to expect the taxpayers of a developing country to further subsidize the harvest operations of a large and profitable multinational timber corporation—as noted, poor Malaysians and Brazilians are already doing too much to maintain profit margins for American and Japanese shareholders and cheap prices for foreign consumers.

More often than not, it is likely to be a combination of groups that share the costs. Some fundamental requirements—such as not clearcutting in riparian zones, not locating roads on steep slopes, and not disturbing endangered species habitat—should probably be mandated by law, thus passing these costs on to consumers. Hunters and anglers will probably continue to pay for managing the species they seek, and in due course, everyone whose recreation focuses on wildlife may be willing to pay something for the privilege. The greatest opportunities may lie with taxpayer-supported incentive programs, and hopefully various interest groups will join to explore these possibilities creatively.

AN INDUSTRIAL/ENVIRONMENTAL FORUM

In November 1986, 28 people gathered at an inn in Tenant's Harbor, Maine to spend three days discussing the future of Maine's forests. Assembled by the Maine Audubon Society, these people represented the forest industry, environmental groups, government agencies, the financial community, and academia. They brought a common desire to improve the forests of Maine as a ecological, economic, and social resource. They voiced grand, visionary ideas: public acquisition of a million acres of forests, constructing a paper mill that would strengthen the demand for hardwood pulp, and developing a research and development park focusing on forest products. And they formed committees.

One of the committees was charged with developing a plan designed to assure that Maine would have a "diverse, productive, and sustainable forest." Initially, the committee constructed three independent visions of how the Maine forest would look in a Panglossian "best of all possible worlds," as seen from the perspectives of wood production, wildlife, and recreation. These three visions were then merged into a single scenario with eight major parts; they can be summarized briefly:

1. *Species composition:* A natural distribution of native tree species should be retained.
2. *Age structure:* A balanced age class distribution should be maintained, including 10–20% stands over 100 years old.
3. *Spatial diversity:* Five scales of management ranging from single tree selection to large clearcuts (40–100 acres) should be used approximately equally.
4. *Regeneration:* Following harvests, steps should be taken to assure the regeneration of a desirable stand.
5. *Productivity:* Site fertility should be maintained by limiting the frequency of whole-tree harvests.
6. *Distribution:* Large-scale management should not take place in ecologically or aesthetically sensitive areas such as riparian zones, steep slopes, deer wintering areas, eagle nest sites, campsites, and trail corridors.
7. *Access:* Public access would be guaranteed; there could be reasonable fees for road use, and limitations on access to sensitive areas, but no exclusive leases for hunting or fishing.
8. *Development:* Development for residential, commercial, or industrial purposes would not be allowed.

During the process of forging a single scenario, the idea for a voluntary incentive program arose; why not give a financial reward to landowners who managed their land in concert with the scenario? An obvious approach was to remove the property tax, approximately a dollar per acre per year, for people who participated in the program. It seemed very reasonable that if the public had access to lands being carefully managed with multiple-use objectives, then these forests would be providing the same public benefits as state-owned forests and should not be taxed.

The process for participation would be fairly simple. First, a landowner would have to decide which lands to include in the program. For large landowners, this would mean identifying management units of at least 10,000 acres, and they would not be allowed to practice "ecological gerrymandering" (stringing together stands of low commercial value to assemble 10,000-acre units). Second, a landowner would: describe the current status of the management unit, project how it would eventually conform to the scenario, and propose a plan for achieving this. It was assumed that some parts of the scenario, especially the balance of age classes, might take as much as 50 years to achieve. Thirdly, this plan would be submitted to a state review board that would approve it or suggest necessary changes. Finally, an annual report of activities and achievements would be submitted to the state; some spot-checks to confirm the accuracy of these reports would be necessary.

On further analysis, it appeared that the incentives would probably have to be higher, at least $2 per acre for landowners to participate. Still, the overall cost of the program appeared very reasonable; with roughly 11 million acres of forest eligible for the program, and an estimated participation rate of 50% or less, the annual cost would probably be under $10 million. Ten million dollars per year is a significant sum, but not when compared to the cost of public land acquisition. At $100 to $300 per acre, it could cost the state roughly $200,000,000 to buy just one million acres of land, and Maine would still have far less public land than the national average. Moreover, any land purchased by the state would still decrease state revenues by about a dollar per acre per year because of lost property taxes. There would be some administrative costs for the program, but again they would probably be less than the administrative costs of owning the land.

Members of the committee from the forest industry felt that there was a hidden danger in proposing a voluntary incentive program to the state legislature. They feared that the legislators, who heard frequent complaints about clearcutting from their constituents, would excise the voluntary aspect of the proposal and make it into a mandatory forest practices act. With this in mind, the committee decided to write its own proposal for a forest practices act, and hopefully forestall an attempt to use the incentive program proposal for this purpose. The committee's proposal for a forest practices act had three major components:

1. *Reporting:* All commercial harvest operations would have to be reported to the state and include a regeneration plan, signed by a professional forester, that would result in adequate regeneration within

5 years, or allow the residual growing stock to fully reoccupy the site within 10 years.

2. *Arrangement of harvest areas in time and space:* Even-aged harvesting would not be allowed to exceed one-third of the area of a parcel during a 15-year period. (A parcel would be any continguous ownership; very large contiguous ownerships would have to be subdivided into management parcels with a maximum area of 25,000 acres.) Individual harvested blocks could not exceed 100 acres and must be separated from one another in space by a minimum of 300 ft, or in time by 15 years. On parcels under 60 acres the size of the harvest area could exceed one-third, up to a maximum block size of 20 acres.

3. *Environmental protection:* Several restrictions on harvesting and road building in riparian areas, steep slopes, high altitude areas, wetlands, and deer wintering areas were included.

When the committee's report was brought to the full Forum, there was considerable controversy over the mandatory restrictions on harvest areas, particularly the restriction on cutting no more than a third of a parcel in 15 years. After lengthy debate, it was decided to advocate legislation that would restrict the size and arrangement of harvest areas in time and space, but to not recommend specific numbers. These would have to be decided in the political arena.

At this writing, the Forum's report, which also proposes many steps that would improve the economic environment for Maine's forest industry, has received considerable attention long before its public release, and the Speaker of the House has agreed to sponsor legislation in the next session of the Maine State Legislature that would enact many of the Forum's recommendations. At this juncture, it is impossible to predict just how the tides of politics will shape the outcome.

SUMMARY

There is enormous scope for managing forest landscapes for biological diversity in a manner that complements, rather than conflicts with, timber management. Nevertheless, it is inevitable that there will be costs associated with such a program. Some of the costs may be fairly small; for example, salary for staff to integrate wildlife objectives into the existing planning process. More substantial costs from lost opportunities and inefficiency could accrue if wildlife considerations lead to a decision to compromise the rate of wood production and harvest in a forest; e.g., if large areas of riparian zones, old-growth stands, and snags were removed from harvest plans and relatively expensive harvest techniques, such as single tree selection, were employed. (Of course, when one considers the costs of regeneration and intermediate treatments that often accompany clearcutting, selection harvesting does not seem so costly.)

There are four groups that might bear the extra costs of forest management for biological diversity: wildlife users, wood consumers, taxpayers, and landowners. Some wildlife users (hunters and anglers) pay for their use of wildlife through property leases, license fees, and equipment taxes, but the majority of people are nonconsumptive users and many seem unwilling to be taxed for recreational use of wildlife. Consumers would pay the costs if governments mandated wildlife management through regulations, because corporations would make these costs internal to the price of wood products. Given the wide social value of wildlife and related qualities such as clean air and water, perhaps taxpayers should bear the bulk of the expense. There are two basic ways this can be done: by providing the most costly elements of a diverse landscape on public land (e.g., old growth); or by subsidizing the extra costs that owners experience when they manage wildlife on their land. Among landowners, only small private owners are likely to carry some of the costs; costs borne by corporate and government owners would effectively be paid by consumers and taxpayers, respectively. In most places, some combination of these four groups will probably pay the cost of wildlife management.

Epilogue

. . . thou canst not stir a flower
Without troubling of a star

Francis Thompson

Are these lines the whimsical expression of a romantic poet, or do they speak to a fundamental truth about the relationships that bind the universe? It exercises the imagination to envision how the welfare of a single flower could affect a star, a huge, impassive, and very distant object; it defies credulity. On the other hand, Newton's law of universal gravitation does assert that the mere stirring of a flower affects every other object in the universe. Closer to home, more down to earth, these lines speak of a commonly held truism of ecology, "Everything is connected to everything else." It is this belief that causes many people to look upon the manipulation of forest ecosystems with great concern. It is this belief that was buttressed at several points in this book—Tana River monkeys and their fig trees, mycorrhizal fungi and their logs—when examples were drawn to demonstrate the importance of acting with great care. However, this belief cannot be a call for inaction. People will manipulate forests; we must to survive and prosper. We need wilderness and tracts of old-growth forests where the flowers are not stirred; and we need forests that are managed for wood production. If we set some forests aside, if we wisely manage the rest, we can enjoy diverse and productive forests for many years to come.

APPENDIX 1

U. S. National Policies Related to Maintaining Biological Diversity in Forests

Sarah S. Stockwell

Federal forests comprise 31% (276 million acres) of all U.S. forest lands. The largest public forest landholder (19% of all U.S. forest land) is the U.S. Department of Agriculture's Forest Service, and they must manage their lands for "outdoor recreation, range, timber, watershed, and wildlife and fish purposes" under the *Multiple Use-Sustained Yield Act* of 1960 (MUSY). Although MUSY does not specifically prescribe managing national forests to maintain biological diversity, it requires the Forest Service to consider the values of fish and wildlife resources equally with other renewable forest resources when developing forest plans. MUSY sets the stage for more specific regulations regarding managing forests for biological diversity set forth in the National Forest Management Act.

The U.S. Department of Interior has three agencies with major forest holdings; of these, the largest forest landholder (16% of all U.S. forest land) is the Bureau of Land Management (BLM). Like the Forest Service, the BLM must manage its lands for multiple purposes and must develop land use plans for review by the Secretary of the Interior as outlined in the *Federal Land Policy and Management Act* of 1976 (FLPMA). The Bureau of Land Management currently is preparing habitat management plans for all BLM lands that outline schedules for maintaining and improving habitat for a variety of fish and wildlife species. Priority is given to projects that protect and restore habitats of threatened, endangered, or declining species and to projects that protect riparian and wetland habitats. The Bureau of Land Management also is attempting to protect examples of all habitats or ecosystems present on BLM lands. Another Interior agency, the National Part Service (NPS), is directed to

291

conserve the scenery, natural and historic objects, and wildlife of national parks in an unimpaired state that can be enjoyed by people for many generations to come. Although NPS has no specific directive to maintain biological diversity, they have a great opportunity to do so and have been encouraged to take a more active role in conserving biological diversity in parklands. Finally, the Fish and Wildlife Service (FWS) has a mission to protect and perpetuate all wildlife species, and contributes to maintaining biological diversity through a wide range of activities including management of National Wildlife Refuges.

In addition to adhering to specific agency directives, there are many Federal Acts that direct various federal agencies to manage forest lands. Two of these, the *Endangered Species Act* of 1973 (ESA), and the *National Forest Management Act* of 1976 (NFMA), specifically address issues relevant to maintaining or restoring biological diversity.

The Endangered Species Act requires all federal agencies to protect and conserve species listed by the U.S. Fish and Wildlife Service as threatened or endangered. ESA specifically requires all federal agencies to identify and protect habitat critical to the survival and recovery of any listed species. ESA also directs all federal agencies with jurisdiction over threatened and endangered species to help prepare recovery plans and develop strategies for implementing the plans. Each recovery plan is designed to lead to the eventual delisting of a threatened or endangered species.

The Endangered Species Act is potentially the single most important piece of legislation in the U.S. that directs conservation of biological diversity in forests. In compliance with this act, a considerable effort is spent on conserving rare, threatened, and endangered species, and far more could be undertaken if sufficient funds were available. Moreover, many nonthreatened and nonendangered species and ecosystems are inadvertently protected under authority of ESA when critical habitat is set aside for rare and endangered species.

The *National Forest Management Act* (NFMA) authorizes various amendments to the *Forest and Rangeland Renewable Resources Planning Act* of 1974 (RPA). RPA requires the USDA Forest Service to analyze long-term (40-year) supplies and demands of all renewable resources on U.S. forestlands (both public and private) and to prepare a detailed program outlining how they will manage those resources in a sustainable manner. Because the Forest Service needs information from state and private forest owners to prepare the required documents, Congress passed the *Cooperative Forestry Assistance Act* in 1978. The Forestry Assistance Act ensures that both federal and state foresters will participate in preparing forest management plans that consider local, regional, and national interests. NFMA goes one step further than RPA by requiring the Forest Service to prepare similar management plans for *each* National Forest.

Under NFMA there are specific directives for managing forestlands to maintain or improve biological diversity at the genetic, species, and ecosystem levels. First, NFMA requires the Forest Service to maintain viable populations of existing native and desirable non-native vertebrate species. Second, NFMA directs the Forest Service to maintain the diversity of tree species present at the onset of the management plan and to protect the resources and habitats upon which vertebrate populations and endangered species depend. Third, NFMA directs the Forest Service to preserve and enhance the diversity of plant and animal communities and tree species within each management area so that diversity is equal to or greater than that of a natural, unmanaged forest. Although NFMA only requires specific habitat management for vertebrate species, the Forest Service has occasionally interpreted the regulations liberally to include plants and invertebrates. The Forest Service monitors populations of several management indicator species to insure that forest management plans meet NFMA requirements. Typically, management indicator species represent the array of species whose population levels should change in response to different management practices outlined in the management plan, and include threatened or endangered species, species with special habitat requirements, popular game species, and nongame species of special interest.

Any management plan drawn up by the Forest Service, the Bureau of Land Management, the National Park Service, or the Fish and Wildlife Service also must comply with requirements and guidelines of the Fish and Wildlife Coordination Act of 1958, the Wilderness Act of 1964, the Wild and Scenic Rivers Act of 1968, the National Environmental Policy Act of 1969, the Sikes Act of 1974, and the Clean Water and Clean Air Acts of 1977.

The *Fish and Wildlife Coordination Act* requires all federal agencies to consult the Fish and Wildlife Service before proceeding with any development project that may affect wildlife. During initial stages of planning, if an agency learns from the USFWS that a particular parcel of land has significant wildlife values, the agency may choose to resite the project to avoid destroying that habitat. At the very least, the Fish and Wildlife Coordination Act works in concert with the Endangered Species Act to help minimize or mitigate loss of critical habitat for threatened and endangered species.

The *Sikes Act* requires all federal agencies within the Departments of Agriculture and Interior to coordinate with state fish and wildlife agencies on plans for conserving and rehabilitating fish and wildlife. Thus, federal and state agencies work together to develop the most effective plans for maintaining and rehabilitating wildlife diversity. Furthermore, the Sikes Act encourages federal agencies to develop specific schedules for managing and improving wildlife habitat as outlined in plans developed under the National Forest Management Act or Federal Land Policy and Management Act.

The *National Environmental Policy Act* (NEPA) requires all federal agencies to prepare a report (EIS, environmental impact statement) on the environmental consequences of any proposed development project or program that may significantly alter the environment. NEPA requires federal agencies to list several alternative strategies for implementing the proposed project or program and to outline the potential environmental effects of each strategy. If rare or endangered species or communities would be affected by the proposed project, this information should be disclosed in the environmental impact statement. NEPA also requires that plans for mitigating adverse environmental effects be addressed in the EIS. Environmental impact statements are then critiqued by other public agencies, private organizations, and concerned citizens. The public review process can significantly affect which development strategy is finally approved and implemented. Ultimately, NEPA acts as a check on programs that might unwittingly lower biological diversity.

The *Wilderness Act* established a national system of wilderness areas, areas untrammeled by humans and set aside for their ecological, geological, educational, scenic, or historical significance. Only primitive recreation is allowed in wilderness areas. The Wilderness Act also requires all federal agencies to review their landholdings to determine what areas would be suitable for inclusion in the wilderness system. Wilderness areas are important reservoirs of biological diversity at all levels (genetic, species, and ecosystem).

The *Wild and Scenic Rivers Act* established a system of free-flowing rivers with outstanding scenic, geologic, recreational, cultural, or other attributes and established a process for reviewing rivers for possible inclusion in the system. Designated wild and scenic rivers harbor an array of riparian species and communities and thus add to the overall biological diversity of a forest.

The *Clean Water Act* requires federal agencies to comply with state clean water requirements and special permitting procedures for any dredge or fill activities in waterways or wetlands. The *Clean Air Act* restricts certain federal lands from allowing any increase in sulfur dioxide and particulate levels. Together, these two Acts protect the resources upon which all life depends, indirectly helping to conserve biological diversity.

State and Private Forest Lands

Nonfederal forest land comprises 61% (409 million acres) of all U.S. forest lands. Moreover, over half of all U.S. forest land is owned by private individuals or corporations that do not own their own wood processing plant. Many of these forest landowners do not manage their land solely for wood products. Collectively, these lands present a tremendous opportunity for conserving biological diversity in forests across the U.S.

A few states have forest practice laws restricting landowners from degrading the soil, water, wildlife, or aesthetics of their land so that neighboring private or public lands are not adversely affected. These state laws are primarily directed towards reducing non-point water pollution from forestry operations in response to an amendment to the 1972 *Federal Water Pollution Control Act*.

The *Cooperative Forestry Assistance Act* of 1978 authorizes the Secretary of Agriculture to assist state foresters to manage nonfederal forest lands. In addition to providing help with timber management, the Forestry Assistance Act authorizes financial and technical assistance to state foresters for protecting soils, water quality and quantity, and for improving fish and wildlife habitat on both state and private forest lands. Although this Act does not directly prescribe tools for managing biological diversity, it does encourage forest landowners to protect fish and wildlife habitat. Ideally, state, federal, and private forest management would all be integrated into regional plans for maintaining biological diversity.

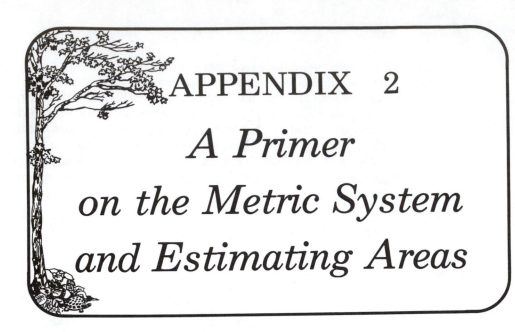

APPENDIX 2

A Primer on the Metric System and Estimating Areas

The usually daunting problem of converting between the English and metric systems of measures is quite simple for the measures of area most relevant to this book: hectares and acres and square miles and kilometers. There are 2.471 acres in a hectare and 2.590 square kilometers in a square mile, making 2.5 a reasonable approximation for both. Remember this number and its reciprocal, 0.4, and you can make many easy conversions: 100 hectares = 250 acres, 100 acres = 40 hectares, 10 km^2 = 4 mi^2, 10 mi^2 = 25 km^2. The only other measure used commonly in this book, meters, can be converted directly to yards with sufficient accuracy for most purposes; e.g., estimating the width of a shoreline buffer forest.

Far more problematic than conversions is estimating area in either system. It probably can't be done with good accuracy, but still it is often useful to be able to roughly guess the area of a unit of land, say plus or minus 25%. Many people should be able to achieve this level of accuracy, especially by familiarizing themselves with the size of some familiar areas; e.g., a city park, a hayfield, a small pond.

The size of some areas are listed that may help readers envision the correct scale while reading this book.

Tennis (singles)	.05 acres	.02 ha
American football field	1.10 acres	.45 ha
Football (soccer) field	1.78 acres	.72 ha
Central Park-New York	840 acres	340 ha
Manhattan	31.2 mi^2	81 km^2
	19,968 acres	8,081 ha

APPENDIX 3

Diversity Indices

Catherine A. Elliott

Diversity indices are used by ecologists to produce quantitative expressions with which the diversity of sets can be measured or compared. Usually the sets being considered are communities and the elements of the sets are species, but diversity indices may also be used to compare the diversity of landscapes in which different community types are the elements and landscapes or regions are the sets (e.g., Hunter 1986). In this discussion we will use the conventional (species as elements in community sets) approach.

The property of a community that diversity indices quantify is the average rarity of species in that community (Patil and Taillie 1982). In diverse communities, most species are relatively rare. Diversity is a combination of two factors, the number of species present, referred to as species richness, and the distribution of individuals among the species, referred to as evenness or equitability. Single species communities are defined as having a diversity of zero, regardless of the index used (Stand E in Table A.3.1).

Let us consider a community of species, numbered $1,2,3,\ldots i \ldots s$, with p_i as the proportional abundance of species i. Species are often ranked in order of decreasing abundance, such that species 1 has the greatest abundance, species s has the least, and, in general $p_1, \geq p_2 \longrightarrow \geq p_s$. A community is considered to be completely even when $p_1 = p_2 = \ldots = p_s = 1/s$.

Patil and Taillie (1982) present a detailed discussion of diversity as a concept and the measurement of diversity. They define the diversity (D) of a community as average species rarity. So $D = \Sigma p_i R(i;p_i)$, where $R(i, p_i)$ is rarity of species i, and D is the diversity index associated with the measure of rarity R.

For a diversity index to be useful, it should meet several criteria:

1. For a given number of species, the index should be at its maximum value when the community is completely even, i.e., the proportion of individuals of each species is $1/s$.
2. For two completely even communities, the one with the greater number of species should have the larger index value.
3. For a community that can be classified in more than one way, the sum of the index values at each level of classification should equal the index value using all levels of classification (see Pielou 1977:pp. 293–294 for a more complete explanation).

Different diversity indices, applied to the same set of communities, may not be consistent in the ordering of the communities from most to least diverse. Patil and Taillie (1982) suggest an intrinsic diversity ordering be defined that does not refer to indices. By their definition, community C′ is intrinsically more diverse than community C provided C leads to C′ by a sequence of the following operations: introducing a species; transferring abundance; or relabelling species. Introducing a species increases diversity indices defined as average rarity. However, depending on the definition of rarity, transferring abundance might not increase such a diversity index. This led Patil and Taillie to impose a fourth criterion:

4. Diversity indices should increase with transfers of abundance from one species to another, less abundant species within the same community.

Let us consider how well some indices meet these four criteria. The two most commonly used diversity indices are Simpson's and Shannon's. Simpson's index:

$$D_s \Sigma p_i\ (1\ -\ p_i)\ =\ 1\ -\ p_i^{2}$$

uses $(1\ -\ p_i)$ as the measure of rarity of species i. D_s is the probability that two individuals, chosen at random from the community, will be of different species. Although it is used as an index of diversity, Simpson's index does not meet the third criterion.

Shannon's index:

$$H'\ =\ -\ p_i \log(\ p_i)$$

was developed from information theory. Its measure of rarity of species i, $-\log(\ p_i)$, reflects the degree of uncertainty of the identity of an individual chosen at random from a community. Shannon's index satisfies all four criteria.

It is important in some instances to have an expression for the evenness of the community. The Stands A and B′ in Table A.3.1 have diversity index values that are very similar. An expression of evenness may serve to further

TABLE A.3.1 Various Diversity Indices for Six Hypothetical Forest Stands with Different Species Richness and Evenness.

Tree Species	Proportional Abundance	*Forest stands*[1]					
		A	B	B'	C	D	E
				Number of Trees per Hectare			
1	$p_1 =$	43	74	62	24	92	100
2	$p_2 =$	32	13	13	17	2	
3	$p_3 =$	25	13	13	17	2	
4	$p_4 =$			12	16	2	
5	$p_5 =$				15	1	
6	$p_6 =$				11	1	
No. species		3	3	4	6	6	1
Cumulative sums	$p_1 =$	.43	.74	.62	(See Fig. A.3.2)		
	$p_1 + p_2 =$	.75	.87	.75			
	$p_1 + p_2 + p_3 =$	1.00	1.00	.88			
	$p_1 + p_2 + p_3 + p_4 =$	1.00	1.00	1.00			
Diversity							
Simpson (D_s)		0.650	0.419	0.567	0.824	0.152	0.000
Shannon (H')		0.466	0.327	0.470	0.766	0.175	0.000
Evenness							
E_s (D_s/D_s max)		0.975	0.629	0.756	0.989	0.182	0.000
J' (H'/H' max)		0.977	0.685	0.781	0.984	0.225	0.000
Dominance							
$1 - E_s$		0.025	0.371	0.244	0.011	0.818	1.000
$1 - J'$		0.023	0.315	0.219	0.016	0.775	1.000

[1]Compositions correspond to those for forests A, B, and C in Table 2.1.

distinguish the stands. Evenness of a community with *s* species can be calculated by comparing its index value to the index value of a completely even community with *s* species.

Maximum index values for a community of *s* species are:

$$D_s \text{ max} = (s - 1)/s$$

$$H' \text{ max} = \log s$$

Evenness is then calculated as the ratio of the observed index value to its maximum:

$$E_s = D_s / D_s \text{ max}$$

$$J' = H' / H' \text{ max}$$

Measures of evenness are dependent on sample size and so they must be used with caution unless the communities being compared have the same or very similar *s* values. When the data set is a sample from a community rather than a complete enumeration, species richness will usually be underestimated because rare species will have been missed. Under these circumstances, maximum diversity will be underestimated and evenness will be overestimated.

Dominance, the opposite of evenness, can be expressed by subtracting evenness values from 1 (see Table A.3.1). Approximately even communities will have lower dominance values the greater the number of species present (Stands A and C, Table A.3.1).

Diversity indices are more sensitive to changes in species richness than they are to evenness. Stands A and B (Table A.3.1) both have 3 species but, because of greater evenness, Stand A ($H' = 0.466$) is more diverse than Stand B ($H' = 0.327$). Adding one species to Stand B, with a minimal change in evenness, is sufficient to make Stand B' ($H' = 0.470$) more diverse than both Stand A and Stand B.

DOMINANCE-DIVERSITY CURVES

Diversity of sets can also be represented graphically using a dominance-diversity curve, also called a relative abundance curve or a species importance curve. Density, coverage, biomass, frequency, productivity, or importance value could be used to rank the species from 1 to *s*, where *s* is the total number of species in the set. The most abundant species (or the species with the greatest density, coverage, biomass, etc.) is given the rank of 1, and the least abundant species is given the rank of *s*. Abundance is then plotted on a logarithmic scale against rank on an arithmetic scale (Figure A.3.1).

Communities with a high species diversity and/or high degree of evenness (low degree of dominance) will tend to have curves that are almost horizontal (Curves A and C). Communities with low species diversity and/or a low degree of evenness (high degree of dominance) will tend to produce curves that are nearly vertical (Curves B and D). Most communities have dominance-diversity curves that are intermediate to the extremes of vertical and horizontal. For further information about comparisons of curves from different communities and environments, and mathematical models that describe the curves, see Whittaker (1965).

COMPARATIVE DIVERSITY PROFILES

Swindel et al. (1987) and Swindel and Grosenbaugh (1988) present another method of graphically comparing communities that incorporates Patil and Taillie's (1982) ideas about intrinsic diversity orderings between communities.

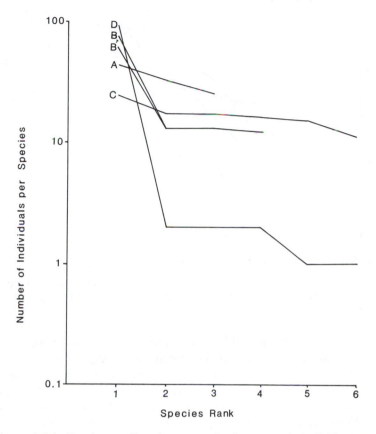

Figure A.3.1 Dominance-diversity curves for forest stands in Table A.3.1. See text for explanation.

The graphs are constructed by first ranking the species in the community in order of decreasing abundance, ($p_1 \geq p_2 \geq \ldots \geq p_s$) and then producing cumulative sums (p_1, $p_1 + p_2$, $p_1 + p_2 + p_3$, . . .) for each community. Pairs of communities are compared by plotting the cumulative sums as has been done with Stands A, B, and B′ in Figure A.3.2. If the resulting curve is entirely below the diagonal line, (Figure A.3.2, Stand A versus Stand B), the community on the y-axis (Stand A) is intrinsically the more diverse. Conversely, if the resulting curve is entirely above the diagonal line, (Figure A.3.2, Stand B versus Stand B′), the community on the y-axis (Stand B) is intrinsically the less diverse. There can also be situations where the curve crosses the diagonal at some point, say d. In figure A.3.2, this is the case with the plot of Stand B′ versus Stand A. The line is at first above the diagonal, then crosses to end below the diagonal; these communities are not intrinsically ordered with respect to diversity. Swindel and Grosenbaugh (1988) describe Stand B′ as

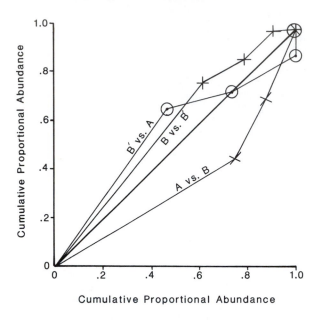

Figure A.3.2 Comparative diversity profiles for Stands A, B, and B′. See text for explanation.

"more diverse in the tail" than Stand A, and present an explanation for this interpretation. Similarly, if the curve were reversed, i.e., below the line, crossing at d, then above the line, the community plotted on the y-axis is less diverse in the tail. The reader is referred to the papers by Patil and Taillie (1982) and those by Swindel et al. (1984 and 1987) and Swindel and Grosenbaugh (1988) for more complete explanations and proofs of these concepts.

Literature Cited and Author Index

ABE, T., and T. MATSUMOTO. 1979. Studies on the distribution and ecological role of termites in a lowland rain forest of West Malaysia. III. Distribution and abundance of termites in Pasoh Forest Reserve. Japan. J. Ecol. 29:337–351. 191

ABER, J. D. 1979. Foliage-height profiles and succession in northern hardwood forests. Ecology 60:18–23. 188

AJAYI, S. S. 1979. Utilization of forest wildlife in West Africa. F.A.O., Rome. (FO: Misc./79/26) 76 pp. 6

ALABACK, P. B. 1982. Dynamics of understory biomass in sitka spruce, western hemlock forests of southeast Alaska. Ecology 63:1932–1948. 78, 184

ALEXANDER, L., and S. R. KELLERT. 1984. Forest landowners' perspectives on wildlife management in New England. Trans. North Am. Wildl. Nat. Resour. Conf. 49:164–173. 278, 285

ALLEN, H. L. 1987. Forest fertilizers. J. For. 85(2):37–46. 206

ALWARD, G. S., B. J. SULLIVAN, and T. W. HOEKSTRA. 1984. Using socioeconomic data in the management of fishing and hunting. Trans. North Am. Wildl. Nat. Resour. Conf. 49:91–103. 277

ALVERSON, W. S. and D. M. WALLER, and S. L. SOLHEIM. 1988. Forests too deer: edge effects in northern Wisconsin. Conserv. Biol. 2:348–358. 107

AMBUEL, B., and S. A. TEMPLE. 1982. Songbird populations in southern Wisconsin forests: 1954 and 1979. J. Field Ornithol. 53:149–158. 122, 123

AMBUEL, B., and S. A. TEMPLE. 1983. Area-dependent changes in the bird communities and vegetation of southern Wisconsin forests. Ecology 64:1057–1068. 125

ANDERSON, H. W., M. D. HOOVER, and K. G. REINHART. 1976. Forests and water:

effects of forest management on floods, sedimentation and water supply. USDA Forest Service Gen. Tech. Rep. PSW-18. 115 pp.   149, 150

ANDERSON, M. T. 1985. Riparian management of coastal Pacific ecosystems. Pages 364–368 *in* Johnson et al. 1985. *op. cit.*   147

ANDERSON, N. H., J. R. SEDELL, L. M. ROBERTS, and F. J. TRISKA. 1978. The role of aquatic invertebrates in processing of wood debris in coniferous forest streams. Am. Midl. Nat. 100:64–82.   141

ANDERSON, R. C., O. L. LOUCKS, and A. M. SWAIN. 1969. Herbaceous response to canopy cover, light intensity, and throughfall precipitation in coniferous forests. Ecology 50:255–263.   186

ANDRE, H. M. 1983. Notes on the ecology of corticolous epiphyte dwellers. 2. Collembola. Pedobiologia 25:271–278.   193

ANDREN, H., and P. ANGELSTAM. 1988. Elevated predation rates as an edge effect in habitat islands: experimental evidence. Ecology 69:544–547.   107

ANTHONY, R. G., and R. KOZLOWSKI. 1982. Heavy metals in tissues of small mammals inhabiting wastewater irrigated habitats. J. Environ. Qual. 11:20–22.   207

ARBUCKLE, J. 1986. Snags: widowmakers or wildlife habitat? Pages 149–157 *in* J. A. Bissonette (ed.). 1986. *op. cit.*   177

ARNETT, JR., R. H. 1985. American insects: a handbook of the insects of America north of Mexico. Van Nostrand Reinhold, New York. 850 pp.

ASKINS, R. A., and M. J. PHILBRICK. 1987. Effect of changes in regional forest abundance on the decline and recovery of a forest bird community. Wilson Bull. 99:7–21.   123, 131

ASKINS, R. A., M. J. PHILBRICK, and D. S. SUGENO. 1987. Relationship between the regional abundance of forest and the composition of forest bird communities. Biol. Conserv. 39:129–152.   122

ATKESON, T. D., and A. S. JOHNSON. 1979. Succession of small mammals on pine plantations in the Georgia Piedmont. Amer. Midl. Nat. 101:385–392.   40, 219

AUDUBON, J. J. 1831. Ornithological biography. Vol. 1. A. Black, Edinburgh.   173

AUGUST, P. V. 1983. The role of habitat complexity and heterogeneity in structuring tropical mammal communities. Ecology 64:1495–1507.   83, 188

AYENSU, E. S., and R. A. DEFILIPPS. 1978. Endangered and threatened plants of the United States. Smithsonian Institution and World Wildlife Fund, Washington, D.C. 301 pp.   4

BAILEY, R. G., R. D. PFISTER, and J. A. HENDERSON. 1978. Nature of land and resource classification. J. For. 76:650–655.   265

BALDA, R. P. 1969. Foliage use by birds of the oak-pine juniper woodland and ponderosa pine forest in southeastern Arizona. Condor 71:399–412.   189

BANASIAK, C. F. 1961. Deer in Maine. Game Div. Bull. No. 6, Dept. Inland Fish and Game, Augusta, Maine.   146

BARKMAN, J. J. 1958. Phytosociology and ecology of cryptogamic epiphytes. Gorcum, Assen, Netherlands. 628 pp.   39, 187

BARKHAM, J. P. 1980a. Population dynamics of the wild daffodil (*Narcissus pseudonar-*

cissus). I. Clonal growth, seed reproduction, morality, and the effects of density. J. Ecol. 68:607–633. 196

BARKHAM, J. P. 1980*b*. Population dynamics of the wild daffodil (*Narcissus pseudonarcissus*). II. Changes in number of shoots and flowers, and the effect of bulb depth on growth and reproduction. J. Ecol. 68:635–664. 196

BARNETT, J. L., R. A. HOW, and W. F. HUMPHREYS. 1978. The use of habitat components by small mammals in eastern Australia. Austral. J. Ecol. 3:277–285. 132

BARROWCLOUGH, G. F., and S. L. COATS. 1985. The demography and population genetics of owls, with special reference to the conservation of the spotted owl (*Strix occidentalis*). Pages 74–85 in Gutierrez and Carey 1985. *op. cit.* 72

BARROWS, C. W. 1980. Roost selection by spotted owls: an adaptation to heat stress. Condor 83:302–309. 72

BARTELS, R., J. D. DELL, R. L. KNIGHT, and G. SCHAEFER. 1985. Dead and down woody material. Pages 171–186 in Browne 1985. *op cit.* 163

BARTON, D. R., W. D. TAYLOR, and R. M. BIETTE. 1985. Dimensions of riparian buffer strips required to maintain trout habitat in southern Ontario streams. North Am. J. Fish. Manage. 5:364–378. 147, 148

BARTON, K., and W. FOSBURGH. 1986. The U.S. Forest Service. Pages 1–156 in A. S. Eno, R. L. DiSilvestro, and W. J. Chandler (eds.). Audubon Wildlife Report 1986. National Audubon Society, New York. 1094 pp. 9, 283

BATZLI, G. O. 1977. Population dynamics of the white-footed mouse in floodplain and upland forests. Am. Midl. Nat. 97:18–32. 143

BEASOM, S., Editor, Journal of Wildlife Management, Texas A&I Univ., Kingsville, Texas. Letter dated 13 December 1985. 3

BENNITT, R., J. S. DIXON, V. H. CAHALANE, W. W. CHASE, and W. L. McATEE. 1937. Statement of policy. J. Wildl. Manage. 1:1–2. 3

BENSON, G. L., and W. F. LAUDENSLAYER, JR. 1986. DYNAST: simulating wildlife responses to forest-management strategies. Pages 351–355 in Verner et al. 1986. *op. cit.* 271

BENZING, D. H. 1983. Vascular epiphytes: a survey with special reference to their interactions with other organisms. Pages 11–24 in Sutton et al. 1983*b*. *op. cit.* 187

BILBY, R. E., and G. E. LIKENS. 1980. Importance of organic debris dams in the structure and function of stream ecosystems. Ecology 61:1107–1113. 142

BINFORD, L. C., B. G. ELLIOTT, and S. W. SINGER. 1975. Discovery of a nest and downy young of the marbled murrelet. Wilson Bull. 87:303–319. 65

BISSONETTE, J. A. (ed.). 1986. Is good forestry good wildlife management? Maine Agric. Expt. Stn. Misc. Publ. No. 689. 377 pp.

BISTROM, O. 1978. Coleoptera and heteroptera in a natural forest in Mantyharju (southern Finland). Not Entomol. 58:95–100. 65

BISWELL, H. H. 1972. Fire ecology in ponderosa pine grassland. Proc. Twelfth Ann. Tall Timbers Fire Ecology Conf., Tall Timbers Res. Stn., Tallahassee, FL. Pages 69–96. 209

BJARVALL, A., E. NILSSON, and L. NORLING. 1977. The importance of virgin forest for capercaillie and marten. Fauna Flora (Stockh.) 72:31–38. 66

BLACK, S., P. BROADHURST, J. HIGHTOWER, and S. SCHAUMAN. 1985. The value of riparian habitat and wildlife to the residents of a rapidly urbanizing community. Pages 413–416 *in* Johnson et al. *op. cit.*   146

BLAIR, R. M., R. ALCANIZ, and A. HARRELL. 1983. Shade intensity influences the nutrient quality and digestibility of southern deer browse leaves. J. Range Manage. 36:257–264.   229

BLAIR, R. M., and D. P. FEDUCCIA. 1977. Midstory hardwoods inhibit deer forage in loblolly pine plantations. J. Wildl. Manage. 41:677–684.   195, 229, 232

BLAKE, J. G., and J. R. KARR. 1987. Breeding birds of isolated woodlots: area and habitat relationships. Ecology 68:1724–1734.   120

BLISS, L. C. (ed.). 1977. Truelove Lowland, Devon Island, Canada: a high Arctic ecosystem. Univ. of Alberta Press, Edmonton. 714 pp.   20

BLOOD, D. A., and G. G. ANWEILER. 1984. Forest nesting habitat of ancient murrelets in the Queen Charlotte Islands. Pages 297–302 *in* Meehan et al. 1984. *op. cit.*   65

BLOOM, A. M. 1978. Sitka black-tailed deer winter range in the Kadashan Bay area, southeast Alaska. J. Wildl. Manage. 42:108–112.   76

BLOUIN, M. S., and E. F. CONNOR. 1985. Is there a best shape for nature reserves? Biol. Cons. 32:277–288.   125

BOECKLEN, W. J. 1986. Effects of habitat heterogeneity on the species-area relationships of forest birds. J. Biogeogr. 13:59–86.   116

BOECKLEN, W. J., and N. J. GOTELLI. 1984. Island biogeographic theory and conservation practice: species-area or specious-area relationships. Biol. Conserv. 29:63–80. 125

BOND, R. R. 1957. Ecological distribution of breeding birds in the upland forests of southern Wisconsin. Ecol. Monogr. 27:351–384.   120

BORMANN, F. H. 1985. Air pollution and forests: an ecosystem perspective. Bioscience 35:434–441.   210

BORMANN, F. H., and G. E. LIKENS. 1979. Pattern and process in a forested ecosystem. Springer-Verlag, New York. 253 pp.   60, 65, 163

BORMANN, F. H., G. E. LIKENS, D. W. FISHER, and R. S. PIERCE. 1968. Nutrient loss accelerated by clear-cutting of a forest ecosystem. Science 159:882–884.   95

BOTKIN, D. B., J. E. ESTES, R. M. MACDONALD, and M. V. WILSON. 1984. Studying the earth's vegetation from space. Bioscience 34:508–514.   267

BOUGERON, P. S. 1983. Spatial aspects of vegetation structure. Pages 29–47 *in* Golley 1983. *op cit.*   182

BOWMAN, G. B., and L. D. HARRIS. 1980. Effect of spatial heterogeneity on ground-nest depredation. J. Wildl. Manage. 44:806–813.   192

BOYCE, J. S. 1961. Forest pathology. 3rd ed. McGraw-Hill, New York. 572 pp.   62

BOYCE, M. S. 1987. A review of the U.S. Forest Service's viability analysis for the spotted owl. Unpublished final report to the National Council of the Paper Industry for Stream and Air Improvement, Inc.   72

BOYCE, S. G. 1977. Management of eastern hardwood forests for multiple benefits (DYNAST-MB). USDA Forest Service Res. Paper SE-168. 116 pp.   270

BOYER, W. D., and D. W. PETERSON. 1983. Longleaf pine. Pages 153–156 *in* R. M.

Burns (tech. comp.). Silvicultural systems of the major forest types of the United States. USDA Forest Service, Washington, D.C. Agriculture Handbook No. 445. 46

BRATTON, S. P. 1974. The effect of the European wild boar (*Sus scrofa*) on the high-elevation vernal flora in Great Smoky Mountains National Park. Bull. Torrey Bot. Club 101:198–206. 185

BRATTON, S. P. 1975. The effect of the European wild boar (*Sus scrofa*) on gray beech forest in the Great Smoky Mountains. Ecology 56:1356–1366. 185

BRAWN, J. D., and R. P. BALDA. 1983. Use of nest boxes in ponderosa pine forests. Pages 159–164 *in* Davis et al. 1983. *op. cit.* 174

BREWBAKER, J. L., and E. M. HUTTON. 1979. Leucaena - versatile tropical tree legume. Pages 207–259 *in* G. A. Ritchie (ed.). New agricultural crops. Am. Assoc. Advance. Sci. Selected Symp. 38, Washington, D.C. 259 pp. 38

BRINSON, M. M., H. D. BRADSHAW, R. N. HOLMES, and J. B. ELKINS, JR. 1980. Litter-fall, stemflow, and throughall nutrient fluxes in an alluvial swamp forest. Ecology 61:827–835. 145

BRINSON, M. M., B. L. SWIFT, R. C. PLANTICO, and J. S. BARCLAY. 1981. Riparian eco-systems: their ecology and status. Eastern Energy and Land Use Team, U.S. Fish and Wildlife Service. FWS/OBS-81/17, Kearneysville, West Virginia. 155 pp. 143, 144, 145, 146, 148, 150

BRITTINGHAM, M. C., and S. A. TEMPLE. 1983. Have cowbirds caused forest songbirds to decline? Bioscience 33:31–35. 107, 122, 132

BROADHEAD, E., and H. WOLDA. 1985. The diversity of Psocoptera in two tropical forests in Panama. J. Anim. Ecol. 54:739–754. 191

BRODO, I. M. 1961. Transplant experiments with corticolous lichens using a new technique. Ecology 42:838:841. 192

BROKAW, N. V. L. 1982. The definition of treefall gap and its effects on measures of forest dynamics. Biotropica 14:158–160. 88, 90

BROKAW, N. V. L. 1985. Treefalls, regrowth, and community structure in tropical forests. Pages 53–69 *in* Pickett and White 1985. *op. cit.* 88

BROUHA, P., and M. G. PARSONS. 1986. Foresters are fish habitat managers. Pages 12–15 *in* Society of American Foresters. Foresters' future: leaders or followers? Society of American Foresters. Bethesda, Maryland. 51 pp. 147

BROWN, J. H. 1981. Two decades of homage to Santa Rosalia: toward a general theory of diversity. Am. Zool. 21:877–888. 26

BROWN, J. H., and A. KODRIC-BROWN. 1977. Turnover rates in insular biogeography: effect of immigration on extinction. Ecology 58:445–449. 118

BROWN, E. R. (ed.). 1985. Management of wildlife and fish habitats in forests and western Oregon and Washington. USDA Forest Service Pub. No. R6-FoWL-192. Portland, Oregon. 322 pp. xi, 248

BRUNIG, E. F. 1983. Vegetation structure and growth. Pages 49–75 *in* Golley 1983. *op. cit.* 182

BRYANT, L. D. 1985. Livestock management in the riparian ecosystem. Pages 285–289 *in* Johnson et al. 1985. *op. cit.* 151

BRYANT, M. D. 1983. The role and management of woody debris in west coast salmonid nursery streams. North Am. J. Fish. Manage. 3:322–330. 142

BUCKMAN, R. E., and R. L. QUINTUS. 1972. Natural areas of the Society of American Foresters. Society of American Foresters, Washington, D.C. 38 pp. 66

BUCKNER, J. L., J. L. LANDERS, J. A. BARKER, and C. J. PERKINS. 1979. Wildlife food plants following preparation of longleaf pine site in southwest Georgia. South J. Appl. For. 3:56–59. 216

BUCKNER, J. L., and L. J. PERKIN. 1974. A plan of forest wildlife habitat evaluation and its use by International Paper Co. Proc. Ann. Conf. Southeast Assoc. Game and Fish Comm. 28:675–682.

BUDD, W. W., P. L. COHEN, P. R. SAUNDERS, and F. R. STEINER. 1987. Stream corridor management in the Pacific Northwest: I. Determination of stream-corridor widths. Environ. Manage 11:587–597. 148

BULL, E. L., and A. D. PARTRIDGE. 1986. Methods of killing trees for use by cavity nesters. Wildl. Soc. Bull. 14:142–146. 175

BULL, E. L., J. W. THOMAS, and K. HORN. 1986. Snag management in national forests in the Pacific Northwest. Western J. Appl. For. 1(2):41–43. 171

BUNNELL, F. L., and G. W. JONES. 1984. Black-tailed deer and old-growth forests—a synthesis. Pages 411–410 *in* Meehan et al. 1984. *op. cit.* 76, 78

BURGESS, R. L., and D. M. SHARPE (eds.). 1981. Forest island dynamics in man-dominated landscapes. Springer-Verlag, New York. 310 pp.

BURY, R. B. 1983. Differences in amphibian populations in logged and old growth redwood forest. Northwest Sci. 57:167–178. 51

CAJANDER, A. K. 1926. The theory of forest types. Acta For. Fenn. 29. 108 pp. 241

CALVERT, W. H., W. ZUCHOWSKI, and L. P. BROWER. 1982. The impact of forest thinning on micro-climate in monarch butterfly over wintering areas of Mexico. Bol. Soc. Bot. Mex. 0(42):11–18. 230

CANHAM, H. O., and J. GRAY. 1986. Federal income tax change and the private forest sector. Grey Towers Press, Milford, Pennsylvania. 41 pp. 284

CANHAM, C. D., and P. L. MARKS. 1985. The response of woody plants to disturbance: patterns of establishment and growth. Pages 197–216 *in* Pickett and White 1985. *op. cit.* 85

CAREY, A. B. 1985. A summary of the scientific basis for spotted owl management. Pages 100–114 *in* Gutierrez and Carey 1985. *op. cit.* 72

CARL, C. M., J. R. DONNELLY, W. J. GABRIEL, L. D. GARRETT, R. A. GREGORY, N. K. HUYLER, W. L. JENKINS, R. E. SENDAK, R. S. WALTERS, and H. W. YAWNEY. 1982. Sugar maple research: sap production, processing, and marketing of maple syrup. USDA Forest Service, Northeastern Forest Experiment Station. Gen. Tech. Rep. NE-72. 109 pp. 228

CARLQUIST, S. 1974. Island biology. Columbia Univ. Press, New York. 660 pp. 119

CARLSON, A. 1986. A comparison of birds inhabiting pine plantation and indigenous forest patches in a tropical mountain area. Biol. Conserv. 35:195–204. 39

CARMICHAEL, JR., D. B., and D. C. GUYNN, JR. 1983. Snag density and utilization by wildlife in the upper Piedmont of South Carolina. Pages 107–110 *in* Davis et al. 1983. *op. cit.* 162

CARSON, R. 1962. Silent spring. Houghton-Mifflin Co., Boston. 368 pp. 211

CAUFIELD, C. 1984. In the rain forest. Knopf, New York. 304 pp. 10, 135

CHADWICK, N. L., D. R. PROGULSKE, and J. T. FINN. 1986. Effects of fuelwood cutting on birds in southern New England. J. Wildl. Manage. 50:398–405. 158

CHANDLER, C., P. CHENEY, P. TTHOMS, L. TRABAUD, and D. WILLIAMS. 1983. Fire in forestry. Vol. 1. Forest fire behavior and effects. Wiley, New York. 450 pp. 185, 209, 210

CHAPMAN, D. W. 1966. The relative contribution of aquatic and terrestrial primary producers to the trophic relations of stream organisms. Spec. Publs. Pymatuning Lab. Fld. Biol. 4:116–130. 141

CHAPMAN, H. H. 1932. Is the longleaf type a climax? Ecology 13:328–334. 44

CHILDERS, E. L., T. L. SHARIK, and C. S. ADKISSON. 1986. Effects of loblolly pine plantations on songbird dynamics in the Virginia Piedmont. J. Wildl. Manage. 50:406–413. 55, 151, 219

CLARK, J. S. 1988. Effect of climate change on fire regimes in northwestern Minnesota. Nature 334:233–235. 61, 96, 209

CLAWSON, M. 1976. The National Forests: a great national asset is poorly managed and unproductive. Science 191:762–767. 284

CLEMENTS, F. E. 1916. Plant succession: an analysis of the development of vegetation. Carnegie Institute, Washington. Publ. 242:1–512. 22

CLEMENTS, J. F. 1981. Birds of the world: a checklist. Facts on File, New York. 562 pp.

CLINE, S. P., A. B. BERG, and H. M. WIGHT. 1980. Snag characteristics and dynamics in Douglas-fir forests, western Oregon. J. Wildl. Manage. 44:773–786. 158, 167

CODY, M. L. 1975. Towards a theory of continental species diversity: Bird distributions over Mediterranean habitat gradients. Pages 214–257 *in* M. L. Cody, and J. M. Diamond (eds.). Ecology and evolution of communities. Harvard Univ. Press, Cambridge, Massachusetts. 545 pp. 31, 39

CODY, M. L. 1985. An introduction to habitat selection in birds. Pages 3–56 *in* M. L. Cody (ed.). Habitat selection in birds. Academic Press, New York. 558 pp. 39

COLEY, P. D., J. P. BRYANT, and F. S. CHAPIN. 1985. Resource availability and plant antiherbivore defense. Science 230:895–899. 39

COLLINS, B. S., K. P. DUNNE, and S. T. A. PICKETT. 1985. Response of forest herbs to canopy gaps. Pages 217–234 *in* Pickett and White 1985. *op. cit.* 186

COLLINS, B. S., and S. T. A. PICKETT. 1988. Response of herb layer cover to experimental canopy gaps. Am. Midl. Natur. 119:281–290. 194

CONLIN, W. M., and R. H. GILES, JR. 1973. Maximizing edge and coverts for quail and small game. Proc. First National Bobwhite Quail Symposium. Oklahoma State University, Stillwater, Oklahoma. 302 pp. 109

CONNER, R. N. 1978. Management for cavity nesting birds. Pages 120–128 *in* R. M. DeGraaf (ed.). Management of southern forests for nongame birds. U.S. Dept. Agric. For. Serv. Gen. Tech. SE-14. 172

CONNER, R. N. 1979. Minimum standards in forest wildlife management. Wildl. Soc. Bull. 7:293–296. 169

CONNER, R. N., and C. S. ADKISSON. 1975. Effects of clearcutting on the diversity of breeding birds. J. For. 73:781–785. 51

CONNER, R. N., J. G. DICKSON, and B. A. LOCKE. 1981. Herbicide-killed trees infected by fungi: potential cavity sites for woodpeckers. Wildl. Soc. Bull. 9:308–310.   175

CONNER, R. N., J. G. DICKSON, B. A. LOCKE, and C. A. SEGELQUIST. 1983*c.* Vegetation characteristics important to common songbirds in east Texas. Wilson Bull. 95:349–361.   227

CONNER, R. N., J. G. DICKSON, and J. H. WILLIAMSON. 1983*a.* Potential woodpecker nest trees through artificial inoculation of heart rots. Pages 68–72 *in* Davis et al. 1983.   175

CONNER, R. N., J. C. KROLL, and D. L. KULHAVY. 1983*b.* The potential of girdled and 2 4-D injected southern red oaks as woodpecker nesting and foraging sites. South. J. Appl. For. 7:125–128.   175, 230

CONNER, R. N., and B. A. LOCKE. 1979. Effects of a prescribed burn on cavity trees of red-cockaded woodpeckers. Wildl. Soc. Bull 7:291–293.   233

CONNER, R. N., and B. A. LOCKE. 1982. Fungi and red-cockaded woodpeckers cavity trees. Wilson Bull. 94:64–70.   161, 179

CONNER, R. N., O. K. MILLER, JR., and C. S. ADKISSON. 1976. Woodpecker dependence on trees infected by fungal heart rots. Wilson Bull. 88:575–581.   161

CONNER, R. N., and K. A. O'HALLORAN. 1987. Cavity-tree selection by red-cockaded woodpeckers as related to growth dynamics of southern pines. Wilson Bull. 99:398–412.   179

CONNOR, E. F., and E. D. McCOY. 1979. The statistics and biology of the species-area relationship. Am. Nat. 113:791–833.   116, 123

CONNOR, E. F., and D. SIMBERLOFF. 1984. Neutral models of species' co-occurrence patterns. Pages 316–331 *in* D. R. Strong, D. Simberloff, L. G. Abele, and A. B. Thistle. Ecological communities. Princeton Univ. Press, Princeton, New Jersey. 613 pp.   125

CONROY, M. J., R. G. ODERWALD, and T. L. SHARIK. 1982. Forage production and nutrient concentrations in thinned loblolly pine plantations. J. Wildl. Manage. 46:719–727.   229

CONSERVATION LAW FOUNDATION OF NEW ENGLAND. (n.d.). Letter soliciting funds to support their suit against the White Mountain National Forest.   262

CORBET, G. B., and J. E. HILL. 1980. A world list of mammalian species. British Museum (National History), London. 226 pp.

COUNCIL ON ENVIRONMENTAL QUALITY. 1980. The global 2000 report to the president. Council on Environmental Quality and Department of State, Washington, D.C. 131

CRANE, E. 1983. The archaeology of beekeeping. Cornell University Press, Ithaca, New York. 360 pp.   173

CRAWFORD, H. S. 1984. Habitat management. Pages 629–646 *in* L. K. Halls (ed.) White-tailed deer. Stackpole, Harrisburg, Pennsylvania. 870 pp.   147, 247

CREPET, W. L. 1983. The role of insect pollination in the evolution of the angiosperms. Pages 31–50 *in* L. Real (ed.). Pollination biology. Academic Press, Orlando, Florida 338 pp.   194

CROKER, T. C., JR. 1979. The longleaf pine story. J. For. Hist. 23:32–43.   43, 44, 45, 46

CROKER, T. C., JR., and W. D. BOYER. 1975. Regenerating longleaf pine naturally.

U.S.D.A. Forest Serv. Res. Pap. SO-105. Southern Forest. Exp. Stn. New Orleans, Louisiana.

CUBBAGE, F. W., and W. C. SIEGEL. 1985. The law regulating private forest practices. J. For. 83:538–545. 280

CUNNINGHAM, D. D. 1907. Plagues and pleasures of life in Bengal. Murray, London. 193

DANIEL, T. W., J. A. HELMS, and F. S. BAKER. 1979. Principles of silviculture. 2nd ed. McGraw-Hill Book Company, New York. 500 pp. 204

DAVIS, J. W., G. A. GOODWIN, and R. A. OCKENFELS. 1983. Snag habitat management. USDA Forest Service Gen. Tech. Rep. RM-99. 226 pp.

DAWSON, W. R., J. D. LIGON, J. R. MURPHY, J. P. MYERS, D. SIMBERLOFF, and J. VERNER. 1987. Report of the Advisory Committee on the Spotted Owl. Condor 89:205–229. 68, 72, 73, 75, 76

DECKER, D. J., G. R. GOFF. 1987. Valuing wildlife. Westview, Boulder, CO. 424 pp. 277

DEGRAAF, R. M. 1987. Breeding birds and gypsy moth defoliation: short-term responses of species and guilds. Wild. Soc. Bull. 15:217–221. 186

DEGRAAF, R. M., and A. L. SHIGO. 1985. Managing cavity trees for wildlife in the northeast. USDA Forest Service Gen. Tech. Rep. NE-101. 175

DELOTTELE, R. S., and R. J. EPTING. 1988. Selection of old trees for cavity excavation by red-cockaded woodpeckers. Wildl. Soc. Bull. 16:48–52. 179

DENNIS, J. V. 1971. Utilization of pine resin by the red-cockaded woodpecker and its effectiveness in protecting roosting and nest sites. Pages 78–95 *in* R. L. Thompson (ed.). The ecology and management of the red-cockaded woodpecker. Bureau of Sport Fisheries and Wildlife, Washington, D.C. 178, 179

DENSLOW, J. S. 1985. Disturbance-mediated coexistence of species. Pages 307–323 *in* Pickett and White. *op. cit.* 27

DESGRANGES, J. L., Y. MAUFFETTE, and G. GAGNON. 1987. Sugar maple forest decline and implications for forest insects and birds. Trans. 52nd N.A. Wildl. and Nat. Res. Conf. 52:677–689. 210

DIAMOND, J. M. 1975. The island dilemma: lessons of modern biogeographical studies for the design of natural preserves. Biol. Conserv. 7:129–146. 115, 123, 124

DIAMOND, J. M. 1976. Island biogeography and conservation: strategies and limitations. Science 193:1027–1029. 125

DIAMOND, J. M., and R. M. MAY. 1981. Island biogeography and the design of natural reserves. Pages 228–252 *in* R. M. May (ed.). Theoretical ecology. (2nd ed.). Sinauer, Sunderland Massachusetts. 489 pp. 124

DIBELLO, F. Unpublished manuscript. 146

DICKSON, J. G. 1981. Impact of forestry practices on wildlife in southern pine forests. Proc. 1981 convention Soc. Am. Foresters. Pages 224–230. 230

DICKSON, J. G., R. N. CONNER, and J. H. WILLIAMSON. 1983. Snag retention increases bird use of a clear cut. J. Wildl. Manage. 47:799–804. 168

DICKSON, J. G., R. N. CONNER, and J. H. WILLIAMSON. 1984. Bird community changes in a young pine plantation in East Texas. South. J. Appl. For. 8:47–51. 55

DICKSON, J. G., and J. C. HUNTLEY. 1985. Streamside managemet zones and wildlife in the southern coastal plain. Pages 263–264 *in* Johnson et al. 1985. *op. cit.* 148

DICKSON, J. G., and R. E. NOBLE. 1978. Vertical distribution of birds in a Louisiana bottomland hardwood forest. Wilson Bull. 90:19–30. 189, 227

DICKSON, J. G., and C. A. SEGELQUIST. 1979. Breeding bird populations in pine and pine-hardwood forest in Texas. J. Wildl. Manage. 43:549–555. 188, 195

DIMOND, J. Dept. Entomology, University of Maine. 53

DISNEY, H. J. S., and A. STOKES. 1976. Birds in pine and native forests. Emu 76:133–138. 39

DIXON, K. R., and T. C. JUELSON. 1987. The political economy of the spotted owl. Ecology 68:772–776. 74, 284

DOBZHANSKY, T., J. R. POWELL, C. E. TAYLOR, and M. ANDREGG. 1979. Ecological variable affecting the dispersal behavior of *Drosophia pseudobscura* and its relatives. Am. Nat. 114:325–334. 94

DOERR, J. G., C. L. BARESCU, J. M. BRIGHENTI, JR., and M. P. MORIN. 1984. Use of clearcutting and old-growth forests by male blue grouse in central southeast Alaska. Pages 309–313 *in* Meehan et al. 1984. *op. cit.* 65

DOLLOF, C. A. 1986. Effects of stream cleaning on juvenile coho salmon and dolly varden in southeast Alaska. Trans. Am. Fish. Soc. 115:743–755. 142

DRESSLER, R. L., G. L. STORM, W. M. TZILKOWSKI, and W. E. SOPPER. 1986. Heavy metals in cottontail rabbits on mined lands treated with sewage sludge. J. Environ. Qual. 15:278–281. 207

DRESSLER, R. C., and G. L. WOOD, 1976. Deer habitat response to irrigation with municipal wastewater. J. Wildl. Manage. 40:639–644. 207

DRISCOLL, P. V. 1977. Comparisons of bird counts from pine forests and indigenous vegetation. Aust. Wildl. Res. 4:281–288. 39

DRITSCHILO, W., H. CORNELL, D. NAFUS, and B. O'CONNER. 1975. Insular biogeography: of mice and mites. Science 190:467–469. 119

DRURY, W. H., and I. C. T. NISBET. 1973. Succession. J. Arnold Arbor. 54:331–368. 20

DUERR, W. A. 1986. Forestry's upheaval: are advances in western civilization redefining the profession? J. For. 841:20–26. 61

DUNN, J. P., T. W. KIMMERER, and G. L. NORDIN. 1986. Attraction of the twolined chestnut borer, *Agrilus bilineatus* (Weber) (Coleoptera: Buprestidae), and associated borers to volatiles of stressed white oak. Can. Entomol. 118:503–509. 161

DUNNE, T., and L. B. LEOPOLD. 1978. Water in environmental planning. Freeman, New York. 818 pp. 142

DYER, M. I., and D. A. CROSSLEY, JR. 1986. Coupling of ecological studies with remote sensing. U.S. Man. and the Biosphere Program, U.S. Dept. of State, Washington, D.C. Publ. 9504. 143 pp. 267

ECKHARDT, R. C. 1979. The adaptive syndromes of two guilds of insectivorous birds in the Colorado Rocky Mountains. Ecol. Monogr. 9:129–149. 36

EGLER, F. E. 1954. Vegetation science concepts. I. Initial floristic composition, a factor in old-field vegetation development. Vegetatio 4:412–417. 21, 22, 50

EHRENFELD, D. 1988. Why put a value on biodiversity? Pages 212–216 *in* Wilson and Peter 1988. 277

EHRLICH, P., and A. EHRLICH. 1981. Extinction. Random House, New York. 305 pp. 9

ELLIOTT, C. A. 1987. Songbird species diversity and habitat use in relation to vegetation structure and size of forest stands and forest-clearcut edges in north-central Maine. Ph.D. Diss., Univ. Maine, Orono. 84 pp. 106, 188, 189

ELTON, C. S. 1958. The ecology of invasions by animals and plants. Methuen, London. 181 pp. 6, 10

ELTON, C. S. 1966. The pattern of animal communities. Methuen, London. 432 pp. 50, 82, 157, 158

ENDLER, J. A. 1977. Geographic variation, speciation, and clines. Princeton Univ. Press, Princeton, New Jersey. 246 pp. 94

ENGE, K. M., and W. R. MARION. 1986. Effects of clearcutting and site preparation on herpetofauna of a north Florida flatwoods. For. Ecol. Manage. 14:177–192. 217

EUBANKS, T. International Paper Co., Augusta, Maine. 279

EVANS, J. 1982. Plantation forestry in the tropics. Clarendon, Oxford. 472 pp. 34

EVANS, K. E., and R. N. CONNER. 1979. Snag management. Pages 214–225 *in* R. M. DeGraaf, and K. E. Evans (eds.). Management of north central and northeastern forests for nongame birds. USDA Forest Service. Gen. Tech. Rep. NC-51. 268 pp. 169, 171, 173

FALINSKA, K. 1979. Modifications of plant populations in forest ecosystems and their ecotones. Pol. Ecol. Stud. 5:89–150. 106

FARNUM, P., R. TIMMIS, and J. L. KULP. 1983. Biotechnology of forest yield. Science 219:694–702. 200, 218

FEDERAL COMMITTEE ON ECOLOGICAL RESERVES. 1977. A directory of research natural areas on federal lands of the United States of America. U.S.D.A. Forest Service, Washington, D.C. 280 pp.

FEINSINGER, P. F., K. G. MURRAY, S. KINSMAN, and W. H. BUSBY. 1986. Floral neighborhood and pollination success in four hummingbird pollinated cloud forest plant species. Ecology 67:449–464. 194

FEINSINGER, P., J. A. WOLFE, and L. A. SWARM. 1982. Island ecology: reduced hummingbird diversity and the pollination biology of plants, Trinadad and Tobago, West Indies. Ecology 63:494–506. 119

FELIX, A. C. III, T. L. SHARIK, and B. S. MCGINNES. 1986. Effects of pine conversion on food plants of northern bobwhite quail, eastern wild turkey, and white-tailed deer in the Virginia piedmont. South J. Appl. For. 10:47–52. 220

FELIX, A. C., III, T. L. SHARIK, B. S. MCGINNES, and W. C. JOHSON. 1983. Succession in loblolly pine plantations converted from second-growth forest in the central piedmont of Virginia. Am. Midl. Nat., 110:365–380. 220

FINEGAN, B. 1984. Forest succession. Nature 312:109–114. 22

FISHER, S. G., and G. E. LIKENS. 1973. Energy flow in Bear Brook, New Hampshire: an integrative approach to stream ecosystem metabolism. Ecol. Monogr. 43:421–439. 141

FOGEL, R., M. OGAWA, and J. M. TRAPPE. 1973. Terrestrial decomposition: a synopsis. Rep. 135, 12pp. Coniferous For. Biome, Seattle, Washington.   159

FOOD and AGRICULTURE ORGANIZATION. 1978. Development and investment in the forestry sector. FAO United Nations, Rome. FO:COFO-78/2.   34

FORMAN, R. T. T., and J. BAUDRY. 1984. Hedgerows and hedgerow networks in landscape ecology. Environ. Manage. 8:495–510.   129

FORMAN, R. T., A. E. GALLI, and C. F. LECK. 1976. Forest size and avian diversity in New Jersey woodlots with some land use implications. Oecologia 26:1–8.   120, 122

FORMAN, R. T. T., and M. GODRON. 1986. Landscape ecology. Wiley, New York. 619 pp.   106, 129

FORSMAN, E. D., E. C. MESLOW, and H. M. WIGHT. 1984. Distribution and biology of the spotted owl in Oregon. Wildl. Monogr. 87:1–64.   65, 72, 82, 259

FOSTER, D. R. 1983. The history and pattern of fire in the boreal forest of southeastern Labrador. Can. J. Bot. 61:2459–2471.   88, 89

FOSTER, D. R. 1988. Species and stand response to catastrophic wind in central New England, U.S.A. J. Ecol. 76:135–151.   88

FOWELLS, H. A. 1965. Silvics of forest trees of the United States. Agriculture Handbook No. 271, USDA Forest Service, Washington, D.C. 762 pp.   63, 265

FRANKEL, O. H., and M. E. SOULE'. 1981. Conservation and evolution. Cambridge Univ. Press, Cambridge, Great Britain. 327 pp.   94, 244

FRANKLIN, J. F., K. CROMACK, JR., W. DENISON, A. McKEE, C. MASER, J. SEDELL, F. SWANSON, and G. JUDAY. 1981. Ecological characteristics of old-growth Douglas-fir forests. U.S.D.A. Forest Service Gen. Tech. Rep. PNW-118. 48 pp.   65

FRANKLIN, J. F., and R. T. T. FORMAN. 1987. Creating landscape patterns by forest cutting: ecological consequences and principles. Landscape Ecol. 1:5–18.   133, 255

FRANKLIN, J. F., and M. A. HEMSTROM. 1981. Aspects of succession in the coniferous forests of the Pacific Northwest. Pages 212–229 *in* West et al. 1981. *op. cit.* 185

FRANZREB, K. E. 1978. Tree species used by birds in logged and unlogged mixed coniferous forests. Wilson Bull. 90:221–238.   158

FRANZREB, K. E., and R. D. OHMART. 1978. The effects of timber harvesting on breeding birds in a mixed-coniferous forest. Condor 80:431–441.   36, 188

FREEMAN, A. M. 1979. The benefits of environmental improvement: theory and practice. Johns Hopkins Univ. Press, Baltimore. 272 pp.   277

FREEMAN, A. M., R. H. HAVEMAN, and A. V. KNEESE. 1973. The economics of environmental policy. Wiley, New York. 184 pp.   280, 284

FRENCH, N. R., R. K. STEINHORST, and D. M. SWIFT. 1979. Grassland biomass trophic pyramids. Pages 59–87 *in* N. R. French (ed.). Perspectives in grassland ecology. Springer-Verlag, New York. 204 pp.   20

FRIEND, G. R. 1982. Bird populations in exotic pine plantations and indigenous eucalypt forests in Gippsland, Victoria. Emu 82:80–91.   39

FROKE, J. B. 1983. The role of nest boxes in bird research and management. Pages 10–13 *in* Davis et al. 1983.   173

GALLAGER, J. L. 1974. Sampling macro-organic matter profiles in salt marsh plant root zones. So. Sci. Am. Proc. 38:154–155. 20

GALLI, A. E., C. F. LECK, and R. T. T. FORMAN. 1976. Avian distribution patterns in forest islands of different sizes in central New Jersey. Auk 93:356–364. 120, 122

GAMLIN, L. 1988. Sweden's factory forests. New Scientist 117:41–47. 200

GANO, L. 1917. A study in physiographic ecology in northern Florida. Bot. Gaz. 63:337–372. 35

GATES, J. E., and L. W. GYSEL. 1978. Avian nest dispersion and fledging success in field forest ecotones. Ecology 59:871–883. 107, 122, 132

GEIST, V. 1988. How markets in wildlife meat and parts, and the sale of hunting privileges, jeopardize wildlife conservation. Conserv. Biol. 2:15–26. 279

GERSON, U., and M. R. D. SEAWARD. 1977. Lichen-invertebrate associations. Pages 69–119 *in* M. R. D. Seaward (ed.). Lichen ecology. Academic Press, London. 550 pp. 193

GILBERT, F. F. 1974. *Parelaphostrongylus tenuis* in Maine: II. Prevalence in moose. J. Wildl. Manage. 38:42–46. 99

GILES, JR. R. H. 1978. Wildlife management. Freeman, San Francisco. 416 pp. 102, 110

GILPIN, M. E. 1987. Spatial structure and population vulnerability. Pp. 125–139 *in* Soule' 1987. *op. cit.* 244

GILPIN, M. E., and J. M. DIAMOND. 1982. Factors contributing to non-randomness in species co-occurrences on islands. Oecologia 52:75–84. 128

GILPIN, M. E., and J. M. DIAMOND. 1984. Are species co-occurrences on islands non-random, and are null hypotheses useful in community ecology? Pages 297–315 *in* D. R. Strong, D. Simberloff, L. G. Abele, and A. B. Thistle. Ecological communities. Princeton Univ. Press, Princeton, New Jersey. 613 pp. 125

GILPIN, M. E., and M. E. SOULE'. 1986. Minimum viable populations: processes of species extinction. Pages 19–34 *in* Soule' 1986. *op. cit.*

GIVNISH, T. J. 1982. On the adaptive significance of leaf height in forest herbs. Am Nat. 120:353–381. 191

GLENN-LEWIN, D. C. 1977. Species diversity in North American temperate forests. Vegetatio 33:153–162. 37

GLINSKI, R. L., T. G. GRUBB, and L. A. FORBIS. 1983. Snag use by selected raptors. Pages 130–133 *in* Davis et al. 1983. *op. cit.* 163

GOLLEY, F. B. (ed.). 1983. Tropical rain forest ecosystems. (Ecosystems of the world: Vol. 14A) Elsevier, Amsterdam. 381 pp.

GOODELL, B. S., A. KIMBALL, and M. L. HUNTER, JR. 1986. Application of wood science to the creation and maintenance of snags for wildlife. Pages 135–140 *in* Bissonette 1986. *op. cit.* 175

GOODMAN, D. 1975. The theory of diversity-stability relationships in ecology. Quart. Rev. Biol. 50:237–266. 10

GOTFRYD, A., and R. I. C. HANSELL. 1986. Predictions of bird-community metrics in urban woodlots. Pages 321–326 *in* Verner et al. 1986. *op. cit.* 112, 132

GOULDING, M. 1980. The fishes and the forest. University California Press, Berkeley. 280 pp.   141

GREENWOOD, P. J., and P. H. HARVEY. 1982. The natal and breeding dispersal of birds. Ann. Rev. Ecol. Syst. 13:1–21.   94

GRIER, C. C., and R. S. LOGAN. 1977. Old-growth *Pseudotsuga menziesii* communities of a western Oregon watershed: biomass distribution and production budgets. Ecol. Monogr. 47:373–400.   157

GRIME, J. P. 1979. Plant strategies and vegetation processes. Wiley, New York. 222 pp.   27, 85, 202, 214

GRUBB, T. C., JR., D. R. PETIT, and D. L. KRUSAC. 1983. Artificial trees for primary cavity users. Pages 151–154 *in* Davis et al. 1983. *op. cit.*   174

GRUNDA, B. 1985. Activity of decomposers and processes of decomposition in soil. Pages 389–414 *in* M. Penka, M. Vyskot, E. Klimo, and F. Vasicek (eds.). Floodplain forest ecosystem. Czechoslovak Academy of Sciences, Prague. 464 pp.   145

GULLION, G. W. n.d. Managing woodlots for fuel and wildlife. Ruffed Grouse Society, Coraopolis, Pennsylvania. 16 pp.   55, 56

GUTIERREZ, R. J. 1985. An overview of recent research on the spotted owl. Pages 39–49 in Gutierrez and Carey 1985. *op. cit.*   72

GUTIERREZ, R. J., and A. B. CAREY (eds.). 1985. Ecology and management of the spotted owl in the Pacific Northwest. USDA Forest Service Gen. Tech. Rep. PNW-185.

HAAPANEN, A., and T. WAARAMAKI. 1977. Changes in the use of wetlands in 2 drainage basins and the effects for example on waterfowl populations. Riistatiet Julkaisuja 36:19–48.   208

HACKETT, W. P. 1985. Juvenility, maturation, and rejuvenation in woody plants. Hort. Rev. 7:109–155.   58

HAILA, Y. 1983. Land birds on northern islands: a sampling metaphor for insular colonization. Oikos 41:334–351.   123

HAILA, Y. 1986. On the semiotic dimension of ecological theory: the case of island biogeography. Biol. and Philosophy 1:377–387.   116, 123

HAILA, Y. 1988. Calculating and miscalculating density: the role of habitat geometry. Ornis Scan. 19:88–92.   117

HAINES, S. G., L. W. HAINES, and G. WHITE. 1979. Nutrient composition of sycamore blades, petioles, and whole leaves. For. Sci. 25:154–160.   220

HALE, M. E., JR. 1965. Vertical distribution of cryptograms in a red maple swamp in Connecticut. Bryologist 68:193–197.   191

HALLE, F., R. A. A. OLDEMAN, and P. B. TOMLINSON. 1978. Tropical trees and forests. Springer-Verlag, Berlin. 441 pp.   181, 182

HALLISEY, D. M., and G. W. WOOD. 1976. Prescribed fire in scrub oak habitat in central Pennsylvania. J. Wildl. Manage. 40:507–516.   234

HAMER, T. E., F. B. SAMPSON, K. O'HALLORAN, and L. W. BREWER. 1987. Activity patterns and habitat use of the barred owl and spotted owl in northwestern Washington. Paper presented at Northern Forest Owl Symposium, Feb. 3–7, 1987, Winnipeg, Canada (Cited from Marcot and Holthausen 1987).   73

HANLEY, T. A. 1983. Black-tailed deer, elk, and forest edge in a western Cascades watershed. J. Wildl. Manage. 47:237–242. 105

HANSEN, R. W., and E. A. OSGOOD. 1984. Effects of split application of Sevin-4-oil on pollinators and fruit set in a spruce-fir forest. Can. Ent. 116:457–464. 212

HANSSON, L. 1983. Bird numbers across edges between mature conifer forest and clear-cuts in central Sweden. Ornis Scand. 14:97–103. 106

HARESTAD, A. S., and F. L. BUNNELL. 1981. Prediction of snow-water equivalents in coniferous forests. Can. J. For. Res. 11:854–857. 76

HARESTAD, A. S., J. A. ROCHELLE, and F. L. BUNNELL. 1982. Old-growth forests and black-tailed deer on Vancouver Island. Trans. North Am. Wildl. Nat. Resour. Conf. 47:343–352. 78

HARLEY, J. L., and S. E. SMITH. 1983. Mycorrhizal symbiosis. Academic Press, London. 483 pp. 164

HARLOW, R. F., and D. C. GUYNN, JR. 1983. Snag densities in managed stands of the South Carolina coastal plain. South. J. Appl. For. 7:224–229. 169, 171

HARLOW, R. F., B. A. SANDERS, J. B. WHELAN, and L. C. CHAPPEL. 1980. Deer habitat on the Ocala National Forest: improvement through forest management. South J. Appl. For. 4:98–102. 227

HARNEY, B. A., and R. D. DUESER. 1987. Vertical stratification of activity of two *Peromyscus* species: an experimental analysis. Ecology 68:1084–1091. 189

HARRIMAN, R., and B. R. S. MORRISON. 1982. Ecology of streams draining forested and non-forested catchments in an area of central Scotland subject to acid precipitation. Hydrobiologia 88:251–263. 150

HARRIS, L. D. 1980. Forest and wildlife dynamics in the southeast. Trans. N. Am. Wildl. Nat. Res. Conf. 45:307–322. 110

HARRIS, L. D. 1984. The fragmented forest, island biogeography theory and the preservation of biotic diversity. Univ. Chicago Press, Chicago. 211 pp. 67, 69, 86, 87 115, 120, 130, 132, 133, 145

HARRIS, L. D. 1985. Conservation corridors: A highway system for wildlife. Environmental Info. Center, Florida Conserv. Found., Winter Park, Florida. ENFO Rept. 855. 130

HARRIS, L. D. Univ. of Florida, Gainesville, FL. Conversation 7 January 1986. 67

HARRIS, L. D., D. H. HIRTH, and W. R. MARION. 1979. The development of silvicultural systems for wildlife. Proceed. Ann. L.A. St. Univ. Forestry Symp. 28:65–81. 45

HARRIS, L. D., and C. MASER. 1984. Animal community characteristics. Pages 44–68 *in* Harris 1984. 53

HARRIS, L. D., and J. D. McELVEEN. 1981. Effect of forest edges on north Florida breeding birds. IMPAC Reports vol. 6, no. 4. Cited from Harris 1984. 110, 111

HARRIS, L. D., and P. SKOOG. 1980. Some wildlife habitat - forestry relationships in the southeastern coastal plain. Pages 103–119 *in* R. H. Chabreck, and R. H. Mills (eds.). Integrating timber and wildlife management in southern forests. Proc. 29th Annual Louisiana State Univ. Forestry Symp. 36, 37, 58, 110

HARRIS, L. D., L. D. WHITE, J. E. JOHNSTON, and D. G. MILCHUNAS. 1974. Impact of forest plantations on north Florida wildlife and habitat. Proc. S.E. Game Fish Comm. Conf. 28:659–667. 45

HARVEY, H. W. 1950. On the production of living matter in the sea off Plymouth. J. Mar. Biol. Assoc., U.K. n.s. 29:97–137.   20

HEEDE, B. 1985. Interactions between streamside vegetation and stream dynamics. Pages 54–58 *in* Johnson et al. 1985. *op. cit.*   142

HEINRICHS, J. 1983. Old growth comes of age. J. For. 81:776–779.   66

HEINRICHS, J. 1984. The winged snail darter. J. For. 81:212–215.   71

HEINSELMAN, M. L. 1973. Fire in the virgin forests of the Boundary Waters Canoe Area, Minnesota. Quat. Res. 3:329–382.   61

HEINSELMAN, M. L. 1981*a*. Fire intensity and frequency as factors in the distribution and structure of northern ecosystems. Pages 7–57 *in* Mooney et al. 1981.   62

HEINSELMAN, M. L. 1981*b*. Fire and succession in the conifer forests of northern North America. Pages 374–405 *in* West et al. 1981. *op. cit.*   88

HELIOVAARA, K., and R. VAISANEN. 1984. Effects of modern forestry on northwestern European forest invertebrates—a synthesis. Acta For. Fenn. 189:1–32.   50, 206

HELLE, P. 1984. Effects of habitat area on breading bird communities in northeastern Finland. Ann. Zool. Fennici 21:421–425.   107, 120

HELLE, P. 1985*a*. Habitat selection of breeding birds in relation to forest succession in northeastern Finland. Ornis Fennica 62:113–123.   53, 55

HELLE, P. 1985*b*. Effects of forest fragmentation on bird densities in northern boreal forests. Ornis Fennica. 62:35–41.   131

HELLE, P. 1986. Bird community dynamics in a boreal forest reserve: the importance of large-scale regional trends. Ann. Zool. Fennici 23:157–166.   131

HELLE, E., and P. HELLE. 1982. Edge effect on forest bird densities on offshore islands in the northern Gulf of Bothnia. Ann. Zool. Fenn. 19:165–170.   106

HELLE, P., and J. MUONA. 1985. Invetebrate numbers in edges between clear-fellings and mature forests in northern Finland. Silva Fenn. 19:281–294.   107

HENDERSON, M. T., G. MERRIAM, and J. WEGNER. 1985. Patchy environments and species survival: chipmunks in an agricultural mosaic. Biol. Conserv. 31:95–105. 129, 130

HEWLETT, J. D., and J. D. HELVEY. 1970. Effects of forest clear-felling on the storm hydrograph. Water Resour. Res. 6:768–782.   150

HIBBERT, A. R. 1967. Forest treatment effects on water yield. Pages 527–543 *in* W. E. Sopper, and H. W. Lull (eds.). Forest hydrology. Pergamon Press, Oxford. 813 pp. 149

HICKEY, J. J. 1974. Some historical phases in wildlife conservation. Wildl. Soc. Bull. 2:164–170.   3

HIGGS, A. J., and M. B. USHER. 1980. Should nature reserves be large or small? Nature 285:568–569.   126

HILL, M. O. 1979. The development of a flora in even-aged plantations. Pages 175–192 *in* E. D. Ford, D. C. Malcolm, and J. Atterson (eds.). The ecology of even-aged forest plantations. Inst. Terrestrial Ecology, Cambridge, Great Britain. 582 pp.   40

HINTON, R. B. 1971. Soil survey of Homer-Ninilchik area, Alaska. USDA Soil Conservation Service, Washington, D.C. 48 pp.   89

HOCKER, H. W., JR. 1979. Introduction to forest biology. Wiley, New York. 467 pp. 23

HOEHNE, L. M. 1981. The groundlayer vegetation of forest islands in an urban-suburban matrix. Pages 41–54 *in* Burgess and Sharpe 1981, *op. cit.* 310 pp. 120

HOLBROOK, H. L. 1974. A system for wildlife habitat management on southern National Forests. Wildl. Soc. Bull. 2:119–123. 236

HOLMES, R. T., and S. K. ROBINSON. 1981. Tree species preferences of foraging insectivorous birds in a northern hardwoods forest. Oecologia 48:31–35. 36

HOLMES, R. T., T. W. SHERRY, and F. W. STURGES. 1986. Bird community dynamics in a temperate forest: long-term trends at Hubbard Brook. Ecol. Monogr. 56:201–220. 36

HOOPER, R. G. 1988. Longleaf pines used for cavities by red-cockaded woodpeckers. J. Wildl. Manage. 52:392–398. 179

HOOVER, R. L., and D. L. WILLIS (eds.). 1984. Managing forested lands for wildlife. Col. Div. Wildl. and U.S.D.A. Forest Service, Rocky Mountain Region, Denver, Colorado. 459 pp. xi

HOOVER, S. L., D. A. KING, and W. J. MATTER. 1985. A wilderness riparian environment: visitor satisfaction, perceptions, reality, and management. Pages 223–226 *in* Johnson et al. 1985. *op. cit.* 146

HORN, H. S. 1971. The adaptive geometry of trees. Princeton Univ. Press, Princeton, New Jersey. 144 pp. 186

HORN, K. M. 1985. The Puerto Rican parrot. Pages 486–492 *in* A. S. Eno, and R. L. Di Silvestro (eds.). Audubon Wildlife Report 1985. National Audubon Society, New York. 671 pp. 162

HORNADAY, W. T. 1913. Our vanishing wild life: its extermination and preservation. New York Zoological Society, New York. 411 pp. 2, 3

HORSLEY, S. B. 1977. Allelopathic inhibition of black cherry. II. Inhibition by woodland grass, ferns, and club moss. Can. J. For. Res. 7:515–519. 186

HORTON, R. E. 1945. Erosional development of streams and their drainage basins; hydrophysical approach to quantitative morphology. Bull. Geol. Soc. Am. 56:275–370. 87

HORTON, S. P., and R. W. MANNAN. 1988. Effects of prescribed fire on snags and cavity-nesting birds in southeastern Arizona pine forests. Wildl. Soc. Bull. 16:37–44. 233

HOWE, H. F. 1986. Seed dispersal by fruit-eating birds and mammals. Pages 123–189 *in* D. R. Wolong (ed.). Seed dispersal. Academic Press, Sydney, Australia. 322 pp. 193, 194

HUBBELL, S. P., and R. B. FOSTER. 1986. Commonness and rarity in a neotropical forest: implications for tropical tree conservation. Pages 205–231 *in* Soule' 1986. *op. cit.* 68, 107, 132

HUFFAKER, C. B. 1980. New technology of pest control. Wiley, New York. 500 pp. 212

HUNT, H. M. 1976. Big game utilization of hardwood cuts in Saskatchewan. Proc. 12th North Am. Moose Conf. and Workshop. Pages 91–126. 100

HUNTER, M. L., JR. 1980. Microhabitat selection for singing and other behaviour in great tits, *Parus major:* some visual and acoustical considerations. Anim. Behav. 28:468–475. 39, 191

HUNTER, M. L., JR. 1986. The diversity of New England forest ecosystems. Pages 35–47 *in* Bissonette 1986. *op. cit.* 81, 253, 297

HUNTER, M. L., JR., G. L. JACOBSON, JR., and T. WEBB III. 1988. Paleoecology and the coarse-filter approach to maintaining biological diversity. Conserv. Biol. 2:375–385 CH3 ISL 25, 130, 266

HUNTER, M. L., JR., and J. W. WITHAM. 1985. Effects of a carbaryl-induced depression of arthropod abundance on the behavior of Parulinae warblers. Can. J. Zool. 63:2612–2616. 212

HUNTER, M. L. JR., J. W. WITHAM, and H. B. DOW. 1984. Effects of a carbaryl-induced depression in invertebrate abundance on the growth and behavior of American black duck and mallard ducklings. Can. J. Zool. 62:452–456. 212

HURST, G. A., J. J. CAMPO, and M. B. BROOKS. 1982. Effects of precommercial thinning and fertilizing on deer *Odocoileus virginianus* forage in a loblolly pine plantation. South J. Appl. For. 6:140–144. 206

HURST, G. A., and R. C. WARREN. 1980. Intensive pine plantation management and white-tailed deer habitat. Proc. Ann. Louisiana State Univ. Forestry Symp. 29:90–102. 234

HUSTON, M., and T. SMITH. 1987. Plant succession: life history and competition. Am. Nat. 130:168–198. 22

HUTCHINSON, G. E. 1959. Homage to Santa Rosalia or why are there so many kinds of animals? Am. Nat. 93:145–159. 163

HUTCHINSON, B. A., and D. R. MATT. 1977. The distribution of solar radiation within a deciduous forest. Ecol. Monogr. 47:185–207. 186

HYDE, W. F. 1980. Timber supply, land allocation, and economic efficiency. Johns Hopkins Univ. Press, Baltimore, Maryland. 224 pp. 201, 259

HYNES, H. B. N. 1970. The ecology of running waters. University of Toronto Press, Toronto. 555 pp. 141

HYNSON, J. R., P. ADAMUS, S. TIBBETTS, and R. DARNELL. 1982. Handbook for protection of fish and wildlife from construction of farm and forest roads. U.S. Fish and Wildlife Service, Washington, DC. FWS/OBS-82/18. 153 pp. 149

JACKSON, J. A. 1979. Tree surfaces as foraging substrates for insectivorous birds. Pages 69-93 *in* J. G. Dickson, R. N. Conner, R. R. Fleet, J. A. Jackson, and J. C. Kroll. The role of insectivorous birds in forest ecosystems. Academic Press, New York. 381 pp. 36

JACKSON, J. A., M. R. LENNARTZ, and R. G. HOOPER. 1979. Tree age and cavity initiation by red-cockaded woodpeckers. J. For. 77:102–103. 179

JACOBSON, G. L., JR., T. WEBB III, and E. C. GRIMM. 1987. Patterns and rates of vegetation change during the deglaciation of eastern North America. Pages 277–288 *in* W. F. Ruddiman and H. E. Wright, Jr. (eds.). North America and adjacent oceans during the last deglaciation. Geological Society of America, Boulder, Colorado. 24, 62

JANZEN, D. H. 1966. Coevolution of mutualism between ants and acacias in Central America. Evolution 20:249–275. 26

JANZEN, D. H. 1970. Herbivores and the number of tree species in tropical forests. Am. Nat. 104:501–528.   27

JANZEN, D. H. 1973. Host plants as islands. II. Competition in evolutionary and contemporary time. Am. Nat. 107:786–790.   119

JANZEN, D. H. 1986. The eternal external threat. Pages 286–303 *in* Soule′ 1986. *op. cit.*   133

JARVINEN, O. 1982. Conservation of endangered plant populations: single large or several small reserves? Oikos 38:301–307.   126

JENNINGS, D. USDA Forest Service entomologist. Orono, Maine   9

JESSOP, R. M. 1982. The impact of forestation on the avifauna of a Scottish moor. Arboric J. 6:107–119.   219

JOHANSEN, C. A. 1977. Pesticides and pollinators. Ann. Rev. Ent.22:177–192.   212

JOHNSON, A. H., and T. G. SICCAMA. 1983. Acidic deposition and forest decline. Environ. Sci. Technol. 17:294–305.   210

JOHNSON, A. S., and J. L. LANDERS. 1978. Fruit production in slash pine plantations in Georgia. J. Wildl. Manage. 42:606–613.   195, 216, 234

JOHNSON, A. S., and J. L. LANDERS. 1982. Habitat relationships of summer resident birds in slash pine flatwoods. J. Wildl. Manage. 46:416–428.   208

JOHNSON, R. R., and C. H. LOWE. 1985. On the development of riparian ecology. Pages 112–116 *in* Johnson et al. 1985. *op. cit.*   139

JOHNSON, R. R., C. D. ZIEBELL, D. R. PATTON, P. F. FFOLLIOTT, and R. H. HAMRE (eds.). 1985. Riparian ecosystems and their management: reconciling conflicting uses. USDA Forest Service. Gen. Tech. Rep. RM-120. 523 pp.   140

JORDANO, P. 1987. Patterns of mutualistic interactions in pollinations and seed dispersal: connectance, dependence, asymmetries, and coevolution. Am. Nat. 129:657–677.   194

KALLMAN, H. 1987. Restoring America's wildlife, 1937–1987. USDI Fish and Wildlife Service, Washington, D.C. 394 pp.   279

KATTEL, B. Nepal Department of National Parks and Wildlife. Pers. comm.   75

KEARNEY, S. R., and F. F. GILBERT. 1976. Habitat use by white-tailed deer and moose on sympatric range. J. Wildl. Manage. 40:645–657.   98

KENNEDY, C. E. J., and T. R. E. SOUTHWOOD. 1984. The number of species of insects associated with British trees: a re-analysis. J. Anim. Ecol. 53:455–478.   37, 39

KIKKAWA, J. 1986. Complexity, diversity, and stability. Pages 41–62 in Kikkawa and Anderson 1986. *op. cit.*   10

KIKKAWA, J. and D. J. ANDERSON (eds.). 1986. Community ecology: pattern and process. Blackwell, Melbourne. 432 pp.

KILGORE, B. M. 1981. Fire in ecosystem distribution and structure: western forests and scrublands. Pages 58–89 *in* Mooney et al. 1981. *op. cit.*   89

KING, C. 1984. Immigrant killers. Oxford Univ. Press, Aukland, New Zealand. 224 pp.   6

KIRCHOFF, M. D., and J. W. SCHOEN. 1987. Forest cover and snow: implication for deer habitat in southeast Alaska. J. Wildl. Manage. 51:28–33.   78

KISTCHINSKI, A. A. 1974. The moose in north-east Siberia. Nat. Can. 101:179–184. 146

KLEIN, R. M., and T. D. PERKINS. 1988. Primary and secondary causes and consequences of contemporary forest decline. Bot. Rev. 54:1–43. 213

KLOPATEK, J. M., R. J. OLSON, C. J. EMERSON, and J. L. JONES. 1979. Land-use conflicts with natural vegetation in the United States. Environ. Conserv. 6:191–200. 42, 146

KNIGHT, F. B. and H. J. HEIKKENEN. 1980. Principles of forest entomology. 5th ed. McGraw-Hill Book Co., New York. 461 pp. 41, 212

KOMAREK, R. 1963. Fire and the changing wildlife habitat. Proc. 2nd Ann. Tall Timbers Fire Ecology Conf., Tall Timbers Res. Sta., Tallahassee, FL. 2:35–43. 232

KOMAREK, R. 1966. A discussion of wildlife management, fire, and the wildlife landscape. Proc. Fifth Annual Tall Timbers Fire Ecology Conf., Tall Timbers Res. Stn., Tallahassee, FL. 5:177–194. 209, 232

KORTE, P. A., and L. H. FREDERICKSON. 1977. Loss of Missouri's lowland hardwood ecosystem. Trans. 42nd N. Am. Wildl. Nat. Resour. Conf. 42:31–41. 146

KOZLOWSKI, T. T. 1984. Reponses of woody plants to flooding. Pages 129–163 *in* T. T. Kozlowski (ed.). Flooding and plant growth. Academic Press, Orlando, Florida. 356 pp. 143

KRISTEK, J. 1985. Structure of insects, spiders, and harvestmen of a floodplain forest. Pages 327–356 *in* M. Penka, M. Vyskot, E. Klimo, and F. Vasicek. Floodplain forest ecosystem. Czechoslovak Academy of Science, Prague. 466 pp. 145

KROODSMA, R. L. 1982. Edge effect on breeding forest birds along a power-line corridor. J. Appl. Ecol. 19:361–370. 106

KROODSMA, R. L. 1984. Effect of edge on breeding forest bird species. Wilson Bull. 96:426–436. 106

KROODSMA, R. L. 1987. Edge effect on breeding birds along power-line corridors in east Tennessee. Am. Midl. Natur. 118:275–283. 106

KRUMMEL, J. R., R. H. GARDNER, G. SUGIHARA, R. V. O'NEILL, and P. R. COLEMAN. 1987. Landscape patterns in a disturbed environment. Oikos 48:321–324. 267

LACK, D. 1976. Island biology. Univ. California Press, Berkeley. 445 pp. 116

LAESSLE, A. M. 1961. A micro-limnological study of Jamaican bromeliads. Ecology 42:499–517. 193

LAMOTHE, L. 1980. Birds of the Araucaria pine plantations and natural forests near Bulolo Papua, New Guinea. Corella 4:127–131. 219

LANDE, R. 1988. Demographic models of the northern spotted owl (*Strix occidentalis caurina*) Oecologia 75:601–607. 72

LANDRES, P. B., J. VERNER, and J. W. THOMAS. 1988. Ecological uses of vertebrate indicator species: A critique. Conserv. Biol. 2:316–328. 242

LANGSAETER, A. 1941. About thinning in even-aged stands of spruce, fir, and pine. Meddel. f. d. Norske Skogforsoksvesen 8:131–216. Cited in Daniel, T. W., J. A. Helms, and F. S. Baker. 1979. Principles of silviculture. 2nd ed. McGraw-Hill Book Company, New York. 500 pp. 227

LANYON, W. E. 1981. Breeding birds and old field succession on fallow Long Island, New York farmland. Bull. Am. Mus. Nat. Hist. 168:1–60.   227

LAUDENSLAYER, W. F., JR., and R. P. BALDA. 1976. Breeding bird use of a pinyon-juniper-ponderosa pine ecotone. Auk 93:571–586.   106

LAUTENSCHLAGER, R. A., and R. T. BOWYER. 1985. Wildlife management by public referendum: when professionals fail to communicate. Wildl. Soc. Bull. 13:564–570. 97

LAWTON, J. H. 1978. Host-plant influences on insect diversity: the effects of time and space. Pages 105–125 *in* L. A. Mound and N. Waloff (eds.). Diversity of insect faunas. Symposium of Royal Entomological Society, 9.   50

LAYCOCK, G. 1966. The alien animals. Natural History Press, Garden City, New York. 240 pp.   6

LEDIG, F. W. 1988. The conservation of diversity in forest trees. Bioscience 38:471–479.   10

LENNARTZ, M. R., and R. F. HARLOW. 1979. The role of parent and helper red-cockaded woodpeckers at the nest. Wilson Bull. 91:331–335.   179

LENNARTZ, M. R., and R. A. LANCIA. In prep. Old-growth wildlife in second-growth forests: opportunities for creative silviculture.   70

LENNERSTEDT, I. 1983. Fodoomraden hos lovsangare *Phylloscopus trochilus* och svart-vit flugsnappare *Ficedula hypoleuca* i fjallsjorkskog. Var. Fagelv 42:11–20.   162

LEOPOLD, A. 1933. Game management. Scribner, New York. 481 pp.   101, 109, 245

LEOPOLD, A. 1936. Deer and dauerwald in Germany. J. For. 34:366–375, 460–466.   35

LEOPOLD, A. 1949. A Sand County almanac and sketcher here and there. Oxford Univ. Press, New York. 226 pp.   xi, 11, 252

LEOPOLD, L. B., M. G. WOLMAN, and J. P. MILLER. 1964. Fluvial processes in geomorphology. Freeman, San Francisco. 522 pp.   140

LEVENSON, J. B. 1981. Woodlots as biogeographic islands in southeastern Wisconsin. Pages 13–39 *in* Burgess and Sharpe 1981. *op. cit.*   120

LEWIS, C. E., G. W. TANNER, and W. S. TERRY. 1985. Double vs. single-row pine plantations for wood and forage production. South J. Appl. For. 9:55–61.   219

LEWIS, S. J., and F. B. SAMSON. 1981. Use of upland forests by birds following spray irrigation with municipal wastewater. Environ. Pollut. Ser. A. Ecol. Biol. 26:267–274.   206

LIEBERMAN, D., M. LIEBERMAN, and C. MARTIN. 1987. Notes on seeds in elephant dung from Bio National Park, Ghana. Biotropica 19:365–369.   241

LIGON, J. D. 1970. Behavior and breeding biology of the red-cockaded woodpecker. Auk 87:255–278.   178

LIGON, J. D., P. B. STACEY, R. N. CONNER, C. E. BOCK, and C. S. ADKISSON. 1986. Report of the American Ornithologists' Union committee for the conservation of the red-cockaded woodpecker. Auk 103:848–855.   179

LIKENS, G. E., F. H. BORMANN, N. M. JOHNSON, D. W. FISHER, and R. S. PIERCE. 1970. Effects of forest cutting and herbicide treatment on nutrient budgets in the Hubbard Brook watershed-ecosystem. Ecol. Monogr. 40:23–47.   150

LINDSEY, A. A., R. O. PETTY, D. K. STERLING, and W. VANASDALL. 1961. Vegetation and environment along the Wabash and Tippecanoe Rivers. Ecol. Monogr. 31:105–156.   144

LONGHURST, W. M., K. K. OH, M. B. JONES, and R. E. KEPNER. 1968. A basis for the palatability of deer forage plants. Trans. North Am. Wildl. Conf. 33:181–192.   37

LORIMER, C. G. 1980. Age structure and disturbance history of a southern Appalachian virgin forest. Ecology 61:1169–1184.   184

LOVEJOY, T. E. 1974. Bird diversity and abundance in Amazon forest communities. Living Bird 13:127–191.   188

LOVEJOY, T. E., R. O. BIERREGAARD, J. M. RANKIN, and H. O. R. SCHUBERT. 1983. Ecological dynamics of tropical forest fragments. Pages 377–384 *in* Sutton et al. 1983*b. op. cit.*   135

LOVEJOY, T. E., R. O. BIERREGAARD, JR., A. B. RYLANDS, J. R. MALCOLM. C. E. QUINTELA, L. H. HARPER, K. S. BROWN, JR., A. H. POWELL, G. V. N. POWELL, H. O. R. SCHUBERT, and M. B. HAYS. 1986. Edge and other effects of isolation on Amazon forest fragments. Pages 257–285 *in* Soule' 1986. *op. cit.*   129, 135

LOVEJOY, T. E., and D. C. OREN. 1981. Minimum critical size of ecosystems. Pages 7–12 *in* Burgess and Sharp 1981. *op. cit.*   123

LOVEJOY, T. E., J. M. RANKIN, R. O. BIERREGAARD, JR., K. S. BROWN, JR., L. H. EMMONS, and M. E. VAN DER VOORT. 1984. Ecosystem decay of Amazon forest remnants. Pages 295–325 *in* M. H. Nitecki (ed.). Extinctions. Univ. Chicago Press. Chicago.   135

LOWMAN, M. D. 1985. Temporal and spatial variability in insect grazing of the canopies of 5 Australian rainforest tree species. Aust. J. Ecol. 10:7–24.   191

LOYN, R. H. 1980. Bird population in a mixed eucalypt forest used for production of wood in Gippsland, Victoria. Emu 80:145–156.   50, 51, 53, 55

LOYN, R. H. 1985. Birds in fragmented forests in Gippsland, Victoria. Pages 323–331 *in* A. Keast, H. F. Recher, H. Ford, and D. Saunders (eds.) Birds of eucalyptus forest and woodlands. Surrey Beatty, Syndney.   120

LYNCH, J. R. and D. F. WHIGHAM. 1984. Effects of forest fragmentation on breeding bird communities in Maryland, U.S.A. Biol. Conserv. 28:287–324.   120

MACARTHUR, R. H. 1958. Population ecology of some warblers of northeastern coniferous forests. Ecology 39:599–619.   189

MACARTHUR, R. H., and J. W. MACARTHUR. 1961. On bird species diversity. Ecology 42:594–598.   188

MACARTHUR, R. H. and E. O. WILSON. 1967. The theory of island biogeography. Princeton Univ. Press, Princeton, New Jersey. 203 pp.   115, 116, 123

MACCLINTOCK, L., R. F. WHITCOMB, and B. L. WHITCOMB. 1977. Island biogeography and habitat islands of eastern forest. II. Evidence for the value of corridors and minimization of isolation in preservations of biotic diversity. Am. Birds 31:6–16.   129

MACKOWSKI, C. M. 1984. The ontogeny of hollows in Blackbutt (*Eucalyptus pilularis*) and its relevance to the management of forests for possums, gliders and timer. Pages 553–567 *in* A. Smith and I. Hume (eds.) Possums and gliders. Australian Mammal Society, Sydney. 598 pp.   167

MACMAHON, J. A. 1981. Successional processes: comparisons among biomes with special reference to probable roles of influences on animals. Pages 277–304 *in* West et al. 1981. *op. cit.* 48

MADER, H. J. 1984. Animal habitat isolation by roads and agricultural fields. Biol. Conserv. 29:81–96. 131, 258

MAGUIRE, D. A., and R. T. T. FORMAN. 1983. Herb cover effects on tree seedling patterns in a mature hemlock-hardwood forest. Ecology 64:1367–1380. 186

MAGURRAN, A. E. 1985. The diversity of macrolepidoptera in two contrasting woodland habitats at Banagher, Northern Ireland. Proc. Royal Irish Acad. Sect. B 85:121–132. 39

MAINE CRITICAL AREAS PROGRAM. 1983. Natural old-growth forest stand in Maine. Maine State Planning Office, Augusta. 254 pp. 61

MANDELBROT, B. B. 1983. The fractal geometry of nature. Freeman, San Francisco. 468 pp. 89

MANNAN, R. W., and E. C. MESLOW. 1984. Bird populations and vegetation characteristics in managed and old-growth forests, northeastern Oregon. J. Wildl. Manage. 48:1219–1238. 70

MANNAN, R. W., M. L. MORRISON, and E. C. MESLOW. 1984. Comment: the use of guilds in forest bird management. Wildl. Soc. Bull. 12:426–430. 246

MARCEAU, J. P. 1981. Mechanical weed control in eastern Canada. Pages 94–101 *in* H. A. Holt and B. C. Fischer (eds.). Weed control in forest management. Purdue Research Foundation, West Lafayette, Indiana. 305 pp. 217

MARCOT, B. G., and R. HOLTHAUSEN. 1987. Analyzing population viability of the spotted owl in the Pacific Northwest. Trans. North Am. Wildl. Nat. Resour. Conf. 52:333–347. 72

MARION, W. R., and L. D. HARRIS. 1982. Relationships between increasing forest productivity and fauna in the flatwoods of the southeastern coastal plain. Proc. Ann. Conf. Soc. Am. For. Pages 215–222. 221, 232

MARK, H. 1943. Elasticity and strength. Pages 991–1052 *in* E. Ott (ed.). Cellulose and cellulose derivitives. Interscience, New York. 1176 pp. 60

MARKOWSKY, J. K. and M. L. HUNTER, JR. (in prep.). Habitat parameters for three sympatric salamander species. 145

MARMELSTEIN, A. (ed.). 1977. Classification, inventory, and analysis of fish and wildlife habitat. USDI Fish and Wildl. Service, Washington, D.C. FWS/OBS-78/76. 604 pp. 265

MARQUISS, M., D. A. RATCLIFFE, and R. ROXBURGH. 1985. The numbers, breeding success, and diet of golden eagles in southern Scotland in relation to changes in land use. Biol. Conserv. 34:121–140. 42, 221

MARSH, C. W., A. D. JOHNS, and J. M. AYERS. 1987. Effects of habitat disturbance on rain forest primates. Pages 83–107 *in* C. W. Marsh and R. A. Mittermeier (eds.) Primate conservation in the tropical rain forest. Alan Liss, New York. 365 pp. 240

MARTIN, C. W., and R. S. PIERCE. 1980. Clearcutting patterns affect nitrate and calcium in streams of New Hampshire. J. For. 78:268–272. 150

MARTIN, J. P. 1981. Mechanical weed control in southern forests. Pages 102–107 *in* H. A. Holt and B. C. Fischer (eds.). Weed control in forest management. Purdue Research Foundation, West Lafayette, Indiana. 305 pp. 217

MARTIN, S. G., D. A. THIEL, J. W. DUNCAN, and W. R. LANCE. 1987. Effects of a paper industry sludge containing dioxin on wildlife in red pine plantations. Proc. 1987 TAPPI Environ. Conf. Pages 363–377.   207

MASER, C. Bureau of Land Management, Forest Science Lab., Corvallis, Oregon. Conversation 18 July, 1986.   164

MASER, C., R. G. ANDERSON, K. CROMACK, JR., J. T. WILLIAMS, and R. E. MARTIN. 1979. Dead and down woody material. Pages 78–95 *in* Thomas 1979. *op. cit.*   159, 163

MASER, C., and J. M. TRAPPE, (eds.). 1984. The seen and unseen world of the fallen tree. USDA Forest Service, Gen. Tech. Rep. PNW-164.   158, 164

MASER, C., J. M. TRAPPE, and R. A. NUSSBAUM. 1978a. Fungal-small mammal interrelationships with emphasis on Oregon coniferous forests. Ecology 59:799–809.   164

MASER, C., J. M. TRAPPE, and D. C. URE. 1978b. Implications of small mammal mycophagy to the management of western confierous forests. Trans. North Am. Wildl. and Nat. Resour. Conf. 43:78–88.   164

MATTHIAE, P. E., and F. STEARNS. 1981. Mammals in forest islands in southeastern Wisconsin. Page 55–66 *in* Burgess and Sharpe 1981. *op. cit.*   120

MAY, R. M. 1974. Stability and complexity in model ecosystems. (2nd ed.). Princeton Univ. Press, Princeton, New Jersey. 265 pp.   10

MAY, R. M., 1986. The search for patterns in the balance of nature: advances and retreats. Ecology 67:1115–1126.   87

MCCOMB, W. C. 1982. Forestry and wildlife habitat management in central hardwoods. J. For. 80:490–492.   230

MCCOMB, W. C., S. A. BONNEY, R. M. SHEFFIELD, and N. D. COST. 1986a. Snag resources in Florida: are they sufficient for average populations of primary cavity-nesters? Wildl. Soc. Bull. 14:40–48.   158, 171

MCCOMB, W. C., S. A. BONNEY, R. M. SHEFFIELD, and N. D. COST. 1986b. Den tree characteristics and abundance in Florida and South Carolina. J. Wildl. Manage. 50:584–591.   161

MCCOMB, W. C., and R. N. MULLER. 1983. Snag densities in old growth and 2nd growth Appalachian forests. J. Wildl. Manage. 47:376–382.   158

MCCOMB, W. C., and R. E. NOBLE. 1980. Effects of single-tree selection cutting upon snag and natural cavity characteristics in Connecticut. Trans. Northeast Sect. Wildl. Soc. 37:50–57.   158

MCCOMB, W. C., and R. E. NOBLE. 1981. Nest box and natural-cavity use in three mid-southern forest habitats. J. Wildl. Manage. 45:93–101.   174

MCCOMB, W. C., and R. E. NOBLE. 1982. Invertebrate use of natural tree cavities and vertebrate nest boxes. Am. Midl. Nat. 107:163–172.   162

MCCOMB, W. C., and R. L. RUMSEY. 1982. Response of small mammals to forest clearings created by herbicides in the central Appalachians. Brimeleyana 8:121–134.   224

MCCOMB, W. C., and R. L. RUMSEY. 1983a. Characteristics and cavity-nesting bird use of picloram-created snags in the central Appalacians. South. J. Appl. For. 7:34–37.   175

MCCOMB, W. C., and R. L. RUMSEY. 1983b. Bird density and habitat use in forest openings created by herbicides and clearcutting in the central Appalachians. Brimleyana 9:83–95.   224

McCORMACK, M. L., JR., and C. H. BANKS. 1983. An aerial technique to adjust spruce-fir density of stocking. Proc. N.E. Weed Sci. Soc. 37:297–300.   224

McCORMACK, M. L., JR., F. B. KNIGHT, and R. A. ROGERS. 1978. Glyphosate as a management tool for increasing forest productivity. Proc. N.E. Weed Sci. Soc. 32:285–286.   224

McFADDEN, M. W., D. L. DAHLSTEN, C. W. BERISFORD, F. B. KNIGHT, and W. W. METTERHOUSE. 1981. Integrated pest management in China's forests. J. For. 79:722–726.   210

McILROY, J. C. 1978. The effects of forestry practices on wildlife in Australia: a review. Aust. For. 41:78–94.   39

McINTOSH, R. P. 1980. The relationship between succession and the recovery process in ecosystems. Pages 11–62 *in* J. Cairns, Jr. (ed.). The recovery process in damaged ecosystems. Ann Arbor Science Publ., Ann Arbor, Michigan. 167 pp.   22

McLAY, C. L. 1974. The species diversity of New Zealand forest birds: some possible consequences of the modification of beech forests. New Zealand J. Zool. 1:179–196.   39

McLELLAN, C. H., A. P. DOBSON, D. S. WILCOVE, and J. F. LYNCH. 1986. Effects of forest fragmentation of New-and Old-world bird communities: empirical observations and theoretical implications. Pages 305–313 *in* Verner et al. 1986. *op. cit.*   122, 128

McMAHON, T. A., and J. T. BONNER. 1983. On size and life. Scientific American Books, New York. 255 pp.   89

McPEEK, G. A., W. C. McCOMB, J. J. MORIARITY, and G. E. JACOBS. 1987. Bark-foraging bird abundance unaffected by increased snag availability in a mixed mesophytic forest. Wilson Bull. 99:253–257.   171

MEALEY, S. P., J. P. LIPSCOMB, and K. J. NORMAN. 1982. Solving the habitat dispersion problem in forest planning. Trans. North Am. Wildl. Nat. Resour. Conf. 47:142–153.   110

MEEHAN, W. R., T. R. MERRELL, JR., and T. A. HANLEY (eds.). 1984. Fish and wildlife relationships in old-growth forests. Amer. Inst. Fish. Res. Biol. (Proc. of symposium Juneau, Alaska, 12–15 April 1982). 425 pp.

MEEHAN, W. R., F. J. SWANSON, and J. R. SEDELL. 1977. Influences of riparian vegetation on aquatic ecosystems with particular reference to salmonid fish and their food supply. Pages 137–145 *in* R. R. Johnson and D. A. Jones (eds.) Importance, preservation, and management of riparian habitat; a symposium. USDA Forest Service Gen. Tech. Rep. RM-43.   141

MENARD, G., R. MCNEIL, and A. BOUCHARD. 1982. Les facterus indicatifs de la diversite des peuplements d'oiseaux forestiers du sud du Quebec. Nat. Can. 109:39–50.   37

MENASCO, K. A. 1983. Providing snag habitat for the future. Pages 205–210 *in* Davis et al. 1983.   176

MENGES, E. S. 1986. Environmental correlates of herb species composition in five southern Wisconsin floodplain forests. Am. Midl. Nat. 115:106–117.   144

METZLER, K. J., and A. W. H. DAMMAN. 1985. Vegetation patterns in the Connecticut River flood plain in relation to frequency and duration of flooding. Nat. Can. 112:535–548. 144

MIDDLETON, J., and G. MERRIAM. 1985. The rationale for conservation: problems from a virgin forest. Biol. Conserv. 33:133–145. 51

MILLER, D. H., and L. L. GETZ. 1977. Factors influencing local distribution and species diversity of forest small mammals in New England. Can. J. Zool. 55:806–814. 143

MITSCH, W. J., C. L. DORGE, and J. R. WIEMHOFF. 1979. Ecosystem dynamics and a phosphorous budget of an alluvial cypress swamp in southern Illinois. Ecology 60:1116–1124. 145

MOEN, A. N. 1973. Wildlife ecology. W. H. Freeman, San Francisco. 458 pp. 37

MOONEY, H. A., T. M. BONNICKSEN, N. L. CHRISTENSEN, J. E. LOTAN, and W. A. REINERS (eds.). 1981. Fire regimes and ecosystem properties. U.S.D.A. Forest Service Gen. Tech. Rep. WO-26. 594 pp.

MOORE, D. M. 1982. Flora Europaea check-list and chromosome index. Cambridge University Press, Cambridge.

MOORE, N. W., and M. D. HOOPER. 1975. On the number of bird species in British woods. Biol. Conserv. 8:239–250. 120

MOORE, W. H., and B. F. SWINDEL. 1981. Effects of site preparation on dry prairie vegetation in south Florida. South. J. Appl. For. 5:89–92. 216

MOORE, W. H., B. F. SWINDEL, and W. S. TERRY. 1982. Vegetative response to prescribed fire in a north Florida flatwoods forest. J. Range. Manage. 35:386–389. 233

MORING, J. R., and R. L. LANTZ. 1975. The Alsea watershed study: effects of logging on the aquatic resources of three headwater streams on the Alsea River, Oregon. Part I - Biological studies. (Also see: Part II - Changes in environmental conditions. Part III - Discussion and recommendations). Fishery Research Rept. No. 9, Oregon Dept. Fish and Wildlife, Corvalis, Oregon. 151

MORRIS, L. A., W. L. PRITCHETT, and B. F. SWINDEL. 1983. Displacement of nutrients into windrows during site preparation of a flatwood forest. Soil Sci. Soc. Am. J. 47:591–594. 215

MORRISON, M. L., M. G. RAPHAEL, and R. C. HEALD. 1983. The use of high-cut stumps by cavity-nesting birds. Pages 73–79 *in* Davis et al. 1983. *op. cit.* 177

MORRISON, M. L., I. C. TIMOSSI, K. A. WITH, and P. N. MANLEY. 1985. Use of tree species by forest birds during winter and summer. J. Wildl. Manage. 49:1098–1102. 36

MORSE, D. H. 1976. Variables affecting the density and territory size of breeding spruce-woods warblers. Ecology 57:290–301. 36

MOSS, D. 1978*a*. Song-bird populations in forestry plantations. Quart. J. For. 72:5–14. 40

MOSS, D. 1978*b*. Diversity of woodland song-bird populations. J. Anim. Ecol. 47:521–527. 188

MOTT, D. G. 1963. The forest and the spruce budworm. Pages 189–202 *in* R. F. Morris (ed.). The dynamics of epidemic spruce budworm populations. Mem. Entomol. Soc. Can. 31. 332 pp. 41

MULLER, R. N. 1978. The phenology, growth, and ecosystem dynamics of *Erythronium americanum* in the northern hardwood forest. Ecol. Monogr. 48:1–20. 186

MURPHY, D. D., and B. A. WILCOX. 1986. Butterfly diversity in natural habitat fragments: a test of the validity of vertebrate-based management. Pages 287–292 *in* Verner et al. 1986. *op. cit.* 120

MURPHY, M. L., C. P. HAWKINS, and N. H. ANDERSON. 1981. Effects of canopy modification and accumulated sediment on stream communities. Trans. Am. Fish. Soc. 110:469–478. 147

MYERS, N. 1979. The sinking ark. Pergamon, Oxford. 307 pp. 9, 33, 281

MYERS, N. 1980. Conversion of tropical moist forests. National Academy of Sciences, Washington, D.C. 205 pp. 135

MYERS, N. 1984. The primary source: tropical forests and our future. Norton, New York. 399 pp. 135

MYERS, N. 1987. The extinction spasm impending: synergisms at work. Conserv. Biol. 1:14–21. 42

NADKARNI, N. 1981. Canopy roots: convergent evoluation in rainforest nutrient cycles. Science 214:1023–1024. 187

NEITRO, W. A., V. W. BINKLEY, S. P. CLINE, R. W. MANNAN, B. G. MARCOT, D. TAYLOR, and F. F. WAGNER. 1985. Snags (wildlife trees). Pages 129–169 *in* Browne 1985. *op. cit.* 168–173, 177, 269

NEWMARK, W. D. 1987. A land-bridge island perspective on mammalian extinctions in western North American parks. Nature 325:430–432. 129

NEWTON, M., and E. C. COLE. 1987. A sustained yield scheme for old-growth Douglas fir. West J. Appl. For. 2:22–25. 70

NILSSON, C., G. GRELSSON, M. JOHANSSON, and U. SPERENS. 1988. Can rarity and diversity be predicted in vegetation along river banks? Biol. Conserv. 44:201–212. 126, 143

NILSSON, S. G., J. BENGTSSON, and S. As. 1988. Habitat diversity or area *per se*? Species richness of woody plants, carabid beetles, and land snails on islands. J. Anim. Ecol. 57:685–704. 126

NOBLE, R. E., and R. B. HAMILTON. 1975. Bird populations in even-aged loblolly pine forests of southeastern Louisiana. Proc. S.E. Game Fish Comm. Conf. 29:441–450. 221

NOON, B. R. 1981. The distribution of an avian guild along a temperate elevational gradient; the importance and expression of competition. Ecol. Monogr. 51:105–124. 266

NORTON, B. G. 1987. Why preserve natural variety? Princeton Univ. Press, Princeton, New Jersey. 281 pp. 277

NOSS, R. F. 1983. A regional landscape approach to maintain diversity. Bioscience 33:700–706. 107

NOSS, R. F. 1987. From plant communities to landscapes in conservative inventories: a look at the Nature Conservancy (USA). Biol. Conserv. 41:11–37. 238

NOSS, R. F. 1987. Corridors in real landscapes: a reply to Simberloff and Cox. Conserv. Biol. 1:159–164. 130

Noss, R. F., and L. D. Harris. 1986. Nodes, networks, and MUM's: preserving diversity at all scales. Environ. Manage. 10:299–309.   130, 133, 258

Noy-Meir, I. 1973. Desert ecosystems: environment and producers. Ann. Rev. Ecol. Syst. 4:25–51.   20

Nyberg, J. B., F. L. Bunnell, D. W. Janz, and R. M. Ellis. 1986. Managing young forests as black-tailed deer winter ranges. British Columbia Ministry of Forests, Land Management Report No. 37. 49 pp.   230, 248

Nyberg, J. B., A. S. Harestad, and F. L. Bunnell. 1987. "Old-growth" by design: managing young forests for old-growth wildlife. Trans. N. Amer. Wildl. Nat. Res. Conf. 52:70–81.   70, 79

O'Connor, R. J. 1986. Biological characteristics of invaders among bird species in Britain. Phil. Tans. R. Soc. Lond. B 314:583–598.   122

O'Connor, R. J., and M. J. Shrubb. 1986. Farming and birds. Cambridge University Press, Cambridge, Great Britain.   253

Odum, E. P. 1969. The strategy of ecosystem development. Science 164:262–270.   22, 49, 50

Odum, H. T., and E. P. Odum. 1955. Trophic structure and productivity of a windward coral reef community on Eniwetok Atoll. Ecol. Monogr. 25:291–320.   20

Oedekoven, K. 1980. The vanishing forest. Environ. Policy Law 6:184–185.   121

Office of Technology Assessment. 1987. Technologies to maintain biological diversity. OTA-F-330 Congress of the United States, Washington, D.C.   9, 12

Oliver, C. D. 1980. Even-aged development of mixed-species stands. J. For. 78:201–203.   183, 195

Oliver, C. D. 1981. Forest development in North America following major disturbances. Ecol. Manage. 3:153–168.   183, 195, 208, 214

O'Meara, T. E., J. B. Haufler, L. H. Stelter, and J. G. Nagy. 1981. Nongame wildlife responses to chaining of pinyon juniper woodlands. J. Wildl. Manage. 45:381–389.   217

O'Meara, T. E., L. A. Rowse, W. R. Marion, and L. D. Harris. 1985. Numerical responses of flatwoods avifauna to clearcutting. Fla. Sci. 48:208–219.   106

Opler, P. A., H. G. Baker, and G. W. Frankie. 1977. Recovery of tropical lowland forest ecosystems. Pages 379 to 421 *in* J. Cairns, K. Dickson, and E. E. Herricks (eds.). Recovery and restoration of damaged ecosystems. Univ. Press Virginia, Virginia.   95

Orians, G. H. 1980. Habitat selection: general theory and application to human behavior. Pages 49–66 *in* J. S. Lockard (ed.). The evolution of human social behavior. Elsevier, New York. P336.   94

Ormerod, S. J., G. W. Mawle, and R. E. Edwards. 1987. The influence of forest on aquatic fauna. Pages 37–49 *in* J. E. G. Good (ed.) Environmental aspects of plantation forestry in Wales. Institute of Terrestrial Ecology, Grange-over-sands, Cumbria, Great Britain. 77 pp.   150

Orwell, G. 1946. Animal farm. Harcourt Brace, New York. 118 pp.   5, 11, 236

Osborne, P. 1982. Some effects of Dutch elm disease on nesting farmland birds. Bird Study 29:2–16.   213

OSBORNE, J. G., and V. L. HARPER. 1937. The effect of seedbed preparation on first-year establishment of longleaf and slash pine. J. For. 35:63–68.   44

OVINGTON, J. D., and J. S. OLSON. 1970. Biomass and chemical content of El Verde lower Montane rain forest plants. Pages 453–477 *in* H. T. Odum (ed.). A tropical rain forest: a study of irradiation and ecology at El Verde, Puerto Rico. Book 3. U.S. Atomic Energy Commission, Office of Information Services.   20

OXLEY, D. J., M. B. FENTON, and G. R. CARMODY. 1974. The effects of roads on populations of small mammals. J. Appl. Ecol. 11:51–59.   258

PAC, D. F., R. J. MACKIE, and H. E. JORGENSEN. 1984. Relationships between mule deer and forest in southwestern Montana—some precautionary observations. Pages 321–328 *in* Meehan et al. 1984. *op. cit.*   79

PACKHAM, J. R., and D. J. L. HARDING. 1982. Ecology of woodland processes. Arnold, London. 262 pp.   19

PAINE, R. T. 1966. Food web complexity and species diversity. Am. Nat. 100:65–75.   27

PAINE, R. T. 1969. A note on trophic complexity and community structure. Am. Nat. 103:91–93.   241

PARKER, K. L., C. T. ROBBINS, and T. A. HANLEY. 1984. Energy expenditures for locomotion by mule deer and elk. J. Wildl. Manage. 48:474–488.   76

PARKER, M., F. J. WOOD, JR., B. H. SMITH, and R. G. ELDER. 1985. Erosional downcutting in lower order riparian ecosystems: have historical changes been caused by removal of beaver? Pages 35–38 *in* Johnson et al. 1985. *op. cit.*   142

PARKER, S. P. (ed.). 1982. Synopsis and classification of living organisms. McGraw-Hill, New York. 1232 pp.   163

PATIL, G. P., and C. TAILLIE. 1982. Diversity as a concept and its measurements. J. Amer. Stat. Assoc. 77:548–561.   8, 297, 298, 302

PATRIC, J. H., and G. M. AUBERTIN. 1977. Long-term effects of repeated logging on an Appalachian stream. J. For. 75:492–494.   150

PATTON, D. R. 1975. A diversity index for quantifying habitat "edge". Wildl. Soc. Bull. 3:171–173.   108

PEARSALL, S. H., D. DURHAM, and D. C. EAGAR. 1986. Evaluation methods in the United States. Pages 111–133 *in* M. B. Usher (ed.). Wildlife conservation evaluation. Chapman Hall, London. 394 pp.   269

PEARSE, P. H. 1967. The optimum forest rotation. Forestry Chronicle 43:178–195.   59

PEARSON, D. L. 1975. The relation of foliage complexity to ecological diversity of three Amazonian bird communities. Condor 77:453–466.   188

PETERKEN, G. F. 1981. Woodland conservation and management. Chapman Hall, London. 328 pp.   42, 195, 196

PETERSON, J. T. 1978. The ecology of social boundaries. Univ. Illinois Press, Urbana. 141 pp.   112, 114

PETERSON, J. T. 1981. Game farming and inter ethnic relations in northeastern Luzon Philippines. Hum. Ecol. 9(1):1–22.   112, 114

PETIT, D. R., K. E. PETIT, T. C. GRUBB, JR., and L. J. REICHHARDT. 1985. Habitat and snag selection by woodpeckers in a clear-cut: an analysis using artificial snags. Wilson Bull. 97:525–533.   174

PHILLIPS, J. C. 1925. A natural history of the ducks. Vol. 3. Houghton Mifflin, Boston. 173

PHILLIPS, R. W., R. L. LANTZ, E. W. CLAIRE, and J. R. MORING. 1975. Some effects of gravel mixtures on emergence of coho salmon and steelhead trout fry. Trans. Am. Fish. Soc. 104:461–466. 142

PICKETT, S. T. A., and J. N. THOMPSON. 1978. Patch dynamics and the design of nature reserves. Biol. Conserv. 13:27–37. 128

PICKETT, S. T. A. and P. S. WHITE (eds.). 1985. The ecology of natural disturbance and patch dynamics. Academic Press, New York. 472 pp. 208

PIELOU, E. C. 1977. Mathematical ecology. John Wiley and Sons, Toronto. 385 pp. 298

PIEROVICH, J. M., E. H. CLARKE, S. G. PICKFORD, and F. R. WARD. 1975. Forest residues management guidelines for the Pacific Northwest. USDA Forest Service Gen. Tech. Rep. PNW-33. 281 pp. 176

PIMM, S. L. 1986. Community stability and structure. Pages 309–329 *in* Soule' 1986. *op. cit.* 10

PITELKA, L. F., D. S. STANTON, and M. O. PECKENHAM. 1980. Effects of light and density on resource allocation in a forest herb, *Aster acuminatus* (compositae). Am. J. Bot. 67:942–948. 186

PITTENDRIGH, C. S. 1948. The bromeliad-*Anopheles*-malaria complex in Trinidad. I. The bromeliad flora. Evolution 2:58–89. 192

PLATT, W. J., G. W. EVANS, and M. M. DAVIS. In press. Effects of fire season on flowering of forbs and shrubs in longleaf pine forests. Oecologia. 233

PLATT, W. J., G. W. EVANS, and S. L. RATHBUN. 1988. The population dynamics of a long-lived conifer (*Pinus palustris*). Am. Nat. 131:491–525. 44

POCS, T. 1982. Tropical forest bryophytes. Pages 59–104 *in* A. J. E. Smith (ed.). Bryophyte ecology. Chapman Hall, London. 511 pp. 187

POORE, M. E. D. 1968. Studies in Malaysian rain forest. I. The forest on Triassic sediments in Jengka Forest Reserve. J. Ecol. 56:143–196. 68

POWELL, D. S., and D. R. DICKSON. 1984. Forest statistics for Maine: 1971 and 1982. U.S. Forest Service Resour. Bull. NE-81. 194 pp. 158

POWER, D. M. 1972. Numbers of bird species on the California islands. Evolution 26:451–463. 119

PRESCOTT-ALLEN, C., and R. PRESCOTT-ALLEN. 1986. The first resource. Yale Univ. Press, New Haven, Conn. 529 pp. 10

PRESCOTT-ALLEN, R., and C. PRECOTT-ALLEN. 1982. What's wildlife worth? Earthscan, London. 92 pp. 10, 41

PRESTON, D. J. 1976. The rediscovery of *Betula uber.* Amer. For. 82(8):16–20. 238

PRINS, H. T. T. and G. R. IASON. 1989. Dangerous lions and nonchalant buffalo. Behaviour 108:262–296.

PRITCHETT, W. L. 1979. Properties and management of forest soils. Wiley, New York. 500 pp. 37

PYNE, S. J. 1984. Introduction to wildland fire. John Wiley and Sons, New York. 455 pp. 214, 232

QUINN, J. F., and S. P. HARRISON. 1988. Effects of habitat fragmentation and isolation on species richness: evidence from biogeographic patterns. Oecologia 75:132–140. 126

RACKHAM, O. 1980. Ancient woodland: its history, vegetation and uses in England. Edward Arnold, London. 402 pp. 63, 102

RADOSEVICH, S. R., and J. S. HOLT. 1984. Weed ecology. Wiley, New York. 265 pp. 217, 224

RADTKE, R., and J. BYELICH. 1963. Kirtland's warbler management. Wilson Bull. 75:208–215, 210

RADVANYI, A. 1970. Small mammals and regeneration of white spruce forests in western Alberta. Ecology 51:1102–1105. 219

RADVANYI, A. 1973. Seed losses to small mammals and birds. Proc. Direct Seeding Symposium. Dept. Environ., Canadian For. Serv. Publ. No. 1339. Pages 67–74. 219

RADVANYI, A. 1974. Small mammal census and control on a hardwood plantation. Proc. 6th Vertebrate Pest Conf., Anaheim, CA. 6:9–19. 219

RAINVILLE, R. P., S. C. RAINVILLE, and E. L. LIDER. 1986. Riparian silvicultural strategies for fish habitat emphasis. Pages 23–33 in Society of American Foresters. Foresters' future: leaders or followers? Society of American Foresters, Bethesda, Maryland. 51 pp. 147

RAND, A. S. 1964. Ecological distribution in anoline lizards of Puerto Rico. Ecology 45:745–752. 190

RANDALL, A. 1988. What mainstream economists have to say about the value of biodiversity. Pages 217–223 in Wilson and Peter. 1988. op. cit. 277

RANNEY, J. W., M. C. BRUNER, and J. B. LEVENSON. 1981. The importance of edge in the structure of dynamics of forest islands. Pages 67–95 in Burgess and Sharpe 1981. op. cit. 106, 107, 133

RAPHAEL, M. G., and R. H. BARRETT. 1984. Diversity and abundance of wildlife in late successional Douglas-fir forest. Proc. Soc. Amer. Foresters, 1983:352–360. 72

RAPHAEL, M. G., and M. WHITE. 1984. Use of snags by cavity nesting birds in the Sierra-Nevada California. Wildl. Monogr. 86:1–66. 158, 162, 166–173

RASMUSSEN, M. 1985. Below-cost timber sales. Amer. For. 91(1):10, 62–64. 284

RATTI, J. T., and K. P. REESE. 1988. Preliminary test of the ecological trap hypothesis. J. Wildl. Manage. 52:484–491. 133

READER, R. J. 1987. Loss of species from deciduous forest understorey immediately following selective tree harvesting. Biol. Conserv. 42:231–244. 69, 194

RECHER, H. F. 1969. Bird species diversity and habitat diversity in Australia and North America. Am. Nat. 103:75–80. 188

RECHER, H. F., D. LUNNEY, P. SMITH, and W. ROHAN-JONES. 1981. Woodchips or wildlife? Aust. Nat. Hist. 20:239–244. 65

RECHER, H. F., W. ROHAN-JONES, and P. SMITH. 1980. Effects of the Eden woodchip industry on terrestrial vertebrates with recommendations for management. Forestry Commission of New South Wales Res. Note. 42. 83 pp. 65, 94

RECHER, H. F., J. SHIELDS, R. KAVANAUGH, and G. WEBB. 1987. Retaining remnant mature forests for nature conservation at Eden, New South Wales: a review of theory and practice. Pages 177–194 *in* D. A. Saunders, G. W. Arnold, A. A. Burbridge, A. J. M. Hopkins (eds.) Nature conservation: the role of remnants of native vegetation. Surrey Beatty, Syndney, Australia. 94, 130, 146

REED, J. M., P. D. DOERR, and J. R. WALTERS. 1988. Minimum viable population size of the red-cockaded woodpecker. J. Wildl. Manage. 52:385–391. 179

REGAL, P. J. 1977. Ecology and evolution of flowering plant dominance. Science 196:622–629. 37

REPENNING, R. W., and R. F. LABISKY. 1985. Effects of even-age timber management on bird communities of the longleaf pine forest in northern Florida. J. Wildl. Manage. 49:1088–1098. 45

RICE, J., B. W. ANDERSON, and R. D. OHMART. 1984. Comparison of the importance of different habitat attributes to avian community organization. J. Wild. Manage. 48:895–911. 36

RICHARDS, K. 1982. Rivers, form and function in alluvial processes. Methuen, New York. 358 pp. 142

RICHARDS, P. W. 1983. The three-dimensional structure of tropical rain forest. Page 3–10 *in* Sutton et al. 1983*a. op. cit.* 182

RICHARDSON, D. H. S., and C. M. YOUNG. 1977. Pages 121–144 *in* M. R. D. Seaward (ed.). Lichen ecology. Academic Press, London. 550 pp. 193

RICKLEFS, R. E. 1987. Community diversity: relative roles of local and regional processes. Science 235:167–171. 26, 27

RISHEL, G. B., J. A. LYNCH, and E. S. CORBETT. 1982. Seasonal stream temperature changes following forest harvesting. J. Environ. Qual. 11:112–116. 150

ROMME, W. H., and D. H. KNIGHT. 1982. Landscape diversity: the concept applied to Yellowstone National Park. Bioscience 32:664–670. 253

ROOT, R. B. 1967. The niche exploitation pattern of the blue-gray gnatcatcher. Ecol. Monogr. 37:317–350. 246

ROSE, F. 1974. The epiphytes of oak. Pages 250–273 *in* M. G. Morris, and F. H. Perring (eds.). The British oak. Classey, Faringdon. 376 pp. 193

ROSENBERG, K. V., and M. G. RAPHAEL. 1986. Effect of forest fragmentation on vertebrates in Douglas-fir forests. Pages 263–272 *in* Verner et al. 1986. *op. cit.* 131

ROTH, R. R. 1976. Spatial heterogeneity and bird species diversity. Ecology 57:773–782. 83

RUFFER, D. G. 1961. Effect of flooding on a population of mice. J. Mammal. 42:494–502. 143

RUGGIERO, L. F., R. S. HOLTHAUSEN, B. G. MARCOT, K. B. AUBRY, J. W. THOMAS, and E. C. MESLOW. 1988. Ecological dependency: the concept and its implications for research and management. Trans. North Am. Wildl. Nat. Resourc. Conf. 53:115–126. 245

RUNKLE, J. R. 1982. Patterns of disturbance in some old-growth mesic forests of eastern North America. Ecology 63:1533–1546.

RUNKLE, J. R. 1985. Disturbance regimes in temperate forests. Pages 17–33 *in* Pickett and White 1985. *op. cit.* 472 pp. 83

RUSTERHOLZ, K. A., and R. HOWE. 1979. Species-area relations of birds on small islands in a Minnesota lake. Evolution 33:468–477. 118

SALWASSER, H. 1986. Modeling habitat relationships of terrestrial vertebrates—the manager's viewpoint. Pages 419–424 *in* Verner et al. 1986. *op. cit.* 269, 272

SALWASSER, H. 1987. Spotted owls: turning a battleground into a blueprint. Ecology 68:776–779. 74

SALWASSER, H., S. P. MEALEY, and K. JOHNSON. 1984. Wildlife population viability: a question of risk. Trans. North Am. Wildl. Nat. Resource Conf. 49:421–434. 244

SALWASSER, H., and J. C. TAPPEINER II. 1981. An ecosystem approach to integrated timber and wildlife habitat management. Trans. North Amer. Wildl. Natur. Resour. Conf. 46:473–487. 255

SALWASSER, H., J. W. THOMAS, and F. B. SAMSON. 1984. Applying the diversity concept to National Forest management. Pages 59–69 *in* J. L. Cooley and J. H. Cooley (eds.) Natural diversity in forest ecosystems. Inst. of Ecology, Univ. Georgia, Athens, Georgia. 290 pp. 254

SAMPSON, F. B., and F. L. KNOPF. 1982. In search of a diversity ethic for wildlife management. Trans. N. Amer. Wildlife Nat. Res. Conf. 47:421–431. 13

SANDBERG, D. V. 1980. Duff reduction by prescribed underburning in Douglas fir. USDA For. Serv. Res. Pap. PNW-272. 18 pp. 215

SANKHALA, K. 1978. Tiger. Collins, London. 220 pp. 242

SANTILLO, D. J. 1987. Response of small mammals and breeding birds to herbicide-induced habitat changes on clearcuts in Maine. M. S. Thesis. Univ. of Maine, Orono. 74 pp. 223, 224

SAUNDERS, D. A., G. T. SMITH, and I. ROWLEY. 1982. The availability and dimensions of tree hollows that provide nest sites for cockatoos (Psittaciformes) in western Australia. Aust. Wildl. Res. 9:541–556. 162

SAVELY, H. E., JR. 1939. Ecological relations of certain animals in dead pine and oak logs. Ecol. Monogr. 9:322–385. 158

SCHAL, C. 1982. Intraspecific vertical stratification as a mate finding mechanism in tropical cockroaches. Science 215:1405–1407. 191

SCHALLER, G. B., H. JUNCHU, P. WENSHI, and Z. JING. 1985. The giant pandas of Wolong. Univ. Chicago Press, Chicago. 298 pp. 82

SCHLESINGER, R. B. 1979. Natural removal mechanisms for chemical pollutants in the environment. Bioscience 29:95–101. 151

SCHMIDT, R. A. 1978. Diseases in forest ecosystems: the importance of functional diversity. Pages 287–315 *in* J. G. Horsfall and E. B. Cowling (eds.). Plant diseases: an advanced treatise; Vol. 2. How disease develops in populations. Academic Press, New York. 436 pp. 41

SCHOEN, J. W., M. D. KIRCHHOFF, and O. C. WALLMO. 1984. Sitka black-tailed deer/old-growth relationships in southeast Alaska: implications for management. Pages 315–319 *in* Meehan et al. 1984. 69

SCHOEN, J. W., O. C. WALMO, and M. D. KIRCHOFF. 1981. Wildlife-forest relationships: is a reevaluation of old growth necessary? Trans. North Am. Wildl. Nat. Resour. Conf. 46:531–544. 65, 79

SCHOENER, T. W. 1967. The ecological significance of sexual dimorphism in size in the lizard *Anolis conspersus*. Science 155:474–477. 190

SCHOENER, T. W. 1986. Resource partitioning. Pages 91–126 *in* Kikkawa and Anderson 1986. *op. cit.* 189

SCHUSTER, E. G., and J. G. JONES. 1985. Below-cost timber sales: analysis of a forest policy issue. USDA Forest Service Gen. Tech. Rept. INT-183. 17 pp. 284

SCOTT, V. E. 1978. Characteristics of ponderosa pine snags used by cavity-nesting birds in Arizona. J. For. 76:26–28. 171

SCOTT, V. E. 1979. Bird response to snag removal in ponderosa pine. J. For. 77:26–28. 162

SCOTT, V. E., J. A. WHELAN, and P. L. SVOBODA. 1980. Cavity-nesting birds and forest management. Pages 311–324 *in* R. M. DeGraff, and N. G. Tilghman (eds.). Management of western forests and grasslands for nongame birds. USDA Forest Service Gen. Tech. Rep. INT-86. 162, 176

SEAGLE, S. W., R. A. LANCIA, D. A. ADAMS, M. R. LENNARTZ, and H. A. DEVINE. 1987. Integrating timber and red-cockaded woodpecker habitat management. Trans. No. Am. Wildl. Nat. Resour. Conf. 52:41–52. 54

SEDELL, J. R., and F. J. SWANSON. 1984. Ecological characteristics of streams in old-growth forests of the Pacific Northwest. Pages 9–16 *in* Meehan et al. 1984. 141

SEDGWICK, J. A., and F. L. KNOPF. 1986. Cavity-nesting birds and the cavity-tree resource in plains cottonwood bottomlands. J. Wildl. Manage. 50:247–252. 158, 171

SEDJO, R. A. 1983. The comparative economics of plantation forestry. Resources for the Future, Washington, DC. 161 pp. 34

SENFT, J. F., B. A. BENDTSEN, and W. L. GALLIGAN. 1985. Weak wood: fast grown trees make problem lumber. J. For. 83:476–484. 60

SEPIK, G. F., R. B. OWEN, JR., and M. W. COULTER. 1981. A landowner's guide to woodcock management in the northeast. Maine Agric. Expt. Stn. Misc. Rep. 253. 82, 248

SERVHEEN, C. 1985. The grizzly bear. Pp. 401–415 *in* A. S. Eno and R. L. DiSilvestro (eds.). Audubon Wildlife Report 1985. National Audubon Society, New York. 671 pp. 16

SEVERINGHAUS, W. D. 1981. Guild theory development as a mechanism for assessing environmental impact. Environ. Manage. 5:187–190. 246

SEYMOUR, R. S., P. R. HANNAH, J. R. GRACE, and D. A. MARQUIS. 1986. Silviculture: the next 30 years, the past 30 years. Part IV. The northeast. J. For. 84(7):31–38. 54

SHAFFER, C. H., and J. V. GWYNN. 1967. Management of the eastern turkey in oak-pine and pine forests of Virginia and the southeast. Pages 303–342 *in* O. H. Hewitt (ed.) The wild turkey and its management. Wildlife Society, Washington, D.C. 589 pp. 237

SHAFFER, M. L. 1981. Minimum population sizes for species conservation. Bioscience 31:131–134. 243

SHAW, W. W., and W. R. MANGUN. 1984. Nonconsumptive use of wildlife in the United States. USDI Fish and Wildlife Service, Resour. Publ. 154. Washington, D.C. 4, 279

SHUGART, H. H. 1984. A theory of forest dynamics. Springer-Verlag, New York. 278 pp. 268

SIMARD, A. J., D. A. HAINES, R. W. BLANK, and J. S. FROST. 1983. The Mack Lake fire. USDA For. Serv. North Central Forest Exp. Sta., Gen. Tech. Rep. NC-83. 36 pp. 232

SIMBERLOFF, D. 1986. Design of nature reserves. Pages 315–337 *in* M. B. Usher (ed.) Wildlife conservation evaluation. Chapman Hall, London. 394 pp. 125

SIMBERLOFF, D. 1987. The spotted owl fracas: mixing academic, applied, and political ecology. Ecology 68:766–772. 72, 74

SIMBERLOFF, D. S., and L. G. ABELE. 1976a. Island biogeography theory and conservation practice. Science 191:285–286. 125, 126

SIMBERLOFF, D. S., and L. G. ABELE. 1976b. Island biogeography and conservation: strategy and limitations. Science 193:1032. 125

SIMBERLOFF, D. S., and L. G. ABELE. 1982. Refuse design and island biogeographic theory: effects of fragmentation. Am. Nat. 120:41–50. 125, 126

SIMBERLOFF, D. S., and L. G. ABELE. 1984. Conservation and obfuscation: subdivision of reserves. Oikos 42:399–401. 125

SIMBERLOFF, D., and J. COX. 1987 Consequences and costs of conservation corridors. Conserv. Biol. 1:63–71. 130

SIMBERLOFF, D. S., and GOTELLI, N. 1984. Effects of insularization on plant species richness in the prairie forest ecotone. Biol. Conserv. 29:27–46. 126

SIMPSON, E. H. 1949. Measurement of diversity. nature 163:688. 298

SISCO, C., and R. J. GUTIERREZ. 1984. Winter ecology of radio-tagged spotted owls in Six Rivers National Forest, Humboldt County, CA. USDA Forest Service Six Rivers National Forest, Final Rep. 140 pp. 72

SLAGSVOLD, T. 1977. Bird population changes after clearance of deciduous scrub. Biol. Conserv. 12:229–243. 223

SLATKIN, M. 1985. Gene flow in natural populations. Ann. Rev. Ecol. Syst. 16:393–430. 94

SMALL, M. F., and M. L. HUNTER. 1988. Forest fragmentation and avian nest predation in forested landscapes. Oecologia 76:62–64. 131, 258

SMALL, M. F. and M. L. HUNTER, JR. 1989. Response of passerines to abrupt forest-river and forest-powerline edges in Maine. Wilson Bull. 101:77–83. 106

SMITH, A. J. E. 1982. Epiphytes and epiliths. Pages 191–227 *in* A. J. E. Smith (ed.). Bryophyte ecology. Chapman Hall, London. 511 pp. 191

SMITH, D. G. 1976. Effect of vegetation on lateral migration of anastomosed channels of a glacier meltwater river. Bull. Geol. Soc. Amer. 87:857–860. 142

SMITH, D. M. 1986. The practice of silviculture. 8th ed. Wiley, New York. 578 pp. 47, 84, 203, 221, 227

SMITH, K. D. 1974. The utilization of gum trees by birds in Africa. Ibis 116:155–164. 39

SOCIETY OF AMERICAN FORESTERS. 1967. Forest cover types of North America. Society of American Foresters, Washington, D.C. 67 pp. 24, 265, 266

SOCIETY OF AMERICAN FORESTERS. 1984. Scheduling the harvest of old growth. Society of American Foresters. Bethesda, Maryland. 44 pp. 68

SOLLINS, P., C. C. GRIER, F. M. MCCORISON, K. CROMACK, JR., R. FOGEL, and R. L. FREDRIKSEN. 1980. The internal element cycles of an old growth Douglas-fir ecosystem in western Oregon. Ecol. Monogr. 50:261–286. 95

SOLLINS, P., and F. M. MCCORISON. 1981. Nitrogen and carbon solution chemistry of an old growth coniferous forest watershed before and after cutting. Water Resour. Res. 17:1409–1418. 95

SORG, C. F., and J. LOOMIS. 1985. An introduction to wildlife valuation techniques. Wildl. Soc. Bull. 13:38–46. 277

SORK, V. L. 1983. Distribution of pignut hickory (*Carya glabra*) along a forest to edge transect, and factors affecting seedling recruitment. Bull. Torrey Bot. Club 110:494–506. 106

SOULÉ, M. E. 1980. Thresholds for survival: maintaining fitness and evolutionary potential. Pages 151–170 *in* M. E. Soule' and B. A. Wilcox (eds.). Conservation biology: an evolutionary-ecological perspective. Sinauer, Sunderland, Massachusetts. 395 pp. 72

SOULÉ, M. E. (ed.) 1986. Conservation biology. Sinauer, Sunderland, Massachusetts. 584 pp.

SOULÉ, M. E. (ed.) 1987. Viable populations for conservation. Cambridge Univ. Press, Cambridge, Great Britain. 189 pp. 94, 118, 127, 243, 244, 245

SOUTHWOOD, T. R. E., V. K. BROWN, and P. M. READER. 1979. The relationships of plant and insect diversities in succession. Biol. J. Linnaean Soc. 12:327–348. 188, 191

SPRADBERRY, J. P., and A. A. KIRK. 1981. Experimental studies on the responses of European siricid woodwasps to host trees. Ann. Appl. Biol. 98:179–186. 210

SPURR, S. H. 1952. Origin of the concept of forest succession. Ecology 33:426–427. 20

SPURR, S. H. 1979. Silviculture. Scientific American 240:76–82, 87–91. 201

STAUFFER, D. F., and L. B. BEST. 1980. Habitat selection by birds of riparian communities evaluating effects of habitat alterations. J. Wildl. Manage. 441:1–15. 145, 148

STEELE, R. C., and R. C. WELCH (eds.). 1973. Monks Wood. A nature reserve record. Natural Environment Research Council, Huntingdon, Great Britain. 9

STEINBLUMS, I. J., H. A. FROEHLICH, and J. K. LYONS. 1984. Designing stable buffer strips for stream protection. J. For. 82:49–52. 148

STEINHART, P. 1981. Leave the dead. Audubon 83(1):6–7. 162

STEVENS, L., B. T. BROWN, J. M. SIMPSON, and R. R. JOHNSON. 1977. The importance of riparian habitat to migrating birds. Pages 156–164 *in* Johnson and Jones (eds.). Importance, preservation, and management of riparian habitat. USDA Forest Service Gen. Tech. Rept. RM-43. 217 pp. 146

STEVENSON, S. K., and J. A. ROCHELLE. 1984. Lichen litterfall—its availability and utilization by black-tailed deer. Pages 391–396 *in* Meehan et al. 1984. *op. cit.* 78

STOUT, A. T. 1985. Below-cost timber sales. Amer. For 91(1):11. 44. 284

STRANSKY, J. J., and L. K. HALLS. 1976. Browse quality affected by pine site preparation in east Texas. Proc. 30th Ann. Conf. S.E. Assn. Game and Fish Comm. Pages 507–512. 216

STRANSKY, J. J., and L. K. HALLS. 1980. Fruiting of woody plants affected by site preparation and prior land use. J. Wildl. Manage. 44:258–263. 216–217

STRANSKY, J. J., and R. F. HARLOW. 1981. Effects of fire on deer habitat in the southeast. Pages 135–142. *In* G. W. Wood (ed.) Prescribed fire and wildlife in southern forests. Belle W. Baruch Forest Science Inst., Clemson Univ., Georgetown, SC. 233–234

STRANSKY, J. J., and J. H. ROESE. 1984. Promoting soft mast for wildlife in intensively managed forests. Wildl. Soc. Bull. 12:234–240. 195

STREETER, D. T. 1974. Ecological aspects of oak woodland conservation. Pages 341–354 in M. G. Morris and F. H. Perring (eds.). The Brish oak. Classey, Faringdon. 376 pp. 120

STRELKE, W. K., and J. G. DICKSON. 1980. Effect of forest clear cut edge on breeding birds in east Texas. J. Wildl. Manage. 44:559–567. 106

STRONG, D. R., J. H. LAWTON, T. R. E. SOUTHWOOD. 1984. Insects on plants. Harvard Univ. Press, Cambridge. 313 pp. 39, 82, 191

STRONG, D. R., JR. 1977. Epiphyte loads, tree falls, and perennial forest disruption: a mechanism for maintaining higher tree species richness in the tropics without animals. J. Biogeogr. 4:215–218. 187

SUTTON, S. L., C. P. J. ASH, and A. GRUNDY. 1983a. The vertical distribution of flying insects in lowland rain forests of Panama, Papua, New Guinea and Brunei. Zool. J. Linnaean Soc. 78:287–297. 191

SUTTON, S. L., and P. J. HUDSON. 1980. The vertical distribution of small flying insects in the lowland rain forest of Zaire. Zool. J. Linnaean Soc. 68:111–123. 189

SUTTON, S. L., T. C. WHITMORE, and A. C. CHADWICK (eds.). 1983b. Tropical rain forest: ecology and management. Blackwell, Oxford. 498 pp.

SWANSON, F. J., S. V. GREGORY, J. R. SEDELL, and A. G. CAMPBELL. 1982. Land-water interactions: the riparian zone. Pages 267–291 *in* R. L. Edmonds (ed.). Analysis of coniferous forest ecosystems in the western United States. Hutchinson Ross, Stroudsberg, Pennsylvania. 419 pp. 140, 143, 144

SWEENEY, J. M., M. E. GARNER, and R. P. BURKERT. 1984. Analysis of white-tailed deer use of forest clear-cuts. J. Wildl. Manage. 48:652–655. 100

SWIFT, B. L., J. S. LARSON, and R. M. DEGRAAF. 1984. Relationship of breeding bird density and diversity to habitat variables in forested wetlands. Wilson Bull. 96:48–59. 208

SWINDEL, B. F., L. F. CONDE, and J. E. SMITH. 1984. Species diversity: concept, measurement, and response to clearcutting and site-preparation. For. Ecol. and Manage. 8:11–22. 302

SWINDEL, B. F., L. F. CONDE, and J. E. SMITH. 1987. Index-free diversity orderings: concept, measurement, and observed response to clearcutting and site-preparation. For. Ecol. and Manage. 20:195–208. 8, 300, 302

SWINDEL, B. F., and L. R. GROSENBAUGH. 1988. Species diversity in young Douglas-fir plantations compared to old growth. For. Ecol. and Manage. 23:227–231.   300, 301, 302

SWINDEL, B. F., W. R. MARION, L. D. HARRIS, L. A. MORRIS, W. L. PRITCHETT, L. F. CONDE, H. RIEKERK, and E. T. SULLIVAN. 1983. Multi-resource effects of harvest site preparation and planting in pine flatwoods. South. J. Appl. For. 7:6–15.   216–217

SZACKI, J. 1987. Ecological corridor as a factor determining the structure and organization of a bank vole population. Acta Theriol. 32:31–44.   129

TEER, J. G., G. V. BURGER, and C. Y. DeKNATEL. 1983. State-supported habitat management and commercial hunting on private lands in the United States. Trans. North Am. Wildl. Nat. Resour. Conf. 48:445–456.   279

TEFLER, E. S. 1970. Winter habitat selection by moose and white-tailed deer. J. Wildl. Manage. 34:553–559.   98

TEMPLE, S. A. 1977. Plant-animal mutualism: coevolution with dodo leads to near extinction of plant. Science 197:885–886.   11

TEMPLE, S. A. 1986. Predicting impacts of habitat fragmentation on forest birds: a comparison of two models. Pages 301–304 in Verner et al. 1986. *op. cit.*   134

TEMPLE, S. A. 1987. Predation on turtle nest increases near ecological edges. Copeia 1987:250–252.   107

TERBORGH, J. 1976. Island biogeography and conservation: strategy and limitations. Science 193:1029–1030.   125, 241

TERBORGH, J. 1983. Five New World primates. Princeton Univ. Press, Princeton, NJ. 241

TERBORGH, J. 1986. Keystone plant resources in the tropical forest. Pages 330–344 *in* Soule' 1986. *op. cit.*   241

TERBORGH, J., and S. ROBINSON. 1986. Guilds and their utility in ecology. Pages 65–90 *in* Kikkawa and Anderson 1986. *op. cit.*   246

TESCH, S. D. 1981. Comparative stand development in an old growth Douglas-fir *Pseudotsuga menziesii*-var-Glauca forest in western Montana. Can. J. For. Res. 11:82–89.   184

THE NATURE CONSERVANCY. 1982. Natural heritage program operations manual. The Nature Conservancy, Arlington, Virginia.   238

THOMAS, J. W. (ed.). 1979. Wildlife habitats in managed forests: the Blue Mountains of Oregon and Washington. U.S.D.A. Forest Service Agricultural Handbook No. 553. Washington, D.C. 512 pp.   xi, xii, 168

THOMAS, J. W., R. G. ANDERSON, C. MASER, and E. L. BULL. 1979c. Snags. Pages 60–77 *in* Thomas 1979. *op. cit.*   159, 163, 167, 169, 171, 176

THOMAS, J. W., H. BLACK, JR., R. J. SCHERZINGER, and R. J. PEDERSEN. 1979d. Deer and elk. Pages 104–127 *in* Thomas 1979. *op. cit.*   146, 192

THOMAS, J. W., J. D. GILL, J. C. PACK, W. M. HEALY, and H. R. SANDERSON. 1976. Influence of forestland characteristics on spatial distribution of hunters. J. Wildl. Manage. 40:500–506.   258

THOMAS, J. W., C. MASER, and J. E. RODIEK. 1979a. Riparian zones. Pages 40–47 *in* Thomas 1979. *op. cit.*   144, 146, 147

THOMAS, J. W., C. MASER, and J. E. RODIEK. 1979*b*. Edges. Pages 48–59 *in* Thomas 1979. *op. cit.* 110

THOMAS, J. W., R. J. MILLER, C. MASSER, R. G. ANDERSON, and B. E. CARTER. 1979*e*. Plant communities and successional stages. Pages 22–39 *in* Thomas 1979. *op. cit.* 246

THOMAS, J. W., L. F. RUGGIERO, R. W. MANNAN, J. W. SCHOEN, and R. A. LANCIA. 1988. Management and conservation of old-growth forests in the United States. 61

THOMAS, J. W., and J. VERNER. 1986. Forests. Pages 73–91 *in* A. Y. Cooperider, R. J. Boyd, and H. R. Stuart (eds.) Inventory and monitoring of wildlife habitat. USDI Bur. Land Manage., Denver, Colorado. 858 pp. 268

THUROW, L. C. 1980. The zero-sum society. Basic, New York. 230 pp. 284

TILMAN, D. 1985. The resource-ratio hypothesis of plant succession. Am. Nat. 125:827–853. 22

TITTERINGTON, R. W., H. S. CRAWFORD, and B. N. BURGASON. 1979. Songbird responses to commercial clearcutting in Maine spruce-fir forests. J. Wildl. Manage. 43:602–609. 49

TOPHAM, P. B. 1977. Colonization, growth, succession, and competition. Pages 31–68 *in* M. R. D. Seaward (ed.). Lichen ecology. Academic Press, London. 550 pp. 191

TOTH, E. F., D. M. SOLIS, and B. G. MARCOT. 1986. A management strategy for habitat diversity: using models of wildlife-habitat relationships. Pages 139–144 *in* Verner et al. 1986. *op. cit.* 269

TRISKA, F. J., J. R. SEDELL, and S. V. GREGORY. 1982. Coniferous forest streams. Pages 292–332 *in* R. L. Edmonds (ed.). Analysis of coniferous forest ecosystems in the western United States. Hutchinson Ross, Stroudsberg, Pennsylvania. 419 pp. 141

U.S. FISH and WILDLIFE SERVICE. 1982. The northern spotted owl: a status review. U.S. Dept. Interior, Fish and Wildlife Service, Portland, Oregon. 29 pp. 72

U.S. FISH and WILDLIFE SERVICE. 1985. Red-cockaded woodpecker recovery plan. U.S. Fish and Wildl. Serv., Atlanta, Georgia. 88 pp. 179

U.S. FISH and WILDLIFE SERVICE. 1988. 1985 Survey of fishing, hunting, and wildlife associated activities. U.S. Fish and Wildlife Service, Washington D.C. 167 pp. 279

U.S. FISH and WILDLIFE SERVICE. 1988. Box score of listings/recovery plans. U.S. Fish and Wildlife Service Endangered Species Tech. Bull. 13(5):12. 4

USDA FOREST SERVICE. 1986a. Draft supplement to the environmental impact statement for an amendment to the Pacific Northwest Regional Guide. USDA Forest Service, Portland, Oregon. 74

USDA FOREST SERVICE. 1986b. Interim definitions for old-growth Douglas-fir and mixed-conifer forests in the Pacific Northwest and California. USDA Forest Service Res. Note PNW-447. 61

USDA FOREST SERVICE. 1986c. Land and resource management plan: White Mountain National Forest. USDA Forest Service, Eastern Region, Upper Darby, Pennsylvania. 261

USDA FOREST SERVICE. 1986d. National forests fire report-1985. USDA Forest Service, Washington, D.C. 88, 91

USDA SOIL CONSERVATION SERVICE. 1982. National list of scientific plant names. Two volumes. USDA Soil Conservation Service SCS-TP-159. 416 pp.; 438 pp.

USHER, M. B. (ed.). 1986. Wildlife conservation evaluation. Chapman Hall, London. 394 pp.   259

USHER, M. B. 1988. Biological invasions of nature reserves a search for generalizations. Biol. Conserv. 44:119–135.   133

VAISANEN, R. A., and P. RAUHALA. 1983. Succession of land bird communities on large areas of peatland drained for forestry. Ann. Zool. Fennici 20:115–127.   207

VAN BALEN, J. H., C. J. H. BOOY, J. A. VAN FRANCKER, and E. R. OSIECK. 1982. Studies on hole-nesting birds in natural nest sites. 1. Availability and occupation of natural nest site. Ardea 70:1–24.   162

VAN HORNE, B. 1983. Density as a misleading indicator of habitat quality. J. Wildl. Manage. 47:893–901.   170, 246

VANNOTE, R. L., G. W. MINSHALL, K. W. CUMMINS, J. R. SEDELL, and C. E. CUSHING. 1980. The river continuum concept. Can. J. Fish. Aquat. Sci. 37:130–137.   140

VERNER, J. 1984. The guild concept applied to management of bird populations. Environ. Manage. 8:1–13.   246

VERNER, J., M. L. MORRISON, and C. J. RALPH (eds.). 1986. Wildlife 2000. Univ. Wisconsin Press, Madison, Wisconsin. 470 pp.   269

VERRY, E. S., J. R. LEWIS, and K. N. BROOKS. 1983. Aspen clearcutting increases snowmelt and storm flow peaks in north central Minnesota. Water Resour. Bull. 19:59–66.   150

VICKERS, C. R., L. D. HARRIS, and B. F. SWINDEL. 1985. Changes in herpetofauna resulting from ditching of cypress ponds in coastal ponds in coastal plains flatwoods. For. Ecol. Manage. 11:17–29.   143

WAHLENBERG, W. G. 1946. Longleaf pine. Charles Lathrop Pack Forestry Foundation, Washington, D.C. 429 pp.   43, 44

WALES, B. A. 1972. Vegetation analysis of north and south edges in a mature oak-hickory forest. Ecol. Monogr. 42:451–471.   106

WALKER, L.C., and H. V. WAINT, JR. 1966. Silviculture of longleaf pine. Stephen F. Austin College School of Forestry. Bull. 11. 105 pp.   46

WALLMO, O. C., and J. W. SCHOEN. 1980. Response of deer to secondary forest succession in southeast Alaska. For. Sci. 26:448–462.   76, 78

WALSTAD, J. D., and P. J. KUCH (eds.). 1987. Forest vegetation management for conifer production. John Wiley and Sons, New York. 523 pp.   216, 224

WALTER, H., and S. W. BRECKLE. 1985. Ecological systems of the geobiosphere. Trans. by S. Gruber. Springer-Verlag, Berlin. 242 pp.   265

WARING, R. H., and J. F. FRANKLIN. 1979. Evergreen coniferous forests of the Pacific Northwest. Science 204:1380–1386.   75

WARING, R. H., and W. H. SCHLESINGER. 1985. Forest ecosystems: concepts and management. Academic Press, Orlando, Florida. 340 pp.   142

WATT, K. E. F. 1968. Ecology and resource management: a quantitative approach. McGraw-Hill, New York. 450 pp.   10

WEGNER, J. F., and G. MERRIAM. 1979. Movements by birds and small mammals between a wood and adjoining farmland habitat. J. Appl. Ecol. 16:349–357.   129

WENGER, K. F. 1984. Forestry handbook. 2nd ed. Wiley, New York. 1335 pp.   60, 163, 219

WEIN, R. W., and D. A. MacLEAN (eds.). 1983. The role of fire in northern circumpolar ecosystems. Scope 18. Wiley, New York. 322 pp.   210

WEST, D. C., H. H. SHUGART, D. B. BOTKIN (eds.). 1981. Forest succession: concepts and application. Springer-Verlag, New York. 517 pp.

WESTOBY, J. 1982. Halting tropical deforestation: The role of technology. Paper commissioned by U.S. Congress, Office of Technology Assessment, Washington, D.C. 281

WHEELWRIGHT, N. T., W. A. HABER, K. G. MURRAY, and C. GUINDON. 1984. Tropical fruit-eating plants: A survey of a Costa Rican lower montane forest. Biotropica 16:173–192.   194

WHIPPLE, S. A., and R. L. DIX. 1979. Age structure and successional dynamics of a Colorado subalpine forest. Am. Midl. Nat. 101:142–158.   184

WHITCOMB, R. F., J. F. LYNCH, P. A. OPLER, and C. S. ROBBINS. 1976. Island biogeography and conservation: strategy and limitations. Science 193:1030–1032.   125

WHITCOMB, R. F., C. F. ROBBINS, J. F. LYNCH, B. L. WHITCOMB, M. K. KLIMKIEWICZ, and D. BYSTRAK. 1981. Effects of forest fragmentation on avifauna of the eastern deciduous forest. Pages 125–205 *in* Burgess and Sharpe 1981. *op. cit.*   120, 122

WHITCOMB, B. L., R. F. WHITCOMB, and D. BYSTRAK. 1977. Island biogeography and habitat islands of eastern forest. III. Long-term turnover effects of selective logging on the avifauna of forest fragments. Am. Birds 31:17–23.   130

WHITE, A. S. 1985. Presettlement regeneration patterns in a southwestern ponderosa pine stand. Ecology 66:589–594.   185

WHITE, A. S. 1986. Prescribed burning for oak savanna restoration in central Minnesota. USDA For. Serv. Res. Pap. NC-266, Stn. 12 pp.   209

WHITE, L. P., L. D. HARRIS, J. E. JOHNSTON, and D. G. MILCHUNAS. 1975. Impact of site preparation on flatwoods wildlife habitat. Proc. 29th Ann. Conf. S.E. Game and Fish Comm. 29:347–353.   217

WHITNEY, G. G. 1984. Fifty years of change in the arboreal vegetation of Heart's Content, an old growth hemlock-white, pine-northern hardwood stand. Ecology 65:403–408.   185

WHITNEY, G. G. 1987. An ecological history of the Great Lakes forest of Michigan. J. Ecol. 75:667–684.   38

WHITNEY, G. G., and J. R. RUNKLE. 1981. Edge versus age effects in the development of a beech-maple forest. Oikos 37:377–381.   106

WHITTAKER, R. H. 1960. Vegetation of the Siskiyou Mountains, Oregon and California. Ecol. Monogr. 30:279–338.   31

WHITTAKER, R. H. 1965. Dominance and diversity in land plant communities. Science 147:250–260.   300

WHITTAKER, R. H., F. H. BORMANN, G. E. LIKENS, and T. C. SICCAMA. 1974. The Hubbard Brook ecosystem study: forest biomass and production. Ecol. Monogr. 44:233–252.   20

WHITTAKER, R. H., and P. P. FEENY. 1971. Allelochemics: chemical interactions between species. Science 171:757–770.   186

WHITTAKER, R. H., and G. M. WOODWELL. 1969. Structure, production and diversity of the oak-pine forest at Brookhaven, New York. J. Ecol. 57:155–174.   20

WICK, H. L., and P. R. CANUTT. 1979. Impacts on wood production. Pages 148–161 in Thomas 1979. *op. cit.*   176, 275

WICK, H. L., D. FAUSS, and R. ZALUNARDO. 1985. Impacts on wood production. Pages 307–313 *in* Browne 1985. *op. cit.*   276

WIENS, J. A. 1974. Habitat heterogeneity and avian community structure in North American grasslands. Am. Midl. Nat. 91:195–213.   83

WIENS, J. A. 1975. Avian communities, energetics, and functions in coniferous forest habitat. Pages 226–265 in D. R. Smith (tech. coor.) Proc. symposium on management of forest and range habitats for nongame birds. U.S.D.A. Forest Service, Washington, D.C. Gen. Tech. Rap. WO-1.   38

WIENS, J. A. 1976. Population responses to patch environments. Ann. Rev. Ecol. Syst. 7:81–120.   83

WIENS, J. A., C. S. CRAWFORD, and J. R. GOSZ. 1985. Boundary dynamics: a conceptual framework for studying landscape ecosystems. Oikos 45:421–427.   102

WILCOVE, D. S. 1985. Nest predation in forest tracts and the decline of migratory songbirds. Ecology 66:1211–1214.   122

WILCOVE, D. S., C. H. MCLELLAN, and A. P. DOBSON. 1986. Habitat fragmentation in the temperate zone. Pages 237–256 *in* Soule' 1986. *op. cit.*   107, 122, 123

WILKINSON, C. F., and H. M. ANDERSON. 1988. Foresters look at below-cost timber sales: Are they legal? J. For. 85(8):20–26.   284

WILLIAMSON, M. 1981. Island populations. Oxford Univ. Press, Oxford. 286. pp.   116

WILLIAMSON, S. J., and D. H. HIRTH. 1985. An evaluation of edge use by white-tailed deer. Wildl. Soc. Bull. 13:252–257.   100

WILLIS, E. O. 1984. Conservation, subdivision of reserves, and the antidismemberment hypothesis. Oikos 42:396–398.   125

WILLSON, M. F. 1970. Foraging behavior of some winter birds of deciduous woods. Condor 72:169–174.   36

WILLSON, M. F. 1974. Avian community organization and habitat structure. Ecology 55:1017–1029.   188

WILLSON, M. F. 1986. Avian frugivory and seed dispersal in eastern North America. Curr. Ornithol. 3:223–279.   194

WILSON, E. O. 1984. Biophilia. Harvard Univ. Press, Cambridge, Massachusetts. 157 pp.   10, 11

WILSON, E. O. 1988. The current state of biological diversity. Pp. 3–18 *in* Wilson and Peter 1988. *op. cit.*   9, 12

WILSON, E. O., and F. M. PETER. 1988. Biodiversity. National Academy Press, Washington, D.C. 521 pp.   10

WILSON, E. O., and E. O. WILLIS. 1975. Applied biogeography. Pages 522–534 *in* M. L. Cody, and J. M. Diamond (eds.). Ecology and evolution of communities. Harvard Univ. Press, Cambridge, Massachusetts. 545 pp.   124

WILSON, W. L., and A. D. JOHNS. 1982. Diversity and abundance of selected animal species in undisturbed forest, selectively logged forest, and plantations in East Kalimantan, Indonesia. Biol. Conser. 24:205–218. 69, 219

WINSTON, P. W. 1956. The acorn microsere, with special reference to arthropods. Ecology 37:120–132. 82

WOLFE, M. L., and F-C. BERG. 1988. Deer and forestry in Germany: half a century after Leopold. J. For. 86(5):25–31. 36

WONG, M. 1985. Understory birds as indicators of regeneration in a patch of selectively logged West Malaysian rainforest. Pages 249–263 *in* A. W. Diamond and T. E. Lovejoy (eds.) Conservation of tropical forest birds. International Council for Bird Preservation Tech. Publ. 4., Cambridge. 318 pp. 69

WOODELL, S. R. J. 1979. The role of unspecialized pollinators in the reproductive success of Aldabran plants. Phil. Trans. R. Soc. Land. B. 286:99–108. 119

WRI and IIED. 1986. World resources 1986. A report by the World Resources Institute and the International Institute for Environment and Development. Basic, New York. 353 pp. 131

WRIGHT, H. A., and A. W. BAILEY. 1982. Fire ecology. Wiley, New York. 501 pp. 210

WYATT-SMITH, J. 1987. Problems and prospects for natural management of tropical moist forests. Pp. 5–22 *in* F. Mergen and J. R. Vincent (eds.) Natural management of tropical moist forests. School of Forestry and Environmental Studies, Yale University, New Haven, Connecticut. 212 pp. 201

YAHNER, R. H. 1988. Changes in wildlife communities near edges. Conserv. Biol. 2:333–339. 105

YONEDA, M. 1984. Comparative studies on vertical separation foraging behavior and traveling mode of saddle-backed tamarins *Saguinus fuscicollis* and red-chested moustached tamarins *Saguinus labiatus* in northern Bolivia. Primates 25:414–422. 190

YOUNG, H. E. 1974. Biomass, nutrient elements, harvesting and chipping in the complete tree concept. J. Assoc. Consult. Foresters 19(4):91–104. 204

ZACKRISSON, O. 1977. Influence of forest fires on the North Swedish boreal forest. Oikos 29:22–32.

ZARNOWITZ, J. E., and D. A. MANUWAL. 1985. The effects of forest management on cavity-nesting birds in northwestern Washington. J. Wildl. Manage. 49:255–263. 171

ZEEDYK, W. D., and K. E. EVANS. 1975. Silvicultural options and habitat values in deciduous forest. Pages 115–127 *in* D. R. Smith (ed.). Management of forest and range habitats for nongame birds. USDA Forest Service Gen. Tech. Rep. WO-1. 543 pp. 173

ZIMMERMAN, B. L., and R. O. BIERREGAARD. 1986. Relevance of the equilibrium theory of island biogeography and species-area relations to conservation with a case from Amazonia. J. Biogeogr. 13:133–143. 116

ZOBEL, B., and J. TALBERT. 1984. Applied forest tree improvement. Wiley, New York. 505 pp. 10, 41

Scientific Names and Taxonomic Index[1]

Acacia	*Acacia* spp.	17
Alder	*Alnus* spp.	145
Alder, red	*Alnus rubra*	23
Algae	Subkingdom: Thallobionta	160, 141, 145
Anaconda	*Eunectes* spp.	143
Angiosperm	Division: Magnoliophyta	17, 35–38
Ant	Family: Formicidae	19, 80, 107
Ant, army	*Eciton burchelli*	137
Ant, lion	Family: Myrmeleontidae	80
Antbird, rufous-necked	*Gymnopithys rufigula*	137
Apple	*Malus* spp.	119
Aracari	*Pteroglossus* spp.	162
Ash	*Fraxinus* spp.	23, 196
Ash, common	*Fraxinus excelsior*	196
Ash, green	*Fraxinus pennsylvanica*	8
Ash, mountain	*Sorbus americana*	266
Ash, white	*Fraxinus americana*	238
Aspen	*Populus* spp.	17, 23, 36, 38, 55, 57, 63, 149, 167, 168, 181, 207, 225, 255
Aspen, quaking	*Populus tremuloides*	63

[1]Taxonomy follows the following references: Corbet and Hill 1980, Clements 1981, Moore 1982, Parker 1982, USDA Soil Conservation Service 1982, and Arnett 1985.

Geographic Index

Subject Index